PHILIP BALL

CRITICAL MASS

Philip Ball majored in chemistry at the University of Oxford and received a Ph.D. in physics from the University of Bristol. He is the author of *Life's Matrix* (FSG, 2000); *Bright Earth* (FSG, 2002), which was a finalist for the National Book Critics Circle Award; and *The Devil's Doctor* (FSG, 2006). He lives in London with his wife.

CRITICAL MASS

CRITICAL MASS

HOW ONE THING LEADS

TO ANOTHER

——

PHILIP BALL

FARRAR, STRAUS AND GIROUX / NEW YORK

Farrar, Straus and Giroux
19 Union Square West, New York 10003

Distributed in Canada by Douglas & McIntyre Ltd.
Printed in the United States of America
Originally published in 2004 by William Heinemann, Great Britain
Published in 2004 in the United States by Farrar, Straus and Giroux
First American paperback edition, 2006

Grateful acknowledgment is made to the following for permission to reproduce paintings and other materials in their possession: *Figure 1.1, Thomas Hobbes* by Isaac Fuller (1606–72), Burghley House Collection, Lincolnshire, UK/Bridgeman Art Library; *1.2*, title page from Hobbes's *Leviathan*, 1651, London, © Bettmann/CORBIS; *2.1*, engraving of James Clerk Maxwell after a photograph by Fergus, © CORBIS; *3.1*, Alphonse Quetelet, courtesy AKG Images; *5.1a*, Manuel Velarde, Universidad Complutense, Madrid; *5.1b, Reviews of Modern Physics* by M. C. Cross & P. Hohenberg (1993), 65, 851; *5.1c*, David Cannell, University of California at Santa Barbara; *5.3*, Mitsugu Matsushita, Chuo University; *5.4*, photographs by Mitsugu Matsushita; *5.5*, Snow Crystals by W. A. Bentley and W. J. Humphreys (1931), reprinted by Dover, New York, 1962; *5.6, 6.1*, Eshel Ben-Jacob, Tel Aviv University; *6.2b*, © Norbert Wu; *6.3*, Tamás Vicsek, Eötvös-Loránd University; *6.8, 6.9*, Dirk Helbing, Technical University of Dresden; *6.11, 6.12*, Michel Batty, University College, London; *6.13*, Hernán Makse, City University of New York; *6.14*, Rui Carvalho and Alan Penn, University College, London; *7.3, 7.4, Physica A* 263 by D. E. Wolf (1999), 438–51; *10.1*, Alastair Bruce, University of Edinburgh; *13.2, 13.3, Computational Economics* 18(1), J. M. Epstein (2001), 9–24; *13.4*, Paul Ormerod, Volterra Consulting, London; *16.3*, Hawoong Jeong, Korea Advanced Institute of Science and Technology; *16.4, 16.6*, Hawoong Jeong; *16.8*, Albert-Laszlo Barabási, University of Notre Dame; *18.5a, b*, Martin Nowak, Oxford University; *18.5c*, from M. A. Nowak and R. M. May (1993), *International Journal of Bifurcation and Chaos 3*, 35.

The Library of Congress has cataloged the hardcover edition as follows:
Ball, Philip, date.
Critical mass : how one thing leads to another / Philip Ball.—
1st American ed.
p.   cm.
Includes bibliographical references and index.
ISBN-13: 978-0-374-28125-0
ISBN-10: 0-374-28125-4 (hc : alk. paper)
1. Sociology.   2. Human behavior—Philosophy.   3. Physics—Social aspects.   I. Title.

HM585.B35 2004
301—dc22

2003064178

Paperback ISBN-13: 978-0-374-53041-9
Paperback ISBN-10: 0-374-53041-6

www.fsgbooks.com

5   7   9   10   8   6   4

CONTENTS

INTRODUCTION: POLITICAL ARITHMETICK   3

1   RAISING LEVIATHAN
The brutish world of Thomas Hobbes   9

2   LESSER FORCES
The mechanical philosophy of matter   33

3   THE LAW OF LARGE NUMBERS
Regularities from randomness   48

4   THE GRAND AH-WHOOM
Why some things happen all at once   80

5   ON GROWTH AND FORM
The emergence of shape and organization   98

6   THE MARCH OF REASON
Chance and necessity in collective motion   118

7   ON THE ROAD
The inexorable dynamics of traffic   156

8   RHYTHMS OF THE MARKETPLACE
The shaky hidden hand of economics   178

9   AGENTS OF FORTUNE
Why interaction matters to the economy   204

10   UNCOMMON PROPORTIONS
Critical states and the power of the straight line   226

11 THE WORK OF MANY HANDS
The growth of firms   250

12 JOIN THE CLUB
Alliances in business and politics   270

13 MULTITUDES IN THE VALLEY OF DECISION
Collective influence and social change   295

14 THE COLONIZATION OF CULTURE
Globalization, diversity, and synthetic societies   337

15 SMALL WORLDS
Networks that bring us together   352

16 WEAVING THE WEB
The shape of cyberspace   372

17 ORDER IN EDEN
Learning to cooperate   402

18 PAVLOV'S VICTORY
Is reciprocity good for us?   429

19 TOWARD UTOPIA?
Heaven, hell, and social planning   449

EPILOGUE: CURTAIN CALL   467

Notes   471
Bibliography   489
Acknowledgments   503
Index   505

CRITICAL MASS

INTRODUCTION:
POLITICAL ARITHMETICK

O n the seventh of November 1690 a manuscript was delivered to England's new king, William III. William, the Prince of Orange, had only the previous year deposed the unpopular Catholic James II in a bloodless coup; and in that time of turmoil the book's message might have provided some solace to the monarch, for it set out to show that England was a solid and secure force in the world.

The book's author was Sir William Petty, a professor of anatomy at Oxford University and physician general to the English army in Ireland. He had died in 1687, but his work was delivered to the royal court by his son, the Earl of Shelburne. Petty claimed to prove

—That a small Country, and few People, may by their Situation, Trade, and Policy, be equivalent in Wealth and Strength, to a far greater People, and Territory . . .

—That France cannot by reason of Natural and Perpetual Impediments, be more powerful at Sea, then the English, or Hollanders.

—That the People, and Territories of the King of England, are Naturally near as considerable, for Wealth, and Strength, as those of France.

—That the Impediments of Englands Greatness, are but contingent and removeable.

—That the Power and Wealth of England, hath increased above this forty years.

—That one tenth part, of the whole Expense, of the King of England's Subjects; is sufficient to maintain one hundred thousand Foot, thirty thousand Horse, and forty thousand Men at Sea, and to defray all other Charges, of the Government: both Ordinary and Extraordinary, if the same were regularly Taxed and Raised.

—That there are spare Hands enough among the King of England's Subjects, to earn two Millions per annum, more than they now do, and there are Employments, ready, proper, and sufficient, for that purpose.

—That there is Mony sufficient to drive the Trade of the Nation.

—That the King of England's Subjects, have Stock, competent, and convenient to drive the Trade of the whole Commercial World.[1]

England, in other words, was styled for greatness. On what grounds did Petty make these bold assertions? His book was called *Political Arithmetick*, and it claimed to make a science of politics. Just as Isaac Newton's law of gravity ultimately rested on the quantitative measurements and deductions of the astronomers Tycho Brahe and Johannes Kepler, so Petty used numbers to derive proofs of the healthy state of English society.

"The Method I take to do this," he explained,

> is not yet very usual; for instead of using only comparative and superlative Words, and intellectual Arguments, I have taken the course (as a Specimen of the Political Arithmetick I have long aimed at) to express my self in Terms of Number, Weight, or Measure; to use only Arguments of Sense, and to consider only such Causes, as have visible Foundations in Nature; leaving those that depend upon the mutable Minds, Opinions, Appetites, and Passions of particular Men, to the Consideration of others: Really professing my self as unable to speak satisfactorily upon those Grounds (if they may be call'd Grounds), as to foretel the cast of a Dye; to play well at Tennis, Billiards, or Bowles, (without long practice,) by virtue of the most elaborate Conceptions that ever have been written *De Projectilibus & Missilibus*, or of the Angles of Incidence and Reflection.[2]

In other words, while Petty professed to know little about mutable human nature, he believed that society could be understood to the extent that it could be measured and quantified. The science of political arithmetic, he argued, could free a nation's leaders from man's irrationality and be used to fashion sound and verifiable principles of governance.

How dismayed Petty would have been to find that three hundred years later, political scientists are still lamenting the fact that human affairs are dominated by whim and prejudice rather than led by reason and logic. In *Man, the State, and War* (1954), Kenneth Waltz voices the hope that dealings between nations might one day be conducted by the use of rational theory rather than by dogma and polemic. "In the absence of an elaborated theory of international politics," he says, "the causes one finds and the remedies one proposes are often more closely related to temper and training than to the objects and events of the world about us."[3]

Waltz certainly does not envisage anything as simple as what Petty had in mind: a kind of Newtonian physics of society. But Petty's efforts, which now look woefully naïve, nevertheless find an echo in contemporary physics.

Over the past two decades, something extraordinary has been happening in this field of science. Tools, methods, and ideas initially developed to understand how the blind material fabric of the universe behaves are now finding application in arenas for which they were never designed, and for which they might at first glance appear ridiculously inappropriate. Physics is finding its place in a science of society.

This book is about how that happened, why it is worth taking seriously, and where it might lead. It is also about the limits and caveats of a physics of society, whose potential for misapplication is considerable.

We have been here before. In the 1970s, the catastrophe theory of René Thom seemed to promise an understanding of how sudden changes in society might be provoked by small effects. This initiative atrophied rather quickly, since Thom's phenomenological and qualitative theory did not really offer fundamental explanations and mechanisms for the processes it described. Chaos theory, which matured in the 1980s, has so far proved rather more robust, supplying insights into how complicated and ever-changing ("dynamical") systems rapidly cease to be precisely predictable even if their initial states are known in great detail. Chaos theory has been advocated as a model for market economics, and its notion of stable dynamical states called attractors seemed to provide some explanation for why certain modes of social behavior or organization remain immune to small perturbations. But this theory has not delivered anything remotely resembling a science of society.

The current vogue is for the third of the three C's: complexity. The buzzwords here are *emergence* and *self-organization*, as complexity theory seeks to understand how order and stability arise from the interactions of many components according to a few simple rules.

The physics I shall discuss in this book is not unrelated to the idea of complexity—indeed, the two often overlap. But very often what passes today for "complexity science" is really something much older, dressed up in fashionable apparel. The main themes in complexity theory have been studied for more than a hundred years by physicists who evolved a tool kit of concepts and techniques to which complexity studies have added barely a handful of new items. At the root of this sort of physics is a phenomenon that immediately explains why the discipline may have something to say about society: it is a science of *collective behavior*. At face value it is not obvious why the bulk properties of insensate particles of matter should bear any relation to how humans behave en masse. Yet physicists have discovered that systems whose component parts have a capacity to act collectively often show recurrent features even though they might seem to have nothing at all in common with one another.

With that in mind, I hope to show that the new physics of society is able to accommodate just those characteristics of humankind that Petty felt it expedient to exclude: the "mutable Minds, Opinions, Appetites, and Passions of particular Men." I want to suggest that even with our woeful ignorance of why humans behave the way they do, it is possible to make some predictions about how they behave collectively. That is to say, we can make predictions about society even in the face of individual free will. It could even be argued that we can illuminate the *limits* of free will.

William Petty thought that quantification alone was enough to qualify his Political Arithmetick as a science. But his contemporary Thomas Hobbes had a rather deeper appreciation of what a science of society should be about. It must look beyond mere numbers, Hobbes implied, and grapple with the difficult question of mechanism. We must ask not just *how* things happen in society, but *why*. In the first part of this book we shall see where Hobbes's mechanistic approach and Petty's arithmetic led in attempts to understand society, and how—most curiously—they fed back into physics itself in the nineteenth century. We shall see how physics deals with systems of many components, all interacting with one another at once, and why regular and predictable behavior emerges in *statistical* form from such seeming chaos.

Treating people as though they were just so much insensate matter (or rather, *appearing* to do so) is a contentious business, which is why we shall approach the physics-based modeling of society with cautious steps, showing first why *life* (I am tempted to say "mere life") need not in itself present a boundary to the application of statistical physics. First the bacterium; later the world.

Yet you should not expect to find a "theory of society" expounded in this book. Indeed, the modern trend toward "unified theories," or grand, overarching frameworks in science, while having its uses, is arguably unhealthy. If there is such a thing as a physics of society, it does not come in the form of some universal equation into which we feed numbers and out of which emerges a deterministic description of social behavior. The case must be constructed by accumulation of examples, and the tools subtly adapted to each specific purpose. This survey is not by any means exhaustive; but we shall look at what physics has to say about how people move around in open spaces; how they make decisions, cast votes, and form alliances, groups, and companies. We shall see physics used to explain some aspects of the behavior of economic markets and to reveal the hidden structure in networks of social and business contacts. We shall uncover physics of a sort in the politics of conflict and cooperation.

Underlying all of this is a more difficult question: Does physics simply help us to explain and understand, or can we use it to anticipate and thereby avoid problems, to improve our societies, to make a better and safer world? Or is that merely another dream destined for the overflowing graveyard of utopias past?

ONE

RAISING LEVIATHAN

THE BRUTISH WORLD OF THOMAS HOBBES

A work on politics, on morals, a piece of criticism, even a manual on the art of public speaking would, other things being equal, be all the better for having been written by a geometrician.

—Bernard Fontenelle, secretary of the Académie Française,
late seventeenth century

"I perceive," says I, "the world has become so mechanical that I fear we shall quickly become ashamed of it; they will have the world be in large what a watch is in small, which is very regular, and depends only on the just disposing of the several parts of the movement. But pray tell me, madame, had you not formerly a more sublime idea of the universe?"   —Bernard Fontenelle (1686)

The most complete exposition of a social myth often comes when the myth itself is waning.   —Robert M. MacIver (1947)

It is no longer very useful to ask the question "Who governs Britain?" Discuss.
—Exercise in Stephen Cotgrove's *Science of Society* (1967)

Brothers will fight
and kill each other . . .
men will know misery . . .
an axe-age, a sword-age,
shields will be cloven,
a wind-age, a wolf-age,
before the world ends.[1]

This is how the Norsemen envisaged Ragnarok, the Twilight of the Gods; but in political exile in France in 1651, Thomas Hobbes must have thought that he had lived through it already. At Naseby and Marston Moor, Newbury and Edgehill, the stout yeomanry of England had hacked one another to bloody ruin. Oliver Cromwell reigned as Lord Protector of a country shocked to find itself a republic, its line of monarchical succession severed by the executioner's ax.

The combatants in the English Civil War, unlike those in France's Revolution or America's blood-soaked battle of North against South, had few clear ideological distinctions. The Royalists fought under the king's banner, but the Roundheads also claimed allegiance to "King and Parliament." For all his presumptuous arrogance, Charles I had no desire to live outside the constitution and laws of the land. Both sides were Anglican and wary of Papists. There were aristocrats in the Parliamentarian ranks and common folk among the Cavaliers. Many of those who slaughtered one another might have found little to dispute had they wielded words instead of swords.

Such a conflict could be nothing but a prescription for confusion once the beheading of the king brought it to an end. Embarrassed by the power with which fate had invested him, Cromwell searched in vain for a constitutional solution that would guarantee stability. Such was the might his troops, the formidable Ironsides, gave him that as Lord Protector he could experiment more freely than any British ruler before or since—although this was a freedom Cromwell would happily have relinquished had he felt able. Time and again he created parliaments on which to shed some burden of authority, only to dissolve them once he found them unworthy of the responsibility.

In the turmoil of those times, none could be certain that friends would not become foes, or old opponents emerge as allies. The Presbyterian Scottish Parliament, whose fierce antagonism to Charles I had precipitated the conflict between Parliament and Crown in the 1630s, was by 1653 fighting against Cromwell with Charles II as its figurehead. Cromwell himself expelled from the House of Commons the Parliament he had fought to instate, and struggled to maintain control of the monstrous army he had created. After Cromwell's death in 1658, this militia restored Parliament and longed for an end to the Protectorate. John Lambert led the troops to victory over a Royalist uprising in 1659 even as he plotted to restore Charles II to the throne (and conveniently make him brother-in-law to Lambert's daughter). Yet in the end it was by defeating Lambert that George Monk, a former Royalist, restored a Parliament in 1660 that he knew would crown the exiled king.

What could the common people possibly have craved more than stability? Twenty years of war and changing fortunes had convinced them that only a monarchy would supply it; and Charles II, who had narrowly escaped the tender mercies of the Ironsides just eight years previously, found a loyal army and a joyous population awaiting his return from France.

There is no way to understand the extraordinary quest on which Thomas Hobbes (1588–1679) embarked without acknowledging its historical setting. Centuries of monarchical rule over a hierarchical society had been graphically dismembered with the fall of the ax on January 30, 1649. A system of governance previously upheld by divine and moral imperatives had been revealed as arbitrary and contingent. Almost every political idea that was to follow in later centuries was voiced then, in seventeenth-century England, and many of them were put into practice. Soldiers and laborers became Levelers and Diggers, advocating socialist principles of equality and an end to individual ownership of land. Cromwell himself seems to have toyed with the notion of a freely elected democratic government, yet he spent much of his Protectorate heading a military dictatorship. Charles I had dissolved Parliament and had instigated an absolute monarchy in the years before the Civil War.

Which of these systems should a society adopt? The issue was a burning one. Although war between nations was regarded almost as a natural state of affairs, it might hardly pain the common person beyond the imposition of new taxes and levies. But internal strife was agonizing. The Civil War in England, conducted on the whole with restraint toward civilians, was bad enough; but the Thirty Years' War that ravaged Europe from the early part of the seventeenth century killed one in every three inhabitants of many German states. For Hobbes and many of his contemporaries, civil peace was worth almost any price.

England's miseries were a symptom of broader changes in the Western world. The feudal system of the Middle Ages was waning before the rise of a prosperous middle class, and from the ranks of this vigorous and ambitious sector came many of the Parliamentarians, who no longer felt obliged to submit to every royal whim. The monarchy, with its councillors and Star Chamber, harked back to the medievalism of Elizabethan society, but the spirit of the age cleaved to something more democratic, however limited in scope. The Reformation of Luther and Calvin had split Europe asunder; no longer did a single Church rule all of Christendom. The backlash to the assault on ecclesiastical tradition—prompted not only by Luther's heresy but by the humanism of the Renaissance—gave birth to the Counter-Reformation, the Council of Trent, the Jesuits, and a ruthless Inquisition. The greater the religious diversity, it seemed, the greater the intolerance.

Emerging from this ferment were ideas about the nature of the world that were ultimately to prove as challenging as any of the proclamations that Luther allegedly nailed to the church door in Wittenberg. Copernicus had been fortunate to develop his heliocentric theory—the idea that the Earth revolves around the sun—in the early sixteenth century, before the Counter-Reformation, and his first manuscript, circulated around 1530, even received papal sanction. But by 1543, when the full treatise was published after his death, it was prefaced with a note through which the editor, Andreas Osiander, hoped to evade ecclesiastical condemnation by indicating that the new view of the celestial bodies should be regarded simply as a convenient mathematical fiction. How Galileo fared against papal authority when he placed the same idea on firmer footing is the stuff of legend. The Inquisition condemned him in 1616 and forced him to recant in 1633. But by the middle of the seventeenth century, with René Descartes revitalizing the ancient Greek atomic theory and Isaac Newton soon to be admitted to Trinity College in Cambridge, the banishment of magic and superstition by mechanistic science seems in retrospect inevitable.

Hobbes's masterwork, *Leviathan*, was an attempt to develop a political theory out of this mechanical worldview. He set himself a goal that today sounds absurdly ambitious, although at the dawn of the Enlightenment it must have seemed a natural marriage. Hobbes wanted to deduce, by logic and reason no less rigorous than that used by Galileo to understand the laws of motion, how humankind should govern itself. Starting with what he believed to be irreducible and self-evident axioms, he aimed to develop a science of human interactions, politics, and society.

It is hard now to appreciate the magnitude not just of this challenge in itself but of the shift in outlook that it embodied. There has never been any shortage of views on the best means of governance and social organization. Almost without exception, proposals before Hobbes (and many subsequently) were designed to give the proposers the greatest (perceived) advantage. Emperors, kings, and queens sought to justify absolute monarchy by appealing to divine covenant. The Roman Catholic Church was hardly the first theocracy to set itself up as the sole conduit of God's authority. In Plato's *Republic*, one of the earliest of utopian models, cool and self-confident reasoning argued for a state in which philosophers were accorded the highest status. The rebellious English Parliament of the early 1640s demanded that the king transfer virtually all governing power to it. One could always find an argument to put oneself at the top of the heap.

Hobbes was different. What he aimed to do was to apply the method of the theoretical scientist: to stipulate fundamental first principles and see where they led him. In theory any conclusion was possible. By analyzing human nature and how people interact, he might conceivably have found that the most stable society was one based on what we would now call communism, or democracy, or fascism. In practice, Hobbes's reasoning led him toward the conclusion that he had probably preferred at the outset, from which we may reasonably suspect that his method was not as objective as he would have had the world believe. Nonetheless, its claim to have dispensed with bias and to rely only on indisputable logic is what makes *Leviathan* a landmark in the history of political theory.

But it is something more. Hobbes's great work is seen today as historically and even philosophically important, but political science has become a very different beast, and no one seriously entertains the notion that Hobbes's arguments remain convincing. Nor should they, in one sense—for as we shall see, his basic postulates are very much a product of their time. Yet *Leviathan* is a direct and in many ways an astonishingly prescient antecedent to a revolutionary development now taking place at the forefront of modern physics. Scientists are beginning to realize that the theoretical framework which underpins contemporary physics can be adapted to describe social structures and behavior, ranging from how traffic flows to how the economy fluctuates and how businesses are organized.

This framework is not as daunting as it might sound. Contrary to what one might imagine from the popular perception of modern physics, we do not have to delve into the imponderable paradoxes of quantum theory, or the mind-stretching revelations of relativity, or the origins of the universe in the Big Bang, in order to understand the basic ideas involved. No, this is an approach rooted in the behavior of everyday substances and objects: water, sand, magnets, crystals. But what can such things possibly have to say about the way societies organize themselves? A great deal, as it happens.

Hobbes had no inkling of any of this, but he shared the faith of today's physicists that human behavior is not after all so complex that it cannot occasionally be understood on the basis of just a few simple postulates, or by the operation of what we might regard as *natural forces*. For Hobbes, contemplating the tumultuous political landscape of his country, the prime force could not be more plain: the lust for power.

THE LEVIATHAN WAKES

Thomas Hobbes (Figure 1.1) had never been able to take anything for granted. His father was a poorly educated and irascible vicar, a drunkard who left his family when Thomas was sixteen and died "in obscurity." This put his son to little inconvenience, since from a young age Thomas was supported and encouraged by his wealthy and altogether more respectable uncle Francis, a glover and alderman of Malmesbury. Francis watched over the boy's education, helping to nurture a clearly prodigious intellect: by the time the fourteen-year-old Thomas won admittance to Magdalen College at Oxford, he had already translated Euripedes' *Medea* from Greek to Latin. He so excelled at the university that when he graduated, he was recommended to the Earl of Devonshire as a tutor to the earl's son (who was only three years younger than Thomas). From such a position Hobbes was free to continue his studies of the classics. In his early twenties he acted as secretary to Francis Bacon (1561–1626), whose interests ranged from natural science and philosophy to politics and ethics. During this time until Bacon's death, Hobbes showed no evident inclinations toward science; but Bacon's rational turn of thought left a clear imprint on his thinking.

It was not until 1629 that the forty-year-old Hobbes, a committed classicist, had his eyes opened to the power of scientific and mathematical reasoning. The story goes that Hobbes happened to glance at a book that lay open in a library, and was transfixed. The book was Euclid's *Elements of*

1.1   Thomas Hobbes was the first to seek a physics of society.

Geometry, and Hobbes began to follow one of the Propositions. "By God, this is impossible!" he exclaimed, but was soon persuaded otherwise. As Hobbes's contemporary, the gossipy biographer John Aubrey, tells it,

> So he reads the Demonstration of it, which referred him back to such a Proposition; which proposition he read: that referred him back to another, which he also read, and *sic deinceps* [so on], that at last he was demonstratively convinced of that trueth. This made him in love with Geometry.[2]

Hobbes was deeply impressed by how this kind of *deductive reasoning*, beginning with elementary propositions, allowed geometers to reach ineluctable conclusions with which all honest and percipient people would be compelled to agree. It was a prescription for certainty.

The axioms of geometry are, by and large, statements that few people would have trouble supposing. They assert such things as "Two straight lines cannot enclose a space." We can typically convince ourselves of their validity with simple sketches. Other fields of inquiry struggle to muster analogous self-evident starting points. "I think, therefore I am" may have convinced Descartes that, as an axiom, it is "so solid and so certain that all the most extravagant suppositions of the skeptics were incapable of upsetting it"; but in fact every word of the sentence is open to debate, and it has none of the compelling visual power of geometry's first principles.

Hobbes was sufficiently enthused to attempt geometry himself, but he was never a master of the subject. Through clumsy errors he persuaded himself that he had solved the old geometric conundrum of "squaring the circle" (something that was later shown to be impossible). But that was not his principal concern. In the 1630s, the tensions between Crown and Commons led Charles I to dissolve Parliament and embark on an eleven-year period of "personal rule." In the midst of an unstable society, Hobbes wanted to find a theory of governance with credentials as unimpeachable as those of Euclid's geometry.

This meant that he needed some fundamental hypothesis about human behavior, which in turn had to be grounded in the deepest soil of science. And there was one man who had dug deeper than any other. In the spring of 1636, Hobbes traveled to Florence to meet Galileo.

The fundamental laws describing how objects move in space are called Newton's laws, since it was Sir Isaac who first formulated them clearly in his *Principia Mathematica* (1687). But the tallest giant from whose shoulders Newton saw afar was Galileo (1564–1642), who laid the foundations of modern mechanics. Galileo taught the world about falling bodies, which, he said, accelerate constantly as they descend (if one ignores the effects of air resistance). And with his law of inertia, Galileo went beyond the "common sense" view of Aristotle that objects must be continually pushed if they are not to slow down: on the contrary, said Galileo, in the absence of any force an object will continue to move indefinitely in a straight line at constant velocity.

Aristotle's view is the commonsense one because it is what we experience in everyday life. If you stop pedaling your bicycle, you will eventually come to a standstill. But Galileo realized that this is because frictional forces act in nature to slow us down. If we can eliminate all the forces acting on a moving body, including gravity and friction, the natural state of the body is motion in an unchanging direction at unchanging speed. This was a truly profound theory, for it saw beyond the practical limitations of Galileo's age to a beautiful and simple truth. (An air pump that could create a good vacuum and thus eliminate air resistance was not invented until 1654.)

Galileo's law of inertia is without doubt one of the deepest laws of nature.

On meeting the great man, Hobbes became convinced that this must be the axiom he was seeking. Constant motion was the natural state of all things, including people. All human sensations and emotions, he concluded, were the result of motion. From this basic principle Hobbes would work upward to a theory of society.

What, precisely, does Hobbes mean by this assumption? It is, to modern eyes, a cold and soulless (not to mention an obscure) description of human nature. He pictured a person as a sophisticated mechanism acted on by external forces. This machine consists of not only the body, with its nerves, muscles, and sense organs, but also the mind, with its imagination, memory, and reason. The mind is purely a kind of calculating machine—a computer, if you will. Such machines were popular in the seventeenth century: the Scottish mathematician John Napier (1550–1617) devised one, as did the French philosopher and mathematician Blaise Pascal (1623–1662). They were mechanical devices for adding and subtracting numbers; and this, said Hobbes, is all the mind does too: "When a man *Reasoneth*, hee does nothing else but conceive a summe totall, from *Addition* of parcels; or conceive a Remainder, from *Subtraction* of one summe from another . . . For REASON . . . is nothing but *Reckoning*."[3] The body, meanwhile, is merely a system of jointed limbs moved by the strings and pulleys of muscles and nerves. Man is an automaton.

Indeed, Hobbes held that the ingenious mechanical automata created by some inventors of the era were truly possessed of a kind of primitive life. To him there was nothing mysterious or upsetting about such an idea. Others were less sanguine: the Spanish Inquisition imprisoned some makers of automata on the grounds that they were dabbling in witchcraft and black magic.

What impelled Hobbes's mechanical people into action was not just external stimuli relayed to the brain by the apparatus of the senses. They were imbued also with an inner compulsion to remain in motion. For what is death but immobility, and which person does not seek to avoid death? "Every man . . ." said Hobbes, "shuns . . . death, and this he doth, by a certain impulsion of nature, no less than that whereby a stone moves downward."[4]

Mankind's volitions, therefore, are divided by Hobbes into "appetites" and "aversions": the desire to seek ways to continue this motion and to avoid things that obstruct it. Some appetites are innate, such as hunger; others are learned through experience. To decide on a course of action, we weigh the relevant appetites and aversions and act accordingly.

What Hobbes means by "motion" is a little vague, for he clearly does not intend to imply that we are forever seeking to run around at full tilt. Motion is rather a kind of liberty—a freedom to move at will. Those things that impede liberty impede motion. Even if a man sits still, the mechanism of his mind may be in furious motion: the freedom to think is an innate desire too.

What room is there in this mechanical description for free will? According to Hobbes, there is none—he was a strict determinist. Humans are puppets whose strings are pulled by the forces at play in the world. Yet Hobbes saw nothing intolerable in this bleak picture. After all, he believed that he had arrived at this basic, indisputable postulate about human nature by *introspection*, by considering his own nature. The first puppet he saw was himself: "Whosoever looketh into himself, and considereth what he doth, when he does *think, opine, reason, hope, feare,* &c, and upon what grounds; he shall thereby read and know, what are the thoughts, and Passions of all other men, upon the like occasions."[5]

THE MECHANISTIC PHILOSOPHY

If we shudder at this concept of humanity today, it is partly because we regard mechanical, clockwork devices as crude and clumsy. There are now many materialist scientists and philosophers who believe that the brain is a kind of vast and squishy computer whose secrets reside in nothing more than the extreme interconnectedness of its billions of biological switches. This view of the brain as a superior version of our most advanced cultural artifact is neither unusual nor eccentric.

To the intellectuals of the seventeenth century the same was true of the clock, which as a reliable timekeeper was still a recent innovation. In that age there was nothing inelegant about a mechanical picture of humanity; on the contrary, it showed just how wonderfully wrought people were. Descartes said:

> And as a clock, composed of wheels and counterweights, observes not the laws of nature when it is ill made, and points out the hours incorrectly, than when it satisfies

the desire of the maker in every respect; so likewise if the body of man be considered as a kind of machine, so made up and composed of bones, nerves, muscles, veins, blood, and skin, that although there were in it no mind, it would still exhibit the same motions which it at present manifests voluntarily.[6]

As above, so below. If mankind was a clockwork mechanism, so too was the universe. The planets and stars revolved like the gears of a clock, contrived by God the cosmic clockmaker. This set in train the debate about whether or not God's skill had left him any cause to intervene in the world once it was "wound," which culminated in an intemperate argument between Gottfried Leibniz and Isaac Newton (who seldom argued temperately).

And if the universe was a clockwork mechanism, the way to understand it was to take it apart, piece by piece: to apply the reductionist methodology of science. It was precisely this approach that Hobbes chose to use to analyze the workings of society; he would resolve it into its constituent parts and descry in their motions the simple causative forces. This was his intention in *Leviathan*'s precursor *De Cive* (On the Citizen), published in 1642, which contained many of the same ideas: "For everything is best understood by its constitutive causes. For as in a watch, or some such small engine, the matter, figure, and motion of the wheels cannot well be known except it be taken asunder and viewed in parts."[7]

By this time Hobbes had joined other Royalist sympathizers in exile in Paris. He sensed what was in the air in England in 1640, when Charles I had been forced to reconvene Parliament in order to gather taxes to finance the suppression of rebellion in Scotland. So antiroyalist was the new "Short" Parliament, which had smoldered in banished discontent for eleven years, that the king rapidly dissolved it again, only to have to resurrect it once more when the Scottish army reached Durham on its march south. From there it was a downhill slide to the outbreak of civil war in 1642. Fearing that his political writings would draw censure (or worse) from the belligerent Short Parliament of 1640, Hobbes left for France.

Thus, although Hobbes had formulated most of his ideas on "civil governments and the duties of subjects" before the war began, its impending prospect lent his efforts some urgency. He had originally intended to write a three-part thesis that began with traditional physics, extended these ideas to the nature of humankind, and only subsequently developed a "scientific" theory of government. But as he later explained, *De Cive* was hastened by circumstances: "My country some years before the civil war did rage, was so boiling hot with questions concerning the rights of dominion and the obedience due from subjects, the true forerunners of an approaching war; and this was the cause which ripened and plucked from me this third part."[8]

In France, Hobbes joined the circle of mechanistic French philosophers whose acquaintance he had made during his earlier European trip in 1634–1637. Among them were Marin Mersenne (1588–1648) and Pierre Gassendi (1592–1655), colleagues of Descartes and two of the most enthusiastic supporters of the mechanical worldview. In this supportive environment Hobbes refined his theory of human nature and carried it through to deduce the consequences for civic structure. *Leviathan* was published in 1651, and Hobbes presented it in exile to the fugitive Charles II, to whom he had once taught mathematics. There was to be no one, Royalist or Roundhead, who was pleased by what it said.

THE UTOPIANS

Hobbes was not the first to imagine a utopia based on scientific reasoning. The governing philosophers of Plato's Republic live simply and own no private property, but they have absolute power over the lower classes of soldiers and common people, with whom Plato is little concerned. His is a utopia for aristocrats only; the mob might as well be living in a totalitarian, if benevolent, state.

But the word *utopia* comes from the imaginary land devised by the scholar and lawyer Thomas More (1478–1535). In More's book *Utopia* (1516), a sailor named Raphael Hythloday describes the eponymous island where he dwelt for five years after sailing there by chance. The meaning of the name is debated, but the common interpretation renders it as either "good place" or "no place."

In More's Utopia everything is ideal. There is no ownership: everyone lives in identical houses, but the houses are exchanged every ten years to dispel any notion that individuals own their homes. All people of the same sex are dressed alike, and their clothing is simple and immune to fashion. Everyone works—enough but not too much—and they are offered noncompulsory educational lectures. All of the many religions are tolerated, and people live moderately and modestly. It is a vision on the one hand refreshingly liberal, equal, and just, and on the other terrifyingly bland and spiritless.

When Francis Bacon drew up his own version of the perfect society, he made science its linchpin. *New Atlantis* was a book he never finished; it was published, incomplete, the year after his death. The title harks back to Plato, who mentions the fabled lost civilization several times in his dialogues. But Bacon employs the same conceit as More: European sailors are driven off course in the Pacific Ocean and find themselves at the previously unknown island of Bensalem (a Hebrew word meaning "son of peace," although the implication is that this is the "New Jerusalem"). It is a Christian society that dwells on Bensalem, welcoming, kind, and compassionate but also fiercely patriarchal and hierarchical. Central to the culture of Bensalem is Salomon's House, an institution devoted to science and the acquisition of new knowledge. The scientists (Fathers) dress and act rather like priests, and have access to vast resources for research. There are laboratories where nature is not only examined but also imitated and manipulated. Artificial environments resembling mines reproduce the conditions in which metals and minerals are formed; new living species are devised and created. "Neither do we this by chance," a Father explains, "but we know beforehand of what matter and commixture what kind of those creatures will arise."[9]

Salomon's House resembles in many ways a modern research institution, albeit one unfettered by any constraints on research ethics. Some might see in it the blueprint for biotechnological laboratories in which the stuff of life is cut up, spliced, and reconstituted. The Fathers take an oath of secrecy and reveal their inventions only if it suits them. One cannot imagine Bacon having much difficulty with the modern concept of gene patenting by private companies.

But Bacon's Bensalem is essentially an arbitrary society—a vision of what its author considered desirable, and one devoted to, rather than derived from, scientific principles. This is why Hobbes's *Leviathan* is original. He does not describe a society ready-made and shaped by his own preferences, but builds it up, with careful logic, from his mechanistic view of how humans behave.

We should take care with what we mean by that. Hobbes was not especially interested in psychology, or in deducing how people will respond to a particular set of circumstances. He was pursuing a moral philosophy by asking whether a course of behavior is *right*. In this respect, the ground was prepared for him (and characteristically for the times, he does not acknowledge it) by the Dutch philosopher Hugo Grotius (1583–1645), whose *Laws of War and Peace* (1625) attempted to find the irreducible characteristics of human social existence. Grotius was not looking for scientific or mathematical laws as we would now understand them, but for "natural laws," which again might be better regarded as natural rights. With ruthless efficiency, Grotius stripped society of its more pleasant features—benevolence, he said, is all very well, but it is not fundamental. There are only two things that people have a natural right to exercise in the company of their fellows: an expectation that they will not be subjected to unwarranted attack, and the freedom to defend themselves if they are. So long as people confine themselves to self-preservation and refrain from injuring others without cause, society is possible. This, said Grotius, is the "State of Nature," the most basic state of social existence. Civilization generally does rather better than that, encouraging courtesies and friendships and learning and the arts and so forth, but these are all optional extras, and society as such can exist without them.

Thus Grotius's "minimal society" was a grim affair, and his concept of natural rights was not, as we might suppose today, a precondition of liberalism. But it was not at all obvious how even such a brusque, unfriendly society might be maintained. For who was to say when aggression was warranted and when it was not? If food is short, are you justified in killing your neighbor to preserve yourself? Are you justified in doing that preemptively, as an insurance policy against possible famine next year? Even if everyone agrees to recognize their fellows' natural rights, social stability doesn't necessarily follow, because there is no consensus about how to exercise them.

In the hierarchical societies of medieval Europe this seldom became a problem, because people were accustomed to the idea that they should do as they were told by their superiors. They might resent this inequality, but it was rarely questioned. By the Renaissance those certainties had broken down—partly because of changes to the structure of society, partly because of religious unrest and the Reformation, partly because humanism had exposed people to new ways of thinking and there was more awareness of the sheer diversity of societies past and present. Society suddenly seemed to lack foundational principles or agreed behavioral norms.

Hobbes realized that this relativism of opinions about how to exercise natural rights meant that in the end a "state of nature" was all about one thing: power.

HOW TO BUILD A COMMONWEALTH

The person without liberty is without power. Even the most humble and self-effacing of us wants a little power—to choose when we eat and sleep, where we live and with whom, what we may say and do. Many millions of people in the world lack some or all of these freedoms, but they are among those acknowledged internationally, in the Universal Declaration of Human Rights, as liberties that everyone deserves simply by virtue of being alive.

Hobbes defined power as the ability to secure well-being or personal advantage, "to obtain some future apparent Good." People, he said, have some "Naturall Power" that enables them to do this, stemming from innate qualities such as strength, eloquence, and prudence. And they may use these qualities to acquire "Instrumentall Power," which is merely "means and Instruments to acquire more": wealth, reputation, influential friends.

So Hobbes's model of society hinged on the assumption that people (if we say "men" we are not, in this context, being inaccurate) seek to accumulate power, up to a personal level of satiety that varies among individuals.

It is a cold-blooded prescription, for sure. The Scottish political scientist Robert MacIver has complained that it neglects all that is good and worthy in man: "Hobbes ignored all the social bonds that spread out from the life of the family, all the traditions and indoctrinations that hold groups of men together, all the customs and innumerable adjustments that reveal the socializing tendencies of human nature."[10]

Doubtless that is so, and we may want to make the same complaint. The social historian Lewis Mumford condemns this kind of abstraction of society, saying that it reduces the individual to "an atom of power, ruthlessly seeking whatever power can command."[11] It has to be admitted that this is precisely what Hobbes intended.

Yet even the nineteenth-century Romantic Ralph Waldo Emerson seems to agree with the Hobbesian interpretation of human nature when he says "Life is a search after power." And in any event, we can agree or disagree with Hobbes's wolfish view of humanity while nevertheless phrasing the valid question: Given these postulates, what follows? *If* men behave this way, what kind of society can arise and be maintained?

Power is relative: the true measure is the amount by which one man's power *exceeds* that of the others around him. It follows, Hobbes said, that the quest for power is in fact a quest for command over the powers of other men. But how does one command the power of another? In the bourgeois market society that had come to dominate the cultural landscape of the mid–seventeenth century, the answer was simple: he buys it. One man pays another to act on his behalf and to submit to his will.

This does not necessarily mean, as it might sound, simply that a powerful man may hire others to act as bullies, henchmen, and mercenaries. Rather, Hobbes had in mind the way a rich merchant employs workers to make and distribute his goods, or a craftsman takes on assistants to execute a contract. Yet his formulation is as icy as his model of man as machine: "The Value, or WORTH of a man, is as of all other things, his Price; that is to say, so much as would be given for the use of his Power."[12] It is the ethic of the free market: buy out the competition.

It is not obvious that a society in which appetites for power vary need in itself be an unsettled one: those with moderate ambitions might be happy enough to work for those with stronger desires. But Hobbes maintained that some men's appetites know no limits. Such power-hungry individuals destabilize a society in which less ambitious men might otherwise labor in harmony. "I put for a generall inclination of all mankind," he said,

> a perpetuall and restlesse desire of Power after power, that ceaseth only in Death. And the cause of this, is not alwayes that a man hopes for a more intensive delight, than he has already attained to; or that he cannot be content with a moderate power: but because he cannot assure the power and means to live well, which he hath present, without the acquisition of more.[13]

And so all are sucked into a perpetual power struggle. Unchecked, this leads to Hobbes's own vision of a State of Nature, in comparison with which Grotius's version—a crabbed, surly society—might sound positively idyllic. It is as bleak and frightening as you can get.

Without any law or law enforcers, every man is open to violent exploitation by others. When everyone seeks to dominate his neighbor without restraint, says Hobbes, there is "no place for Industry . . . no Culture of the Earth . . . no Knowledge of the face of the Earth . . . no Arts; no Letters; no Society; and which is worst of all, continuall feare, and danger of violent death; And the life of man, solitary, poore, nasty, brutish, and short."[14]

Who would not do all he could to escape such a state? But to proceed logically to a better way, Hobbes found it necessary to introduce two more postulates, which he elevated to the status of Laws of Nature. The first says that a man will not seek actively to harm himself or endanger his life, or to overlook ways of making it safer. Reasonable enough at first glance, this in fact accords us extraordinary percipience in seeing the consequences of the actions we choose, so that we will always make the choice most favorable to our self-preservation. But the second law is still more debatable: "That a man be willing, when others are so too, as farre-forth, as for Peace, and defence of himselfe he shall think it necessary, to lay down this right to all things; and be contented with so much liberty against other men, as he would allow other men against himselfe."[15]

In other words, men will, as a corollary of their instinct for self-preservation, be prepared to suppress their exploitative impulses and cooperate with one another. Only thus can peace and stability be brought to the State of Nature.

But cooperation is not enough. For men's unceasing appetite for power will make them liable to defect from this contract the moment they see any advantage in doing so. We shall see later that Hobbes here essentially formulates, three hundred years ahead of time, one of the most influential behavioral dilemmas of the modern era.

The solution, he reasons, is for men not simply to give up some of their natural rights to do as they please, but to *transfer* these rights to some authority that is then granted the mandate to impose the contract—by force if necessary.

In whom should this authority reside? Hobbes felt that it did not greatly matter, so long as the authority was there. His fundamental postulates assume a degree of equality among men rarely voiced in seventeenth-century Europe: in the State of Nature, no man's status is greater than another's, although some have the advantage of greater "Naturall Power." But then the community *elects* some individual and confers on him absolute power. In effect they choose a monarch and thenceforth defer to him or her without question.

This resolution is a peculiar mixture, for it amounts to the creation of a despotism by democratic means out of an anarchic state. Hobbes admits that the supreme authority could conceivably be an elected body, not an individual—a Parliament, in effect. But he suspects (and who can dispute it?) that with more than one head of state, internal power conflicts will arise sooner or later.

The powers of Hobbes's elected monarch are absolute, stopping only at the right of individuals always to preserve their own lives. It is up to the sovereign, once elected, to decide how much of each man's power he must enlist to maintain the social contract. Even to a tyranny, says Hobbes, citizens owe an obligation of duty and submission.

At the same time, this absolutism unites people into a cohesive unit, a Commonwealth: the Leviathan. It was a curious name to give to a supposedly desirable state of society—almost as though Hobbes positively wanted his readers to envisage a dreadful, oppressive regime. Leviathan is a fearsome sea creature mentioned in the Book of Job:

> If you lay a hand on him,
> You will remember the struggle and never do it again!
> Any hope of subduing him is false;
> The mere sight of him is overpowering . . .
> When he rises up, the mighty are terrified;
> They retreat before his thrashing . . .
> Nothing on earth is his equal—
> A creature without fear.
> He looks down on all that are haughty;
> He is king over all that are proud.[16]

The message is plain—you disobey Leviathan's laws at your peril.

Yet because it has freely elected to be governed this way, the population in some sense shares in the political structure that results. Leviathan is thus "one person, of whose acts a great multitude . . . have made themselves every one the author"[17]—an image reinforced by the dramatic frontispiece to the first edition of the book, probably prepared by the artist Wenceslas Hollar (Figure 1.2). In personifying the State in this way, Hobbes was following a long tradition: in the fourteenth century, the Bishop of Rochester, Thomas Brinton, identified the prince as the head of the "body politic" and the laborers as the feet. Others took delight in anatomizing every member of society, from priests (chest or ears) to merchants (thighs) to judges (ribs).

The justification for the Leviathan, says Hobbes, is "the Convenience, or Aptitude to Produce the Peace, and Security of the People." One can deplore his proposed means of achieving these aims, but the objectives themselves are nevertheless enshrined in all democracies today. In explaining how a mass of selfish individuals can unite to create a sovereign nation, Hobbes gave form to the modern idea of the State. More than this, even: according to historian Frederick Nussbaum, "Hobbes discovered society."[18]

And thus Thomas Hobbes believed he had proved monarchy to be the best system of rule, using science and reason alone. He felt that those nations that had enjoyed prolonged civic stability, such as Imperial Rome, had by good luck or judgment hit on the ideal solution that science now revealed with inexorable logic. "The skill of making, and maintaining Commonwealths," he said, "consisteth in certain Rules, as doth Arithmetique and Geometry; not (as Tennis-play) on Practise onely."[19]

1.2 Hobbes's Leviathan is a sovereign who makes society cohere into a Commonwealth—by rule of the sword, if necessary. "The only way," says Hobbes, "to erect such a Common Power, as may be able to defend [people] from the invasion of Forraigners, and the injuries of one another . . . is, to conferre all their power and strength upon one Man . . . that may reduce all their Wills, by plurality of voices, unto one Will . . . This done, the Multitude so united in one Person, is called a COMMONWEALTH, in latine CIVITAS. This is the Generation of that great LEVIATHAN, or rather (to speak more reverently) of that *Mortall God*, to which wee owe under the *Immortall God*, our peace and defence."

A CALCULUS OF SOCIETY

One might think that Charles II would have been pleased with a treatise claiming to prove scientifically that kings were the best rulers. But he was not at all happy with *Leviathan*, for it proposed that the king comes from the ranks of ordinary men and is instated arbitrarily by election of the masses, like a common parliament! Whereas it was well known that kings ruled by divine decree, deriving their authority not from some social contract but from a heavenly one. To the Royalists, the book was pure treason.

There was no comfort here for supporters of the parliamentary system either. Hobbes's supreme authority, whether an individual or a collective body, subsequently had the right to decide who would succeed it—democracy is exercised once and then relinquished. And to make matters worse, *Leviathan* offended the devout by lambasting those nations which "acquiesce in the great Mysteries of Christian Religion, which are above Reason."[20] This was deemed by many to be a declaration of atheism. Hobbes endeared himself to no one.

So it was a dangerous game that Hobbes now played. In the winter of 1651–1652, shortly after his book appeared, he retreated from the hostility of the exiled Royalists and returned to Cromwell's England, where the desire for peace and stability under the Protectorate had introduced a degree of tolerance. Hobbes made friends within the new regime, and he fitted in quietly enough until Charles II was restored to the throne in 1660. If there was one thing the Royalists, new and old, disliked more than Hobbes's political philosophy, it was his views on religion. He had become widely regarded as an atheist, especially by the dominant Anglican Royalists, and he might well have faced imprisonment if the bill to make Christian heresy a criminal offense had been passed by Parliament in 1666. The threat was ever present for the remainder of Hobbes's lifetime; but in spite of this, and decades of ill health notwithstanding, he survived to the truly venerable age of ninety-one.

No nation chose to put the advice in *Leviathan* into practice. Indeed, according to the historian Richard Olson, "because they seemed to inspire both immorality and revolution, Hobbes's theories were generally feared and detested by all respectable persons."[21] To the Scottish philosopher David Hume, "Hobbes's politics are fitted only to promote tyranny, and his ethics to encourage licentiousness."[22] But because his ideas were argued with such compelling force and precision, they posed a challenge to all subsequent political philosophers. You could be appalled by Hobbes, but you could not ignore him.

Above all, *Leviathan* established the idea that there was room for reason in politics. Previous utopias were not deductive; their validity was simply asserted. In general they sought either to shore up the status quo or to portray a society conjured into existence from the author's imagination, with no explanation of how things got to be that way. The Leviathan, on the other hand, was at least ostensibly the product of mechanistic science. It was not something to be celebrated, but was a necessary evil, the only alternative to grim anarchy.

The social contract proposed by Hobbes might sound like a forerunner of those advocated by John Locke (1623–1704) and Jean-Jacques Rousseau (1712–1778), but it is instead the reverse. To Locke and Rousseau, the power conferred upon the head of state comes with an obligation to serve the interests of the populace; for Hobbes, the common people are contracted to serve their ruler. For Hobbes, the principal fear was of anarchy; for Locke it was the abuse of power, which is why he saw the need for safeguards to avoid absolutism.

But although apparently a proponent of autocracy, Hobbes also provides arguments that can be used to support both bourgeois capitalism and liberalism. Although he expressed an aversion to the way the mercantile society bred men whose "only glory [is] to grow excessively rich by the wisdom of buying and selling," which they do "by making poor people sell their labour to them at their own prices,"[23] he saw bourgeois culture as largely inevitable, and sought a system that would accommodate its selfish tendencies without conflict. To this end he left it to the market to assign the value of everything, people included: "The value of all things contracted for, is measured by the Appetite of the Contractors: and therefore the just value, is that which they be contracted to give."[24] This free-market philosophy found voice in Adam Smith's *Wealth of Nations* in the following century. Those in Britain and the United States (and indeed elsewhere) who lived through the 1980s will recognize it as an attitude that did not wane with the Age of Enlightenment.

A political scientist taking a chronological approach would chart the trajectory of Hobbes's thought via Locke to later thinkers who believed there could be such a thing as a "calculus of society." Along this path we would uncover Jeremy Bentham's utilitarianism in the late eighteenth century, an attempt to harmonize the individual's pursuit of personal happiness with the interests of society. Bentham, like Locke, believed that reason alone could show how this might be achieved. His solution was the "greatest happiness" principle, which would guide society to an optimal state in which the sum total of human happiness was as large as it could possibly be, given the conflicts of interest that inevitably arise when each person seeks his or her own advantage. Bentham's utopia was quite different from Hobbes's: a democracy with equality for all, including votes for women. Bentham and the Philosophical Radicals, who included John Stuart Mill, paved the way for the socialism of Karl Marx. Marx, of course, was also determined to formulate a "scientific" political theory, one that in his case was strongly (and misguidedly) influenced by Darwinism.

And so we might go on. But I shall not. These theories indeed seek a foundation in rationality, and we shall revisit them from time to time. But they are not scientific in the way that the real topic of this book is scientific. There are few political thinkers who have defined a social model with the logical precision of Hobbes, and none who have carried those precepts through to their conclusions in a truly scientific, rather than suppositional way. This is not by any means to denigrate such models; rather, it is simply to say that their approach is different. Political theorists tend to concern themselves with what they think *ought* to be; scientists concentrate on the way things *are*. The same is true of the new physics of society: it seeks to find descriptions of observed social phenomena and to understand how they might arise from simple assumptions. Equipped with such models, one can then ask what we would need to do in order to obtain a different result instead. Decisions about what is *desirable* should properly be in the realm of public debate—they are no longer scientific questions. In this sense, the science becomes, as it should be, a servant and guide, and not a dictator.

How is it that physics has come to have the confidence, perhaps even the arrogance, to venture into social science? No one in recent decades has set out to construct a physics that would be capable of this. It just so happens that physicists have realized they have at their disposal tools that can be applied to this new task. These tools were not developed for that purpose; they were first developed to understand atoms.

Carolyn Merchant, in her book *The Death of Nature* (1983), argues that the rise of mechanistic, atomistic philosophy in the seventeenth century sanctioned the manipulations and violations of nature that continue to blight the world today. The utopian society envisaged by Thomas Hobbes, in which people are little more than automata impelled this way and that by mechanical forces and where scientific reasoning is the arbiter of social justice, sounds like a chilling place to live. It is hard to imagine how any model of society that regards the behavior of individuals as governed by rigid mathematical rules can offer us a vision of a better way to live, rather than a nightmarish Brave New World.

That, I suspect, is the instinctive objection that many will have to the notion of a "physics of society." But I hope to show that the new incursion of physics into the social, political, and economic sciences is not like this. It is not an attempt to prescribe systems of control and governance, still less to bolster with scientific reasoning prejudices about how society ought to be run. Neither does it really imagine that people are so many soulless, homogeneous effigies to be shuffled this way and that according to blind mathematics. Instead, what physicists are now trying to do is to gain some understanding of how patterns of behavior emerge—and patterns undoubtedly *do* emerge—from the statistical melée of many individuals doing their own idiosyncratic thing: helping or swindling one another, cooperating or conflicting, following the crowd or blazing their own trail. By gaining such knowledge, we might hope to adapt our social structures to the way things *are* rather than the way some architect or politician or town planner thinks they ought to be. We might identify modes of organization that fit with the way we actually and instinctively behave.

These are potential practical benefits of a genuinely inquisitive physics of society, but from such efforts emerges a broader message too. The message is this: collective actions and effects are inevitable. No matter how individualistic we like to think we are, our deeds are often the invisible details of a larger picture. This is not necessarily a description of impotence. Environmentalists and other activists like to entreat us to "think globally, act locally." But the physics of society shows that the reverse can take effect too: by concerning ourselves with nothing more than how we interact with our immediate neighbors, by "thinking locally," we can collectively acquire a coherent, global influence. The consequences of that—good or bad—are worth knowing.

No scientific theory will show us how to build a utopia, but the search for a physics of society will benefit from our acknowledging the lessons of those quixotic attempts, like that of Thomas Hobbes, to do so in the past. These efforts to create a rational utopia show us the dangers of such a rigid program. Science provides not prescriptions but descriptions. With such understanding, we might hope to make our choices with clearer vision.

TWO

LESSER FORCES

THE MECHANICAL PHILOSOPHY OF MATTER

> Nature, it seems, is the popular game
> for milliards and milliards and milliards
> of particles playing their infinite game
> of billiards and billiards and billiards.
> —Piet Hein (1966)

> The Boltzmann is magnificent. I have almost finished it. He is a masterly expounder. I am firmly convinced that the principles of the theory are right, which means that I am convinced that in the case of gases we are really dealing with discrete point masses of definite size, which are moving according to certain conditions . . . This is a step forward in the dynamical explanation of physical phenomena.
> —Albert Einstein (1900)

> I have endeavoured to show that it is the peculiar function of physical science to lead us to the confines of the incomprehensible, and to bid us behold it and receive it in faith, till such time as the mystery shall open.   —James Clerk Maxwell (1856)

L aws make life simpler, and that can be a liberating thing. Immanuel Kant realized this when he said, "Man is free if he needs to obey no person but solely the laws."[1]

It is not a trivial matter that science has come to use legal terminology to describe regularities in nature. "I'm arresting you for breaking the laws of physics," says the policeman to the levitating man in a cartoon. Like many good jokes, this one reveals the snares that language sets. We can break society's laws if we dare; but the laws of physics do not need enforcing, for they are inviolable.

If the Enlightenment enthusiasm for a mechanistic philosophy looks naïve to us now, let us not forget what it offered. Such "natural laws" as Aristotle divined were hardly simplifications; often they were mere tautologies. Objects fell to earth because they had a downward tendency. The sun and moon followed their arcs across the sky because heavenly bodies had a circulating tendency. In contrast, Newton's law of gravity rationalized both why cannonballs fall and why the moon does not. It condensed pages of astronomical data into a concise, simple formula. It helped to fit disparate observations into a single framework. And beyond all this, it suggested that humankind can understand, and not just experience, the hows and whys of existence.

The mechanical laws of Galileo and Newton hold true for planetary orbits and for motes of dust, for a falling apple and a falling star. They are deep and elegant truths, so far as truth can ever be discerned, about the way the universe works. Maybe we can therefore forgive Hobbes and his contemporaries their propensity to use mechanics to explain everything—even the mysteries of the human mind. But in the two centuries that followed the publication of *Leviathan*, delight in mechanics did not diminish. On the contrary, scientists saw ever more reason to believe that they had grasped the central governing principles of all matter and that explanations for all phenomena simply required the right mechanical description.

It is this account of matter at the fundamental level, hatched in the nineteenth century, that underpins the physics of society. In this chapter we shall see where it came from and what it consists of. It is a theory that invokes many players, and each of them is too small to see. That the whole world can be constructed from a small variety of atoms is an astonishing thing. Understanding what they do when they get together is one of science's greatest triumphs. But no one could have expected this understanding to lead where it has.

PIECES OF EVERYTHING

As the fundamental, irreducible constituent particles of all things, atoms (meaning "uncuttable") were postulated around 440 B.C. by the Greek philosopher Leucippus. His pupil Democritus worked out the implications of the hypothesis in great detail. The idea of atoms led to controversy about whether or not there was space (void) between the particles. Anaxagoras (c. 500–428 B.C.) dismissed the notion of void, but Epicurus (341–270 B.C.) questioned how anything could move if all space were packed full of atoms.

Democritus's atomism fell out of favor for two millennia, largely because Aristotle did not like it. Medieval theologians rejected the hypothesis because it could not be accommodated with the Christian notion of transubstantiation. Interest was revived in the fifteenth century by the rediscovery of the poem *De rerum natura* (On the Nature of Things) by the Roman philosopher Lucretius (99–55 B.C.), a follower of the atomistic doctrine of Epicurus.

Galileo, Francis Bacon, Pierre Gassendi, and Isaac Newton believed in atoms, but many other great thinkers did not. While accepting that matter might be made up of small particles, René Descartes saw no reason to assume they could not be divided indefinitely. He asserted that they were borne along like grains of dust in swirling vortices of some all-pervading fluid.

It was generally agreed that the microscopic realm was a world in motion, which implied that mechanics could be used to understand the everyday properties of matter. The idea was first enunciated clearly by Daniel Bernoulli (1700–1782), a mathematician of Flemish descent born in Basel, Switzerland. In 1738 he proposed that gases are composed of tiny particles rushing around and colliding. The pressure exerted by a gas—on the side of an inflated balloon, say—was the result of all the little impacts of the particles hitting the surface.

In 1763 a Serbian Jesuit named Roger Joseph Boscovich (1711–1787) identified the ultimate implication of this mechanical atomic theory. One of the crucial aspects of Isaac Newton's laws of motion is their predictive capability. If we know how an object is moving at any instant—how fast, and in which direction—and if, furthermore, we know the forces acting on it, we can calculate its future trajectory exactly. This predictability made it possible for astronomers to use Newton's laws of motion and gravity to calculate, for example, when future lunar and solar eclipses would happen.

Boscovich realized that if all the world is just atoms in motion and collision, then an all-seeing mind

could, from a continuous arc described in an interval of time, no matter how small, by all points of matter, derive the law [that is, a universal map] of forces itself . . . Now, if the law of forces were known, & the position, velocity & direction of all the points at any given instant, it would be possible for a mind of this type to foresee all the necessary subsequent motions & states, & to predict all the phenomena that necessarily followed from them.[2]

That is to say, a mathematician with godlike omniscience could deduce the rest of history, for ever and ever, from a mere moment in time. Compared with Hobbes's version of determinism, in which people are automata moving at the insistence of mechanical forces, this is an altogether more constraining straitjacket for the world. Nothing is unknown or uncertain and there is no deviation from the inevitable play of forces. The fact that no human mind could possibly make such an astronomical calculation is irrelevant: in Boscovich's view, the future was already defined by the present. The eminent French mathematician Pierre-Simon Laplace (1749–1827) made a similar statement in 1814, which, like its author, is far better known. For such an awesome intelligence, said Laplace, "the future, like the past, would be present before its eyes."[3] Mechanism, it seemed, had banished free will.

DISSIPATION AND DEATH

The implications of a mechanical universe were not just philosophical. With the Industrial Revolution in full swing at the dawn of the nineteenth century, there were pressing practical matters for scientists to address. In a short life terminated prematurely by cholera, the French scientist Nicholas-Léonard-Sadi Carnot (1796–1832) busied himself with one of the most important of these problems: how to optimize the fuel efficiency of a steam engine.

What was true of power generation in Carnot's time is largely true today: extracting work from an engine means generating heat and letting it flow. Think of a coal-fired gas turbine. Heat produced by burning fuel is transferred from the burner to the gas. The hot gas expands, the pressure rises, and a jet is released that drives the rotating blades of the turbine. The rotation turns an electromagnet, creating electricity in the coils. The steam engine, workhorse of the Industrial Revolution, likewise used the expansion of a hot gas: water vapor.

But what, exactly, is heat? In the late eighteenth century many eminent scientists agreed that it was a physical substance called caloric, which flows from hot to cold. The American scientist Benjamin Thompson (later Count Rumford, 1753–1814) thought otherwise. Heat, he suggested, is the random motion of atoms in collision. It is not the *product* of such motions—the frictional heating caused by the atoms' surfaces rubbing together. No, it must be identified with these motions themselves. A substance heats up when its atoms are made to juggle more furiously—for example, as a result of atomic collisions when the substance comes into physical contact with another material in which the motions are already very lively.* Carnot agreed with this proposal: "Heat is then the result of a motion,"[4] he wrote in 1824. The mechanical world of atoms had rationalized an old mystery.

Engineers needed to capture some of this microscopic motion and turn it into the motion of railway carriages, factory machinery, and pumps. Carnot realized that this was contingent on getting heat to *flow* from a hot body to a cooler one. He deduced a general theory for calculating how much of this heat flow can be converted to useful work—the conversion is never perfect, because some heat energy is inevitably squandered—and how this depends on the difference in temperature between the hot heat source and the cold heat sink.

To develop his argument, Carnot considered an engine in which heat flow allowed a gas to expand (when heated) and contract (when cooled), driving a piston in a cyclic process now known as the Carnot cycle. His analysis laid the cornerstone of a new discipline called thermodynamics—literally, "heat movement."

Most people who have encountered thermodynamics blanch at its mention, because it is an awesomely tedious discipline both to learn theoretically and to investigate experimentally. This is a shame, because it is also one of the most astonishing theories in science. Think of it: here is a field of study initiated to help nineteenth-century engineers make better engines, and it turns out to produce some of the grandest and most fundamental statements about the way the entire universe works.† Thermodynamics is the science of change, and without change there is nothing to be said.

Thermodynamics, like Newton's theory of motion, has three laws. The third is hardly worth knowing unless you are a physicist; the first two should be engraved in the mind of anyone who wants to understand science.

The First Law is the easiest: energy is never destroyed but only trans-

---

*This was not an entirely unprecedented notion, for a mechanical theory of heat was first proposed by the Anglo-Irish scientist Robert Boyle in 1675.

†The physicist Erwin Schrödinger was probably right to point out that thermodynamics owes more to steam engines than steam engines do to thermodynamics. The theory has practical implications, without doubt; but it soon leads us into discussions verging on the metaphysical.

formed. Photovoltaic panels gobble up the energy of sunlight, and turn it—some of it, never all—into electrical energy. (Most, alas, is wasted as heat.) In a turbine, heat is transformed into the energy of motion (called kinetic energy) of the turbine blades, and then to electrical energy. Thus the universe conserves its energy. Only when heat was identified as the movement (kinetic energy) of atoms could this law be properly formulated.

The Second Law is more remarkable, and some scientists believe we still don't fully understand it. A testament to its importance is C. P. Snow's famous (albeit perhaps overblown) complaint in his book *The Two Cultures*:

> A good many times I have been present at gatherings of people who, by the standards of the traditional culture, are thought highly educated and who have with considerable gusto been expressing their incredulity at the illiteracy of scientists. Once or twice I have been provoked and have asked the company how many of them could describe the Second Law of Thermodynamics. The response was cold: it was also negative. Yet I was asking something which is about the scientific equivalent of: *Have you read a work of Shakespeare's?*[5]

There are several ways of expressing this law. When the German physicist Rudolph Clausius (1822–1888) first did so in 1850, he said something along the lines that heat always flows from hot to cold. As anticlimaxes go, this is a pretty damp squib. But what he really meant was that there exist processes that go only one way—which are, in other words, *irreversible*. Water does not flow uphill, and neither, figuratively speaking, does heat.

This seemingly innocuous statement is really the secret of all change. If there are irreversible processes, then time has an arrow, a singular direction defined by such processes. The Second Law connects to our perception that we are always moving forward in time, never back.

But Clausius did not let it go at that. He contrived the concept that enabled a mathematical theory of change and irreversibility: *entropy*. Entropy crops up in thermodynamics as a rather abstract quantity, but one can in fact measure it just as one can measure the heat released during a chemical reaction. Crudely speaking, entropy is a measure of the amount of disorder in a system; I shall sharpen this definition shortly. The Second Law reduces to the statement that in all processes of spontaneous change (such as heat flowing from hot to cold), entropy increases.

In 1852, William Thomson (later Lord Kelvin, 1824–1907) noticed something peculiar about the way energy gets transformed. There is, he said, "a universal tendency in nature to the dissipation of mechanical energy."[6] What he meant was that some energy is always "wasted" as heat (that is, random atomic motion). Think of that rotating turbine, in which friction warms up the bearings. It is hard to win back any useful energy* from this heat, which leaks away into the surroundings. In 1854 the German physicist Hermann von Helmholtz (1821–1894) perceived the consequences of this inevitable dissipation: the universe would end up as a uniform, tepid reservoir of heat. No further change would then be possible because there was nowhere colder for the heat to flow. Thus, he said, the universe would ultimately die a "heat death." In the behavior of steam engines we can read the fate of all creation.

THE DANCE OF PROBABILITY

Right from the inception of thermodynamics, scientists wanted to know where its rules came from. If all the world is just atoms in motion, each of them obeying Newton's laws, should it not be possible to deduce the thermodynamics laws just by considering all those invisible collisions?

Daniel Bernoulli began that quest with his explanation of gas pressure. An Englishman named John Herapath (1790–1869) wondered what manner of motions would be required to account for the pressure a gas exerts, and he calculated that the gas particles (atoms or molecules, which are small clusters of atoms) have to be traveling at speeds of something like two kilometers per second.

The pressure of a gas can be altered by changing its temperature. If you heat up a gas in a sealed vessel—that is, in a fixed volume of space—its pressure increases. This is why aerosol cans explode if thrown into a fire. If, on the other hand, the volume isn't fixed—if the walls of the vessel are movable—then a hot gas expands. This is what drives the piston in Carnot's cycle. In other words, the three characteristics of a gas—its temperature, pressure, and volume—are rather like the notorious trio one encounters in

---

*That is, energy that can be used to conduct some mechanical task, such as lifting a weight or moving a wheel. Scientists call useful energy "work."

engineering or business: cost, speed, and quality. That's to say, if you specify any two of them, you have no say over the third: it is decided for you. We
can arrange for a gas to have a particular temperature and pressure; but then
the volume (or equivalently the density—the number of molecules in a given
volume of space) is preordained. Another way of saying this is that if we
keep one of the trio constant, there is a mathematical relationship between
the other two. At fixed volume, for instance, the pressure of a gas is proportional to its temperature.

These relationships between the temperature, pressure, and volume of a
gas—the so-called gas laws—were studied in the seventeenth century by
Robert Boyle. Nearly a century later, Boyle's investigations were refined by
the Frenchmen Jacques Charles (who made the first hydrogen balloon flight,
in 1783) and Joseph-Louis Gay-Lussac.

The challenge was to see whether the gas laws could be derived from a
mechanical model in which atoms are like billiard balls moving in straight
lines until they collide with one another. Rudolph Clausius laid much of the
groundwork for this so-called *kinetic theory of gases* in the 1850s; but it was
brought to fruition mostly by one man, the Scottish physicist James Clerk
Maxwell (1831–1879) (Figure 2.1).

2.1   James Clerk Maxwell, whose introduction
of statistical ideas into the atomic theory of gases
was just one of his major contributions to science. He also clarified the nature of color,
pioneered color photography, and unified all elec-
tromagnetic phenomena in a single theory.

When a snooker player strikes a ball, it is not difficult to calculate what its motion will do to the other balls on the table. But in a single thimbleful of air there are about 10 billion billion atoms. We cannot possibly know how they are all moving at any instant; and even if we did, the task of calculating how the motion would be altered by collisions in the next instant, and the next, is imponderable. So how can we hope to account for everyday behavior, as described by the gas laws, starting from "first principles," from atomic motions?

Maxwell's key insight was that we do not need to know all the details. What is important is not the precise trajectory of every gas particle but their *average* behavior. He pictured a swarm of bees, all buzzing about furiously in the air while the swarm itself hovers as a stationary mass, because on average the bees are no more likely to be flying in one direction than another.

All that matters about the motions of the gas particles, said Maxwell, are two things: how fast, on average, each particle is moving—which determines their average motional (kinetic) energy—and how broad is the spread in speeds on either side of that average. Maxwell intuited that the *distribution* of speeds resembles the kind of bell-shaped curve you see in statistical surveys of the population—for example, of the average wage. We shall see in the next chapter that this intuition owed a great deal to a nascent science of society.

Maxwell's curve, indicating how many gas particles are moving at each speed, rises smoothly from low speeds, hits a peak at the average, and then tails off smoothly toward high speeds (Figure 2.2). This distribution shows that rather few particles have speeds *much* higher than the average. As the Welsh physical chemist Emyr Alun Moelwyn-Hughes once prophetically

2.2 Maxwell's probability distribution for the speeds of particles in a gas. As the gas gets hotter, the peak shifts to higher velocities and gets broader and flatter.

put it, "Energy among molecules is like money among men. The rich are few, the poor numerous."[7]

The average speed of the particles depends on how much kinetic energy the gas as a whole contains. Pump in more energy—heat the gas up—and the average rises: the peak of Maxwell's distribution shifts to higher speeds. But another thing happens too: the bell curve gets flatter and broader, transforming from a tall, steep-sided pinnacle to a lower, more gently sloping hillock. That is to say, the spread of speeds gets wider. (Whether pumping more "energy" into an economy has the same effect on the distribution of wealth is another matter.)

Maxwell's gas does not in fact behave quite like a swarm of bees staying stationary in the air. The particles, unlike the bees, are constantly colliding. This means that their direction of motion is constantly changing, essentially at random. Yet even though each particle moves at random and there is no overall preference for movement in any direction, this does not mean that the particles stay clumped in a swarm. Particles moving at random do actually get somewhere, rather than forever meandering about a fixed position. Their erratic paths take them gradually farther from this point, but in a random direction. This is called a random walk, and physicists like to compare it to the path of a drunken man staggering uncontrollably toward no particular destination (Figure 2.3). A particle moving this way is said to be *diffusing*.

Because of diffusion, a cluster of particles released into the air will gradually expand outward in all directions, rather like an ink droplet dispersing

2.3   A particle bouncing between collisions in a gas executes what is called a random walk, drifting gradually farther from where it began.

in a glass of water. In the same way, two different gases filling adjacent compartments in a box will gradually mix if the partition between them is removed. Maxwell's mathematical analysis allows one to calculate how rapidly a particle moves by diffusion. This is considerably slower than the particle's actual speed because it takes a highly circuitous route to get from A to B.

It was by observing random walks that scientists finally reached a consensus on the existence of atoms. Maxwell's theory was predicated on the idea that gases are composed of atoms and molecules; he even used it, in 1873, to calculate the sizes of molecules, predicting that a hydrogen molecule is 0.0000006 millimeters across. (His estimate was off only by a factor of about three.) But no one had actually *seen* atoms, and even at the end of the nineteenth century some scientists still refused to countenance them. Ernst Mach, a highly influential German physicist, regarded it as poor science to accept the existence of anything inaccessible to direct experience, and so preferred to withhold judgment on the atomistic theory.

But in 1905 Albert Einstein published a seminal paper in which he showed that by assuming gases to be composed of invisible particles (atoms or molecules) executing random walks, one could explain the hitherto mysterious phenomenon of Brownian motion.

Robert Brown was a great botanist, but he had no intention of pronouncing on physical theory. When in 1828 he first saw pollen grains dancing wildly under the microscope as they hung suspended in water, he thought at first that this activity revealed the fundamental "active force" of life, embodied in the old hypothesis of vitalism. He later discovered that unambiguously "dead" grains, including (bizarrely) fragments of the Egyptian Sphinx, behaved the same way, and the various explanations for the movement proposed in the nineteenth century dispensed with these vitalistic notions. Einstein's theory was, however, the first to account convincingly for Brown's observations. Einstein supposed that the tiny pollen grains were small enough to be deflected by collisions with individual molecules of water,* even though the grains were microscopically visible and the molecules were not. Einstein's paper was the first thorough treatment of diffusion, and it

---

*More strictly, the pollen grains experience an imbalance in the number of collisions from different directions—the deviations of their trajectories happen when more molecules strike them from one side than from the other.

made several predictions about Brownian motion that the physicist Jean Perrin verified in a series of extremely precise experiments in 1908. Perrin won the Nobel Prize for physics in 1926, and for validating a theory over two millennia old, that seems fair reward.

FAITH IN NUMBERS

Maxwell's contribution to the kinetic theory of gases is central to the kind of physics with which most of this book will be concerned. He had made physics *statistical*, saying that what matters when we are dealing with huge numbers of virtually identical moving objects is not the detailed behavior of individuals but the average motions, as well as the extent of deviation from those averages. Anyone interested in broadscale human behavior will be familiar with this notion. Demographers do not need to know that Eric Baggins was born on March 6, 1969, but merely what the annual birthrate was. Traffic planners don't care that Mary Parker drove to the Pocono supermarket on Tuesday morning, but ask simply how many cars use Route 209 in the course of a typical day. These statistics become more reliable as the size of the census increases. If we are asking about the behavior of a lump of matter you can hold in your hand, we are typically dealing with billions of billions of molecules, and the statistical behavior is utterly reproducible from one experiment to another. In other words, for example, for two jars of identical gas at the same temperature the Maxwellian distribution of velocities is absolutely identical.

Along with the introduction of statistics comes the notion of probability. Maxwell's distribution tells us nothing exact about the speed of any particular gas molecule. Instead, it tells us the *probability* that a particle selected at random will have a particular speed. The most probable speed is the average speed;* there is a low probability that it will be much faster or much slower than this. It is indeed extremely convenient that statistics are enough for us to account for the behavior of gases, for even with modern instruments we could not gather detailed information on every gas particle.

---

*In fact the *mean* and *most probable* speeds in the distribution are not identical—they differ by a small factor. The peak in the distribution is the most probable speed; the mean speed is slightly greater than this because the curve tails off more slowly than it rises.

Maxwell evinced a certain uneasiness about his kinetic theory, since he recognized that it broke with the mechanistic tradition of using Newton's laws of motion to deduce the exact trajectories of a system's components, as one does, for example, to explain planetary motions. This was, in other words, a new way of doing science. Maxwell realized that the theory had profound philosophical implications, and as we shall see later, he may not have risked publishing it had there not already been good precedent.

Maxwell's "probability distribution" of the speeds of gas particles was a seminal contribution to the kinetic theory, but the truth was that he deduced it using a strong dose of informed guesswork rather than exact mathematics. The job was done more rigorously by a troubled Austrian physicist named Ludwig Boltzmann (1844–1906).

As someone employed regularly to scan the scientific literature for news stories, I have come to appreciate that any paper with a title that begins "Further researches on . . ." should be passed over with alacrity. It tends to be science-speak for "the odds and ends we did not think worth pursuing in our last paper." So it is humbling to be reminded that had I taken this attitude in 1872, I would have missed one of the most explosive papers of the century. In "Further researches on the thermal equilibrium of gas molecules," Boltzmann not only made Maxwell's case watertight but also proved that there truly exist irreversible processes, as the Second Law of Thermodynamics stipulates—and showed why.

Maxwell proved that gas particles, once they achieve a Maxwellian distribution of speeds, will stay that way. But he did not show how they get to that state in the first place. This is what Boltzmann did, by developing a way to calculate how probability distributions change over time. He demonstrated that for particles moving at random, "whatever the initial distribution of kinetic energy may have been, it must always necessarily approach the Maxwellian form after a very long time has elapsed."[8]

Thus Boltzmann put *change* under the lens of the kinetic theory, which at once brought the Second Law into focus. Clausius had proposed that entropy always increases during an irreversible process; Boltzmann clarified what this meant for the probabilities of molecular motions. He showed that entropy can be equated with the number of different arrangements of molecules, which, at the everyday scale, look identical.

Picture a child's balloon on the end of a string. It is full of gas molecules moving at random, and the collisions of these particles with the elastic wall create the pressure that keeps the balloon inflated. At any instant, each molecule is moving on a particular path with a particular speed. If we had a camera so powerful that it could take snapshots showing all the particles, then two snapshots taken an hour—or a minute, or a second—apart would show very different arrangements. Because of the huge number of particles, we could take a billion snapshots and never see the same picture twice. But on the scale at which we typically make laboratory measurements, the gas is just the same in every case: it still has the same temperature, pressure, and volume.

The number of possible arrangements of the molecules here is truly astronomical. But nevertheless it is finite. We can imagine arrangements that would *not* be equivalent to those above—for example, with all the particles in one half of the balloon. In that case, the empty half would deflate. Because the particles are moving at random, there is absolutely nothing in the laws of physics to prevent this arrangement from arising by pure chance. But the likelihood that all the particles would suddenly happen to acquire velocities that took them into the same half of the balloon is so tiny that it is hard to distinguish from zero. The same is true for just about any arrangement other than ones in which all the particles are distributed evenly throughout the balloon.

Thus the balloon stays fully inflated not because Newton's laws of motion say it must but because the arrangements of particles that ensure this are *overwhelmingly* more probable than any others, simply because there are many, many more of them than there are of any arrangements that are not equivalent. By equating the entropy of a state with the number of equivalent molecular arrangements to which it corresponds, Boltzmann was thus saying that the fully inflated state of the balloon has the highest entropy. He deemed the mathematical equation relating entropy to the number of "microstates" of a system to be the apogee of his life's work, and this recondite formula, $S = k \log W$, is engraved on Boltzmann's tombstone.

When change happens in any system, entropy increases because the new arrangement of the constituent particles is more probable than the old. To put it another way, the direction of change—the arrow of time—is determined by probabilities. An ink drop diffuses and disperses in water because it is vastly more probable that the random motions of the ink particles will carry them away from the original droplet in all directions than that they will all conspire to make the droplet move coherently sideways, say, or shrink.

The crucial point about this explanation of the Second Law is that it shows how the irreversibility of time can come about through the operation of mechanical laws which have no preferred direction in time. Picture a movie of two billiard balls coming together, colliding, and moving apart. Played in reverse, the movie would not look at all odd: the reverse collision also obeys Newton's laws.* But the coalescence of a droplet of ink within a glass of initially pale blue water would obviously be time-reversed footage, even though each of the individual collisions between particles that "created" this arrangement would look like those billiard balls hitting in reverse. This is simply due to the effect of very large numbers on the probability of certain processes happening. Entropy does not *have* to increase by cosmic decree; it simply does so because that is overwhelmingly probable.

The theory of Maxwell and Boltzmann was derived from nothing more than the application of Newton's laws of motion to vast numbers of moving molecules—from so-called classical mechanics. It marks the beginnings of the field of *statistical mechanics*. This is the field that provides modern physics with its central organizing framework. By connecting thermodynamics with the properties of atoms in motion, statistical mechanics describes the behavior of matter from the bottom up.

The shift from Newtonian determinism to statistical science is what makes a physics of society possible. It was not a smooth ride; but as we shall now see, it may have been bumpier still if scientists and philosophers had not already begun to appreciate that society itself is fundamentally a statistical phenomenon.

---

*This assumes that no kinetic energy is dissipated in the collision as frictional heat or sound. In the real world these dissipative processes do occur, of course: even the striking of billiard balls is an irreversible process that results in an increase in entropy.

THREE

THE LAW OF LARGE NUMBERS

REGULARITIES FROM RANDOMNESS

It can be stated without exaggeration that more psychology can be learned from statistical averages than from all philosophers, except Aristotle.

—Wilhelm Wundt (1862)

Taken in the mass, and in reference both to the physical and moral laws of his existence, the boasted freedom of man disappears; and hardly an action of his life can be named which usages, conventions, and the stern necessities of his being, do not appear to enjoin on him as inevitable, rather than to leave to the free determination of his choice. —John Herschel (1850)

If there is some precision, there is some science. —Herbert Spencer (1880)

Ludwig Boltzmann was seldom a happy man. His poem "Beethoven in Heaven," written five or six years before he died, expresses an anguish that is all too evidently the physicist's own, as well as a presentiment of what was to come:

> With torment that I'd rather not recall
> My soul at last escaped my mortal body.
> Ascent through space! What happy floating
> For one who suffered such distress and pain.[1]

His scientific achievements perhaps brought him more "distress and pain" than satisfaction and joy, for his ideas were vigorously attacked by several of his contemporaries. Although today most scientists concur with his explanation for the arrow of time, it left open some important questions into which his opponents sunk their teeth. Boltzmann responded robustly to these attacks, but they greatly disheartened him.

Boltzmann was by nature a self-doubting and hesitant man, and his gloom deepened in 1889 when his eldest son died from appendicitis. He grew restless and was unable to settle comfortably at any of the several Austrian and German universities he joined. On September 5, 1906, while on holiday with his family in Duino, near Trieste, he hanged himself, at the age of sixty-two.

Theoretical physics scarcely sounds like a life-threatening activity. But Boltzmann's self-inflicted death was echoed by that of his brilliant successor, the Viennese Paul Ehrenfest, who shot himself in 1933. (After recounting these wretched episodes, the physicist David Goodstein tartly informs the reader of his modern textbook, "Now it is our turn to study statistical mechanics."[2])

It would be unwise to draw any conclusions about the psychological hazards of early-twentieth-century physics, however, before asking whether the suicide rate among physicists was any greater than it was among the population as a whole. Vienna at the fin de siècle was a furnace of intellectual debate, fired by the likes of Sigmund Freud, Arnold Schoenberg, Ludwig Wittgenstein, and Robert Musil; but its citizens were, as Musil observed, nevertheless a soulless, tight-lipped crowd in thrall to convention. "The notion," he says in *The Man Without Qualities*, "that people who live like that could ever get together for the rationally planned navigation of their spiritual life and destiny was simply unrealistic; it was preposterous."[3]

In this rigid and materialistic society, suicides were disturbingly widespread. They claimed the lives of three of Wittgenstein's brothers, Gustav Mahler's brother, and, in 1889, Crown Prince Rudolf of Austria (who killed his mistress first). Boltzmann's sad death does not speak to any broader context until it is seen in the light of the relevant demographic statistics.

To us this seems obvious, but before the nineteenth century hardly anyone would have thought this way. Assessing individual events in the context of their average rate of occurrence is a relatively modern practice. Without it, the world is ripe for magic, superstition, miracles, and conspiracy theories. A few chance events can become evidence for supernatural influence. Even now the relevance of statistics is routinely overlooked in subjective assessments of risk and coincidence. When the "psychic" Uri Geller apparently stopped a few watches among his TV audience in the 1970s, no mention was made of the likelihood of such a thing happening by chance given the very large number of viewers.

Whenever one is trying to make sense of mass behavior, whether of atoms or of people, statistics are indispensable. This now seems so beyond question that it is hard to comprehend the urgency of the philosophical arguments that surrounded the use of statistics in nineteenth-century science. At that time, it seemed that God and human free will were being held hostage to numbers. The roots of a physics of society are enmeshed in this debate, so we shall find some of the moral issues raised by the new discoveries described in this book prefigured by soul-searching from more than a hundred years ago.

The history of statistical mechanics outlined in the previous chapter is the orthodox one that physicists tell routinely. Very rarely is any hint given of the way it really began—not just among the insensate gases of the laboratory but in the behavior of people in society. Speaking of a physics of society may sound like a very postmodern thing to do, but truly there is nothing new under the sun.

MEASURING SOCIETY

In *Leviathan* Thomas Hobbes was arguably taking to its logical conclusion the analogy drawn by his mentor Francis Bacon between the "Body Naturall" and the "Body Politick." This notion implied that politics might be a kind of natural science, with an anatomy waiting to be dissected by the scalpel of systematic and rational inquiry. In attempting to create such a scientific political theory, Hobbes chose mechanical physics as his framework.

Today we think of physics as a supremely quantitative, not to say mathematical, science. Physicists measure the fundamental numbers of nature down to the tenth decimal place. Their formal literature is dense with symbols, equations, and graphs. Even if things were not quite like this in Hobbes's day, it is still striking that *Leviathan* is wholly discursive—there is not a number or an equation in sight. Hobbes liked to make use of physical analogies, but he had no intention of making political science mathematical.

That inevitably had to happen, however, if the endeavor was not just to borrow the ideas but to share the demonstrative force of natural science. William Petty, a disciple of Hobbes, seemed to recognize as much when he called for a "political arithmetic." "To practice upon the Politick," said Petty, "without knowing the Symmetry, Fabrick, and Proportion of it, is as casual as the practice of Old-women and Empyricks."[4]

What numbers was this arithmetic to manipulate? Why naturally, those that measured society. In the 1660s, John Graunt (1620–1674), a London haberdasher and a friend of Petty's, introduced the study of "social numbers" as a means to guide political policy. Chief among the numbers with which he concerned himself were death rates. In *Observations upon the Bills of Mortality* (1662) he drew up tables of mortality figures "whereby all men may both correct my Positions, and raise others of their own."[5] How could one reasonably legislate and govern the people, he asked, without knowing the numbers in which they come and go?

Graunt's statistics were hardly a model of methodological finesse. As he freely admitted, those humble souls responsible for recording deaths were all too easily induced, "after the mist of a Cup of Ale, and the bribe of a two-groat fee, instead of one," to list the cause of death as something anodyne (consumption, say) when the truth was more scandalous (such as syphilis). Yet his tables of causes and ages of death were seen as a bountiful resource for those seeking to understand the flux of society. Graunt, although a mere businessman, was elected a Fellow of the Royal Society, and Charles II averred that "if they found any more such Tradesmen, they should be sure to admit them without any more ado."[6]

William Petty continued to revise the *Observations* after Graunt's death. He was the first to study political economy by means of such social statistics, which he argued could provide a rational basis for formulating policy. In this respect he was an empiricist, working with observations of social aggregates rather than trying to derive theories based on assumptions about the fundamental psychology of individuals in the manner of Hobbes. Petty enjoyed the favor of Charles I, Charles II, and James II (and managed pragmatically to serve Cromwell too), and he was a founding member of the Royal Society. Yet his policy recommendations were largely ignored, and frankly this was often just as well. Petty often exemplifies the dangers of a hyper-rational, analytical approach to social policy that takes no account of its human costs.

Population measures—birthrates and death rates—were the major pre-occupation of early quantifiers of society. It was thought to be of paramount importance for a nation to multiply its subjects, an injunction that was, after all, sanctified in the Bible. The power and glory of a country were believed to be reflected in the size of its population, so much so that some savants proposed that wars of conquest were driven largely by a desire to increase it. In the mid–eighteenth century Johann Peter Süssmilch (1707–1767), a German army chaplain, argued that war could be avoided by removing all checks to the growth of population, obviating the need for kings to gather new subjects from outside their realm.

A focus on mortality was understandable in an age that knew so much of it. Masses died in noisome cities, the "Places of the Waste and Destruction of Mankind,"[7] according to Thomas Short in 1767. Famine and starvation were endemic in the countryside. Few conflicts were quite as devastating as the Thirty Years' War, but war still seemed to be an ever-present part of human affairs and a steady source of attrition in the population. Procreation was the only remedy. Ironically from today's perspective, Protestants in England and Germany denounced Catholicism because its advocacy of celibacy compromised population growth.

By 1826, when the English economist Thomas Robert Malthus (1766–1834) wrote his *Essay on the Principle of Population*—a compelling critique of unchecked population growth which had a profound influence on both Darwin and Marx—governments in Europe and the United States had begun to appreciate the wisdom of counting their citizens. Censuses in fact date back to the Norman efforts to record in the *Domesday Book* the population of England in the eleventh century, although this was not so much an exercise in quantification as the establishment of a bureaucratic basis for the exploitation of a conquered population. By the eighteenth century such social numbers were considered to encode insights into how society functioned. Süssmilch, for example, argued that the differences in birthrates and death rates of boys and girls balanced perfectly so as to provide marriage prospects for them all. In other words, from the chaos of human life arose a kind of law of the masses that stabilized society.

Süssmilch's observations helped establish the idea that society observed rules that were ordained by no government yet could be revealed by counting. This led Immanuel Kant in 1784 to speak of "universal laws" which, however obscure their causes, [permit] us to hope that if we attend to the play of freedom of human will in the large, we may be able to discern a regular movement in it, and that what seems complex and chaotic in the single individual may be seen from the standpoint of the human race as a whole to be a steady and progressive though slow evolution of its original endowment.[8]

On the one hand, this belief in "laws" of society that lay beyond the reach of governments was a product of the Enlightenment faith in the orderliness of the universe. On the other, it is not hard to see within it the specter of the Industrial Revolution, with its faceless masses of toiling humanity like so many swarming insects. Before the nineteenth century, the laws that applied to Graunt's "social numbers" were regarded as evidence of divine wisdom and planning. To later commentators they were the preconditions for catastrophe and revolution.

This study of social numbers needed a name. In 1749 the German scholar Gottfried Achenwall suggested that since this "science" dealt with the natural "states" of society, it should be called *Statistik*. John Sinclair, a Scottish Presbyterian minister, liked the term well enough to introduce it into the English language in his epic *Statistical Account of Scotland*, the first of the twenty-one volumes of which appeared in 1791. The purveyors of this discipline were not mathematicians, however, nor were they "scientists"; they were tabulators of numbers, and they called themselves statists.

THE CHURCH OF NEWTON

Those who collected statistical data soon realized that these data tell us something not just about what has occurred but about the general *probability* of things happening. Statistics thus began to attract the interest of mathematicians who concerned themselves with one of the most philosophically recondite and sometimes counterintuitive branches of their subject: probability theory.

This discipline has its origins not in social numbers but in games. If you play a game of chance, you would be wise to know the odds beforehand. Anyone who does not appreciate that betting on black or betting on number 15 in roulette are quite different things, with different odds, is not destined to spend long in the casino. Games involving dice and other agents of chance have a long history, and in the eighteenth century mathematicians began to investigate the rules that govern them. In this seemingly frivolous area the French mathematician Marie-Jean-Antoine-Nicolas de Caritat, Marquis de Condorcet (1743–1794), found the tools he needed to build one of the most optimistic of all science-based utopias.

Condorcet described his vision in a book written shortly before the architects of the French Revolution condemned him to the guillotine. His *Esquisse d'un tableau historique des progrès de l'esprit humain* (Sketch for a Historical Picture of the Progress of the Human Mind), written in 1793, is a hymn to a kind of rationality that Condorcet must have found in short supply during the Reign of Terror. Society, he believed, must be founded upon reason and guided by its great formalization: science. The eighteenth century was a truly radical time, when belief in liberty and equality (Robespierre's Terror notwithstanding) was more than a matter of good intentions. Many Enlightenment philosophers genuinely trusted that these principles, allied to reason, would usher in a glorious age of freedom. Condorcet argued for the equality of women, and in 1792 he proposed that all patents of nobility be burned, including his own. He became a friend of Voltaire, whose utopian writings were, however, rather more cynical.

Condorcet was a precocious mathematical genius, and his early efforts came to the attention of the eminent French academician Jean le Rond d'Alembert. Under d'Alembert's influence he turned from the pure mathematics of probability to consider social and economic issues. In his study of democratic decision making, *Essai sur l'application de l'analyse à la probabilité des décisions rendues à la plurité des voix* (Essay on the Application of Analysis to the Probability of Majority Decisions) (1785), Condorcet concurred with his fellow "statists" that if (as he maintained) there is indeed a science of human affairs, with its own axioms and laws, then it must be a statistical science. To read these laws, one must gather enough numerical data: "All that is necessary to reduce the whole of nature to laws similar to those which Newton discovered with the aid of calculus, is to have a sufficient number of observations and a mathematics that is complex enough."[9]

Condorcet foresaw that the scientist so equipped would be able to predict the outcome of democratic decision making, so that history itself could become a true science. Then we would indeed be on the threshold of a true utopia, as he indicated in his *Esquisse*:

How consoling for the philosopher who laments the errors, the crimes, the injustices which still pollute the earth and of which he is often the victim is this view of the human race, emancipated from its shackles, released from the empire of fate and from that of the enemies of its progress, advancing with a firm and sure step along the path of truth, virtue and happiness![10]

It is not hard to read in this passage an attempt by the author to console himself in the face of a bleak future. The *Esquisse* was written hurriedly in hiding while Robespierre's agents hunted for its author. His was a dramatic fall from favor, and his story tells us much about the nature of revolutions.

In 1792, Condorcet's intellectual reputation and his support for the Republican cause earned him a seat on the Committee of Nine, which was charged with drafting the new French constitution. Among Condorcet's colleagues was Thomas Paine, a French citizen after his exile from Britain following the publication of his book *The Rights of Man*. The draft constitution was blocked by Robespierre, who resented having been excluded from the committee. When a new version, hastily redrafted by another, makeshift committee and full of loopholes, was accepted, Condorcet published an anonymous letter urging the people to reject it. His authorship did not stay secret for long, and he was convicted of treason.

The *Esquisse*, penned in a safe house in Paris run by one Madame Vernet, is, particularly under the circumstances, strikingly optimistic. Condorcet regards humankind as having "evolved" from the level of beasts to a state of higher intelligence in which people have acquired an innate altruism. He sees no reason why this evolution (anticipating Darwin) cannot continue until people are "perfected"—an idea in stark contrast to Jean-Jacques Rousseau's view that civilized man is corrupted. In the future utopia, says Condorcet, medical science will conquer all disease, and people will be too enlightened to go to war. Education will abolish social inequality, and all people will speak the same language. "Are we not arrived at the point," he asked, "where there is no longer any thing to fear, either from new errors, or the return of old ones? . . . Everything tells us that we are approaching the era of one of the grand revolutions of the human race . . . The present state of knowledge assures us that it will be happy."[11]

This soaring vision was not that of a worldly man. Although Condorcet eluded his captors when they came to arrest him, his refined manner aroused suspicion in the country inn to which he fled, and he was swiftly apprehended. He was taken to prison at Bourg-la-Reine, near Paris. With the guillotine his likely fate, he seems to have committed suicide by poisoning himself in his cell. Had he remained hidden for just several months more, he would have escaped his persecutor forever: Robespierre himself went to his death in July 1794.

The *Esquisse* became posthumously celebrated. Malthus read it but did not share its rosy outlook. Condorcet was aware that population growth could eventually overwhelm available resources and threaten the stability of civilization, but he had a simple remedy—birth control. Malthus did not think it would be so easy. He reckoned that the "passions of mankind" put population outside the control of governments, whether they sought either to encourage or to limit its rise. It was, he believed, a "law of nature" that the populace would multiply exponentially, while society could not increase the means of feeding itself at the same rate. Thus nations must succumb sooner or later to overcrowding, misery, poor health, and social unrest—which would bring a stark choice between repression and revolution.

To escape this fate, people would do well to accept that government alone, however good, cannot steer them clear of catastrophe. Rather, said Malthus in his profoundly influential *Essay*, one needed to know the irrevocable laws, the "internal structure of human society."

Others, while less pessimistic, concurred with Malthus's view that there *was* an internal structure, a set of laws, that dictated the way society behaved and evolved. These laws were deemed to stand in relation to society as Newton's mechanics stood in relation to the motion of bodies. The idea was particularly popular in France; the Baron de la Brède et de Montesquieu (Charles-Louis de Secondat) (1689–1755) adduced it in *The Spirit of the Laws* (1748), which preceded Condorcet by several decades. Claude-Henri de Rouvroy, the Comte de Saint-Simon (1760–1825), shared Condorcet's dream of a society governed by scientific reason, and he imagined that it might lead to the founding of a "Religion of Newton." The vision was particularly explicit in Jean Théophile Desaguliers's *Newtonian System of the World: The Best Model of Government, an Allegorical Poem* (1728), in which he wrote that the notion of a force of attraction "is now as universal in the political, as the philosophical world."[12] David Hume (1711–1776) expressed a desire in his *Treatise of Human Nature* (1739–1740) to become the Newton of the moral sciences by reducing human nature to first principles through empiricism rather than Cartesian axioms. When, in 1741, Hume proposed "that politics may be reduced to a science,"[13] the idea had already become so common-place as to be the subject of satire: Jonathan Swift's Gulliver berates the Brobdingnagians for "not having hitherto reduced *Politicks* into a *Science*, as the more acute wits of Europe have done."[14]

It was Hume who in the 1760s introduced Adam Smith, on a European Grand Tour with his pupil, the young Duke of Buccleuch, to François Quesnay (1694–1774), physician to Louis XV at Versailles. In his sixties Quesnay had begun to take an interest in economics and was collecting facts and figures in the hope of discerning among them laws and "social forces" akin to the physical forces of Newton. Quesnay's *Tableau économique* (Economic Table) (1758) was one of the first works of economic theory, and indeed his followers were the first to be classified as *les économistes*. His work leaves its mark clearly in Smith's *Wealth of Nations* (1776), which the author would have dedicated to Quesnay had the Frenchman not died two years before it was published (see page 178).

Belief in a "scientific" political theory has tended to flourish mostly at the liberal end of the political spectrum. (It is partly for this reason that Thomas Hobbes can rightly be regarded as a liberal.) When Thomas Jefferson wrote the Declaration of Independence, his vision of a free and happy nation was that of a man in love with Newtonian mechanics and the ideals of the Enlightenment who believed that humans are compelled toward happiness just as apples are pulled by gravity toward the earth. Jefferson considered that happiness could be measured as quantitatively as matter could be weighed. Irish-born British statesman Edmund Burke (1729–1797), often considered the "Father of Conservatism," abhorred this sort of thing, holding that a state's laws and institutions cannot be deduced from first principles but can only emerge empirically from particular historical processes. For that reason, one should regard those laws that existed as tried and tested, honed by the whetstone of experience and tradition, and not to be meddled with on account of abstract, "rational" theories. After all, Burke argued, people are too complex to permit any "scientific" analysis of the histories they produce: "In the gross and complicated mass of human passions and concerns, the primitive rights of men undergo such a variety of refractions and reflections, that it becomes absurd to talk of them as if they continued in the simplicity of their original direction."[15] Their "trajectories," in

other words, become randomized. But how telling that even Burke felt compelled to phrase his argument in the terminology of Newtonian mechanics
and optics!

The spirit of the Enlightenment position was, however, brought to its
zenith by the French positivist philosopher Auguste Comte (1798–1857),
who grounded his thinking in a rational religion motivated by the advance-
ment and well-being of humanity. Like Adam Smith, he believed that these
ends could be attained by uncovering the natural laws of society rather than
by political interference. Although he did not share the statists' enthusiasm
for quantification, Comte coined a term that encapsulated his desire for and
faith in a science of civilization: *physique sociale*, social physics. In his
Cours de philosophie positive (1830–1842), he argued that his would complete the scientific description of the world that Galileo, Newton, and others
had begun:

> Now that the human mind has grasped celestial and terrestrial physics, mechanical
> and chemical, organic physics, both vegetable and animal, there remains one sci-
> ence, to fill up the series of sciences of observation—social physics. This is what
> men have now most need of; and this it is the principal aim of the present work to es-
> tablish.[16]

ORDER FROM CHAOS

No one did more to propagate the notion of a scientific understanding of society than the Belgian astronomer Adolphe Quetelet (1796–1874) (Figure
3.1). In Quetelet's work all the developing strands of social inquiry—
Hobbes's mechanistic politics, the value of statistical quantification, and the
belief in natural laws of society—came together, and for a heady half century or so there seemed to be no demarcations between physics, mathematics, economics, politics, and sociology.

Like Hobbes, Quetelet had a vested interest in proving that a scientific
view of society could promote stability. He pursued his career at a time of
great political upheaval in Belgium. Most of the country had become part
of France in the late eighteenth century, but the southern provinces were
absorbed into the Netherlands. In 1830 the Belgians revolted in a bid for
independence, and the resulting conflict made academic study all but im-

3.1 Adolphe Quetelet. His *mé-canique sociale* was an attempt to find "laws of society" analogous to Newton's laws, which governed the motions of inanimate bodies.

possible. Scientists left their posts to enlist in the army, and the universities and colleges were disrupted. The Royal Observatory in Brussels, which Quetelet directed and helped to construct, was occupied by soldiers and, in his own despairing words, "converted into a fortress . . . surrounded with ditches and ramparts."[17] Just months after the revolt, Quetelet published his first paper on *mécanique sociale*, a discipline allied explicitly with Comte's social physics. By making direct analogies between the organizing forces of a solar system and those of an orderly social system, he aimed to show that the study of society was as rule-based as astronomy.

There was good precedent for such comparisons. Indeed, Quetelet felt that astronomers were to thank for introducing statistics to social studies. The first table of mortality figures, he pointed out, was drawn up by Newton's contemporary and friend, the astronomer Edmond Halley, and published in 1693. Quetelet argued that it was natural that astronomers should perceive order in the social sphere:

> The laws that concern man, and those that govern social development, have always had a special attraction for the philosopher, and perhaps most especially for those who

have directed their attention to the system of the universe. Accustomed to considering the laws of the material world, and struck with the admirable harmony that reigns
there, they can not be persuaded that similar laws do not exist in the animate world.[18]

All this became evident to Quetelet when in 1823 he was sent to the
French Royal Observatory in Paris to expand his astronomical knowledge
in preparation for his directorship at the planned observatory in Brussels.
Quetelet indeed discovered much about celestial science, but he also found
that the stars of the French astronomical firmament had a broad interest in
statistics.

THE SHAPE OF ERROR

The dominant figure of French astronomy at the time was Pierre-Simon
Laplace, who extended Newton's mechanics and used it to discover new aspects of planetary motion. Laplace was a formidable mathematician, and
he and his contemporaries knew that their measurements of celestial motions rarely reflected an exact adherence to the mathematical regularity that
Newton's laws demanded. All measurements incur errors that can cause apparent deviations from the smooth relationships predicted by the laws of
mechanics.

The French astronomers developed methods for dealing with these errors
that involved finding the smooth curves or lines that best fitted a scattering
of data points. Laplace and his pupil Siméon-Denis Poisson (1781–1840)
assumed that measurement errors were essentially random. They could be of
any magnitude, but not with equal probability. Rather, the likelihood of an
error of a particular size—a deviation of that magnitude from the "true law"
governing the observations—decreased as its size increased. In the same
way, if you measure the length of your foot using a ruler, you are more likely
to be off by a millimeter than a centimeter. The error will not be the same for
every measurement, even if it is made using the same instruments and techniques. If you measure the feet of all the members of your family, you might
sometimes be off by just half a millimeter, sometimes by two millimeters. It
depends not only on how accurate your ruler is, but on how careful you are
each time. The error is largely a matter of chance. This, the astronomers
realized, connected error rates with probability theory.

To know the likelihood of a particular error, we need to know how *often*, in a large and representative sample of measurements, it occurs. We need to collect statistics about errors. The French scientists found that errors were always distributed in the same way. Not only were there always more small errors than large ones, but this decrease in number with increasing size was predictable. Plotted on a graph, the statistics of errors fitted onto a particular curve which became known as the error curve (Figure 3.2). Its ubiquity also gained it the name *normal distribution*. The German physicist Carl Friedrich Gauss (1777–1855) bequeathed another name to this bell-shaped curve when he analyzed its properties in 1807: measurements that fall on this curve are now most commonly said, by physicists at least, to obey *Gaussian* statistics. Whatever one calls it, the error curve is basically the probability distribution of the outcomes of a *random* process.

This curve was already known to mathematicians interested in probability because in 1733 Abraham de Moivre showed that it described the distribution of outcomes from tossing a coin. There is an equal probability of the coin falling head up or tail up. The result depends (if the toss is fair) on pure chance. Nevertheless, for a large number of throws there will be roughly as many heads as tails.

3.2 The error curve. This bell-shaped curve describes the statistics of all random processes. (Strictly speaking, mathematicians call these *stochastic* processes, meaning that each outcome or observation is independent of the others.)

That this predictable result arises from a series of chance events is not terribly surprising, for it simply reflects the canceling out of equally probable deviations in either direction. A series of several heads in a row is likely to be balanced eventually by a temporary chance prevalence of tails. The equal numbers that result merely remind us of the 50:50 probability for each throw. At the beginning of the eighteenth century Jacob Bernoulli, uncle of Daniel (see page 35), pointed out that when the outcome of an event is governed by a fixed probability ratio (here 1:1), the distribution of actual outcomes will converge on this same ratio if the event is repeated often enough. Poisson enshrined this idea in 1835 in his "law of large numbers," a way of saying that pure randomness gives way to determinism if the number of random events is large. Thus randomness need not, in itself, rule out things happening in a predictable way.

But a 50:50 balance is not guaranteed, nor always observed. If we toss a coin ten times, a ratio of four heads to six tails should not surprise us: chance has shifted the balance in favor of tails so that there is a 20 percent deviation from equality. The more tosses we make, the more closely the numbers converge on a 50:50 distribution. For 100 tosses, 49 heads and 51 tails is not an unlikely outcome—again, a difference of 2, but this time the deviation is a smaller proportion (2 percent) of the whole. Another round of 100 tosses might produce a different outcome—say, 52 heads and 48 tails. De Moivre showed that for many rounds of a fixed number of tosses, the bell-shaped error curve always provides a good fit for the distribution of outcomes.

For coin tossing, it is possible in theory to calculate the shape of the error curve, that is, to write an exact equation describing the probability of each outcome in a single round of tosses. But this calculation is very laborious, especially with eighteenth-century techniques. De Moivre showed instead that the curve could be accurately approximated by a simpler mathematical equation.

We could regard the deviations from a 1:1 ratio of heads and tails as "errors" that shift the "measurement" away from its "true" value. It might seem perverse to do so, since if the coin tossing is done fairly and the counting is accurate, there is no "error" as such but just the operation of chance. However, in the 1770s Laplace realized that errors in measurement were also the outcome of factors that, being too complex to quantify, caused random divergences from true values. So he began to use de Moivre's approximate equation to quantify errors in astronomical measurements.

In the early part of the nineteenth century the French mathematician Joseph Fourier (1768–1830) also began to apply the error curve more widely. As director of the Bureau de Statistique of the Département de la Seine, Fourier published several papers on demographic statistics, and he helped introduce the curve into social science. Laplace too sought to put de Moivre's equation to work in social statistics. In 1781 he argued that the near equality of male and female births in Paris, which others had previously read as a sign of divine providence, was merely the expected result of a random process with two equally probable outcomes, whose variations were consistent with the error curve.

Quetelet was impressed by such examples when he encountered Laplace's work in Paris, and he began to suspect that the error curve was the fundamental leitmotif of human demography. In 1844 he showed that human dimensions—height and girth—were distributed with this same humpbacked profile. To Quetelet this was a sign of regularity and order in nature. Look at a crowded street and you will see people of many sizes. It may well seem at first that (within obvious limits) there is no predictability about the dimensions of a human being. But collect enough data and the bell-shaped curve will emerge.

ORDERLY BEHAVIOR

Quetelet acquired from the French scientists the idea that variation is linked with error. Instead of regarding height differences as a characteristic feature of nature, he saw them as departures from an ideal form. These "errors" become less prominent as greater numbers of people are taken into account, just as Poisson's law of large numbers says they should. This is true, Quetelet decided, not only for physical characteristics but for *behavior*, since the foibles of the individual temperament average out among the tendencies of the mass. Quetelet wrote in 1832 that "whatever concerns the human species, considered *en masse*, belongs to the domain of physical facts; the greater the number of individuals, the more the individual will is submerged beneath the series of general facts which depend on the general causes according to which society exists and is conserved."[19]

Since it was clearly a desirable thing that society should "exist and be conserved," this implied that average behavior was the "right" behavior.

And so Quetelet's social physics became founded on the concept of the "average man" (*l'homme moyen*), whose dimensions and physical features, and also moral and aesthetic attributes, represented a perfect mean to which all should aspire. To be great was to be average—"an individual who epitomized in himself, at a given time, all the qualities of the average man, would represent at once all the greatness, beauty and goodness of that being."[20]

This disturbing worship of uniformity has as its corollary an abhorrence of all variation: "Deviations more or less great from the mean have constituted . . . ugliness in body as well as vice in morals and a state of sickness with regard to the constitution."[21]

The idea of a physical and moral perfection of humankind that is reflected in the conformity to a mathematical ideal dates back to the Renaissance; but now there existed the tools to quantify what perfection was. In retrospect it is easy to read into Quetelet's theory of the average man overtones of racial purity and stifling social conformity; but in an age that believed in physiognomy, such views might have been deemed perfectly rational, although that does not excuse their sinister aspect.

In any event, the French government was soon to discover that *l'homme moyen* had his uses. In 1844, Quetelet compared the distribution of men's heights in the general population with that of 100,000 conscripts drafted into the French army, and found a discrepancy that led him to conclude that about 2,000 men had lied about their height so as to fall below the minimum and evade conscription.

Many of Quetelet's contemporaries were deeply impressed by and enthusiastic about the regularities that he perceived in the statistics of human affairs. Reviewing Quetelet's work in 1850, the eminent English astronomer John Herschel wrote, "No one has exerted himself to better effect in the collection and scientific combination of physical data in those departments which depend for their progress on the accumulation of such data in vast and voluminous masses, spreading over many succeeding years, and gathered from extensive geographical districts."[22]

Florence Nightingale urged that Quetelet's social physics, which she deemed to be an indicator of God's design, be taught at Oxford.* Karl Marx used

---

*Longfellow's romanticized image of Florence Nightingale as "the Lady of the Lamp" does her achievements few favors. She was more formidable than angelic, and would not otherwise have had

Quetelet's statistical laws in developing his labor theory of value. And John Stuart Mill, the utilitarian successor of Jeremy Bentham, felt that Quetelet's work lent support to his conviction that society and history were bound by laws as absolute (if harder to discern) as those of the natural sciences. In *A System of Logic* (1862), Mill had the universal error curve in mind when he wrote that the "very events which in their own nature appear most capricious and uncertain, and which in any individual case no attainable degree of knowledge would enable us to foresee, occur, when considerable numbers are taken into the account, with a degree of regularity approaching to mathematical."[23]

THE SCIENCE OF HISTORY

Awareness of Quetelet's work was promoted in Britain by one of the most avid proponents of a law-bound social physics, Henry Thomas Buckle (1821–1862). Like many adherents of Comte's positivist philosophy, Buckle wanted to fortify the world of human affairs against the meddling influence of governments. In Quetelet's view, governments had little effect on his statistical laws, which transcended human intervention. But Buckle maintained that one shouldn't even try to tamper with them; like Adam Smith, he argued actively for the principle of laissez-faire and for the need for people to be allowed to govern themselves. Left to their own devices, he believed, societies automatically produced "order, symmetry, and law," while "lawgivers are nearly always the obstructors of society, instead of its helpers."[24]

Metaphysical philosophers, in Buckle's view, had in the past pursued the futile goal of trying to unravel the way society works by worrying about what makes individuals tick. The empirical science of social statistics avoided such imponderables by discovering laws within the numbers. To make his case, Buckle felt he needed to show the play of these laws throughout his-

---

such an impact on medical conditions in the Crimean War. Less well known is her love of mathematics, in which as a girl she had begged her parents to let her be tutored. In the Crimea she collected statistics on mortality rates and devised a way of displaying them as wedge-shaped graphs. Quetelet's *Social Physics* was said to be the devout Nightingale's second Bible, and her copy was covered with her annotations.

tory. Traditionally, history had been a narrative about kings and queens. Buckle's history would be different: it would be a science.

Kant anticipated this quest for "historical laws" in alluding to the way that collective behavior smooths out individual unpredictability. In his essay "Idea of a Universal History from a Cosmopolitan Point of View" (1784), he said, "Individual men, and even whole nations, little think, while they are pursuing their own purposes . . . that they are advancing unconsciously under the guidance of a purpose of nature which is unknown to them."[25]

Buckle agreed; for him, history was ruled by the "great truth" that "the actions of men . . . are in reality never inconsistent, but however capricious they may appear only form part of one vast system of universal order . . . the underlying regularity of the modern world."[26]

Buckle set out the case for this universal order in his book *History of Civilization in England*. The first two volumes of this ambitious work were published between 1857 and 1861, but exhausted by his efforts, he died before he could complete it. As a consequence, the *History* says more about the rest of the world than about the author's homeland. Buckle had initially planned a world history and could not resist dispensing the fruits of this great vision in the initial volumes, even though he thought that England showed the operation of the laws of history more clearly than did any other country.

In this great compilation of regularities in social statistics, Buckle drew on many of the same examples as had Quetelet: rates of birth and death, crime, suicide, and marriage. The *History of Civilization in England* helped shape the British intellectual climate of the mid–nineteenth century around the idea of a liberal laissez-faire that held government to be largely unnecessary and consequently unwelcome. According to William Newmarch of the Statistical Society of London in 1860,

> Men are gradually finding out that all attempts at making or administering laws which do not rest upon an accurate view of the social circumstance of the case, are neither more nor less than imposture in one of its most gigantic and perilous forms . . . Every topic from the greatest to the least which the old legislators dealt with according to . . . caprice . . . have all been found to have laws of their own, complete and irrefragable.[27]

In the same year, the British economist Nassau Senior summarized the zeitgeist this way: "The human will obeys laws nearly as certain as those which regulate matter."[28]

To some observers, this new way of looking at human behavior was decidedly strange. In an 1850 issue of *Household Words*, a periodical edited by Charles Dickens, Frederick Hunt's amusement seems to sugarcoat a degree of skepticism: "the savants are superseding the astrologers of old days, and the gipsies and wise women of modern ones, by finding out and revealing the hitherto hidden laws which rule that charming mystery of mysteries—that lode star of young maidens and gay bachelors—matrimony."[29]

Ralph Waldo Emerson was also wary of the supposed certainty of statistical laws. In his 1860 essay "Fate," he explained the central claim of "the new science of Statistics": "that the most casual and extraordinary events— if the basis of population is broad enough—become matter of fixed calculation." He went on to mock what he saw as the rigidity of such an idea: "In a large city, the most casual things, and things whose beauty lies in their casualty, are produced as punctually and to order as the baker's muffin for breakfast. *Punch* makes exactly one capital joke per week; and the journals contrive to furnish one good piece of news every day."[30]

Mark Twain, meanwhile (in a remark he attributed to Benjamin Disraeli), dealt statistics an enduring wound by claiming that "there are three kinds of lies: lies, damned lies, and statistics."[31] And Friedrich Nietzsche, whose belief in the shaping of history by a few "great men" was second to none, was characteristically acerbic: "So far as there are laws in history, laws are worth nothing and history is worth nothing."[32]

FROM PEOPLE TO ATOMS

Others found inspiration in these manifestations of regularity within apparent randomness. The reliability of the statistical laws of society invited scientists to use them as analogies for similarly random processes in the natural world. Among these scientists was James Clerk Maxwell.

A few months after the publication of Buckle's great work, Maxwell wrote to his friend Lewis Campbell: "One night I read 160 pages of Buckle's *History of Civilization*—a bumptious book, strong positivism, emancipation from exploded notions and that style of thing, but a great deal of actually original matter, the result of fertile study, and not mere brainspinning."[33]

When Maxwell came to study the problem of gases in which the constituent particles were constantly engaging in collisions that none could hope

to follow, he recognized this as a problem of the same class as those that
Buckle had pondered in society, in which the immediate causes of individual
behavior must forever be inscrutable:

> The smallest portion of matter which we can subject to experiment consists of millions
> of molecules, not one of which ever becomes individually sensible to us. We cannot,
> therefore, ascertain the actual motions of any one of these molecules; so that we are
> obliged to abandon the strict historical [Newtonian] method, and to adopt the statis-
> tical method of dealing with large groups of molecules . . . In studying the relations
> between quantities of this kind, we meet with a new kind of regularity, the regularity
> of averages, which we can depend upon quite sufficiently for all practical purposes.[34]

As Maxwell indicated in 1873, the experiences of social statisticians lent
him confidence that this statistical approach could extract order from the
microscopic chaos:

> Those uniformities which we observe in our experiments with quantities of matter
> containing millions of molecules are uniformities of the same kind as those ex-
> plained by Laplace and wondered at by Buckle arising from the slumping together of
> multitudes of causes each of which is by no means uniform with the others.[35]

Would Maxwell have dared abandon the "strict historical method," the
obligation to explain everything in terms of the Newtonian mechanics of individual particles, if studies of society had not revealed laws at work even in
complex systems where the direct causes of behavior were obscure? How
otherwise might he have found the faith to look for laws in the face of woefully incomplete knowledge about motions?

Maxwell began his work on the kinetic theory of gases shortly after reading Buckle. But in his early work he also drew on the more analytical studies
of Quetelet, whose wide application of the error curve came to his attention
through John Herschel's description of the Belgian scientist's work in 1850.
There, Herschel had already alluded to connections between social physics
and the early kinetic theory of gases.

Maxwell knew that Rudolf Clausius had used probability laws in 1857 to
deduce the role of molecular collisions in the pressure exerted by a gas on
the walls that confined it. But Clausius was interested only in the average velocity of the particles. Maxwell wanted to know how the velocities were distributed around this average. If the error curve worked so well for describing

variations from the average in social physics, then it should work for him too. In 1859 he proceeded on the assumption that particle motions could be described by Quetelet's error curve, and was able to show what this implied for the measurable properties of a gas.

As we saw in the previous chapter, Maxwell's velocity distribution was merely an assumption until Ludwig Boltzmann showed in 1872 that any group of moving particles in a gas *must* converge on this distribution. Boltzmann also knew of Buckle's work and was not slow to draw analogies between his particles and the individuals in the social censuses that furnished Buckle's statistics: "The molecules are like to many individuals, having the most various states of motion, and the properties of gases only remain unaltered because the number of these molecules which on the average have a given state of motion is constant."[36]

Boltzmann likened the gas laws, a statement of the invariance of statistical averages, to the uniform profits of insurance companies. In 1886, Maxwell's friend Peter Cuthrie Tait compared the statistical approach of the kinetic theory with "the extraordinary steadiness with which the numbers of such totally unpredictable, though not uncommon phenomena as suicides, twin or triple births, dead letters,* &c., in any populous country, are maintained year after year."[37]

Today physicists regard the application of statistical mechanics to social phenomena as a new and risky venture. Few, it seems, recall how the process originated the other way around, in the days when physical science and social science were the twin siblings of a mechanistic philosophy and when it was not in the least disreputable to invoke the habits of people to explain the habits of inanimate particles.

The limitations, not to say the dangers, of reducing human affairs to statistical laws were, however, amply illustrated in other spheres. When Charles Darwin apparently turned humans into highly evolved apes, first in *The Origin of Species* (1859) and then more explicitly in *The Descent of Man* (1871), he appealed to chance and randomness as the engine of nature's variation. The analogy with the kinetic theory of gases was quickly appreciated. In 1877, Charles Peirce wrote:

---

*That is, those which remain in the postal system because they are badly addressed. Laplace had commented on how this was a constant fraction of the total turnover of the postal service.

> Mr. Darwin proposed to apply the statistical method to biology. The same thing has been done in a widely different branch of science, the theory of gases. Though unable to say what the movements of any particular molecule of gas would be, Clausius and Maxwell were yet able . . . by the application of the doctrine of probabilities . . . to deduce certain properties of gases, especially in regard to their heat-relations. In a like manner, Darwin, though unable to say what the operation of variation and natural selection in any individual case may be, demonstrates that in the long run they will, or would, adapt animals to their circumstances.[38]

Darwin's cousin Francis Galton saw that, as natural selection was basically a statistical theory, natural variation within a species could be tamed by Quetelet's error law. Galton's investigations of the statistical distributions of human features and behavior led him to conclude that there was "better" and there was "worse"—that such a distribution implied that men are not "of equal value, as social units, equally capable of voting, and the rest."[39] It was then but a short step to the idea of selective breeding to improve the distribution, as he argued in *Hereditary Genius* (1869). Galton's insistence on the need for statistics in studies of inheritance led him to establish the central mathematical basis of biometrics, the measurement of biological variation; but he is also notorious now as the progenitor of eugenics, an enthusiasm of the socialist left until its acceptance by the fascist right alerted the world to its real implications.

Herbert Spencer, whose infamous phrase "the survival of the fittest" has created much confusion about Darwin's theory, regarded Quetelet's findings as justifying an evolutionary approach to sociology. He also took from Quetelet's concept of *l'homme moyen* the dubious idea that natural selection in nature weeded out flawed variations and resulted in the survival of one "perfect specimen." More generally, the statistical aspect of Darwinism led many in the late nineteenth century to regard it as a theory as "mechanical" as Maxwell's, with effects arising from definite, if unknown, "forces." That was how Boltzmann, for one, saw it.

WILL AND DESTINY

Before the 1850s, "statists" were generally people who collected data about social habits and trends, and statistics itself was best regarded as the empirical arm of political economics. After that time, statistics became less a form of social science and more a *method*, a means of handling quantification in all scientific disciplines. Condorcet and Laplace were among those who foresaw this broad role of the mathematics of probability, but as the economist Antoine-Augustin Cournot admitted in his book on probability theory in 1843, many (himself not included) had more limited visions: "Statistics . . . is principally understood as the collection of facts to which the aggregations of men in political society give rise."[40]

It was the Englishman J. J. Fox who, in a paper presented to the Statistical Society of London in 1860, first clearly enunciated the idea that statistics was not a scientific discipline but a technique. Statistics, he said,

> has no facts of its own; in so far as it is a science, it belongs to the domain of Mathematics. Its great and inestimable value is, that it is a "method" for the prosecution of the other sciences. It is a "method of investigation" founded upon the laws of abstract science; founded on the mathematical theory of probabilities; founded upon that which had been happily termed the "logic of large numbers."[41]

This might make statistics seem a most modest thing, an instrument like Euclid's geometry or the calculus of Newton and Leibniz. Are tools not there simply to be taken up and used?

That the truth was so much otherwise reminds us not only how deeply nineteenth-century scientists examined the philosophical basis of their work but also how much they remained influenced by religious thinking. What statistical thought held up to the looking glass was nothing less than the concept of human free will.

From the outset, statistical approaches to social science were controversial. As statistics looked ever more likely to reveal the supposed natural laws of society, the question of what that implied for *individual human behavior* became impossible to ignore. This may quite reasonably be the first question in the minds of those encountering the new "physics of society" for the first time. The debate that raged (and it did rage) in the nineteenth century can usefully inform us about the arguments.

First, there is the issue of cause and effect—the conclusions we might draw by looking *back* to derive causes from their effects. Many statists believed, reasonably enough, that there was little point in assembling numbers if one was not going to interpret them. But interpretations immediately become politically charged. One of the central concerns of statists of the early nineteenth century (and it still preoccupies sociologists today) was crime. Could statistics tell one how to reduce it and thus achieve the great goal of the times, social improvement? For that, one needed to deduce the causes of crime.

The warning that a correlation between numbers does not necessarily reveal cause and effect ("correlation is not causation") is now almost a mantra among statisticians (though it is not always heeded). That is to say, if two sets of statistics show the same trend, it need not follow that one is caused by the other. Yet in the early days of statistics, many had no qualms about leaping to cause-and-effect conclusions that confirmed their prejudices. When the Frenchman A. Taillandier found in 1828 that 67 percent of prisoners were illiterate, the conclusion seemed clear to him: "What stronger proof could there be that ignorance, like idleness, is a mother of all vices?"[42] (Taillandier did not even bother to state, and presumably did not know, the illiteracy rate in the population as a whole.)

In the face of such abuses the council of the Statistical Society of London, whose cofounders in 1834 included such luminaries as Malthus, Charles Babbage, and William Whewell, tried to keep its practitioners within the proper bounds by announcing that the "Science of Statistics differs from Political Economy, because, although it has the same end in view, it does not discuss causes nor reason upon probable effects; it seeks only to collect, arrange, and compare."[43]

William Farr, of the British General Register Office, echoed the exhortations of the society to "exclude all opinions," telling Florence Nightingale in 1861, "The statistician has nothing to do with causation . . . Statistics should be the dryest of all reading."[44] This was, all the same, frequently honored only in the breach. As Alphonse de Candolle remarked in 1830, in the hands of policy makers statistics could become "an inexhaustible arsenal of double-edged weapons."[45]

Second, there is the matter of what statistics say about future probabilities—the conclusions we might draw by looking *forward*.

As long as social numbers stay in their tables—a mere record of events—they are uncontroversial. But statistics took on an entirely new significance when it became appreciated that these numbers held within them the potential for prediction. If, say, six in every hundred people in England who were alive at the beginning of 1790 had died by the end of the year, and if the same was true for 1791 and 1792, then would we not be justified in expecting, as 1793 dawned, that 6 percent of the current population would be dead by Christmas? That seems eminently reasonable, yet making the extrapolation proved furiously contentious.

It is one thing to know what has happened; it is quite another to claim to know what *will* happen. Of course, death rates are not rigidly constant—it may be that five people in a hundred died in 1791, and seven in a hundred in 1792. All the same, it would seem valid then to suppose that the deaths in 1793 will be close to six in a hundred, as opposed to twenty or fifty.

Yet such a suggestion would have provoked outraged opposition in 1793. How can one possibly know what the death rate will be? What if some fatal and infectious disease ravaged Europe that year? Or if another war broke out? Or even if it was simply an exceptionally healthy year and only one in a hundred died? No one, surely, could be certain that these things would not happen.

The distinction here is between statistics, which are numbers that represent facts, and probabilities. One deals in certainties, as long as good counting methods are employed; the other deals in the unknown. To some philosophers and scientists, these were chalk and cheese and should not be conflated. To do so was not just mathematically or logically bad form; it was heresy, promoting fatalism and undermining free will.

Kant had recognized in 1784 that the regularities evident in Johann Peter Süssmilch's tables of births and deaths seemed to confront belief in free will with a kind of determinism: "Whatever concept one may hold, from a metaphysical point of view, concerning the freedom of the will, certainly its appearances, which are human actions, like every other natural event are determined by universal laws."[46]

Statistical regularities were, however, not difficult to explain away in theological terms: they were evidence of God's wisdom. By engineering, for example, roughly the same number of baby boys as girls, He ensured that in principle there were marriage prospects for all and thereby maintained the stability of society.

But other constancies of statistics could hardly be rationalized so blithely. For what kind of God would arrange for the number of suicides or murders and other crimes to remain constant year after year? Moreover, these records were enough to trouble the unbeliever too. An atheistic biologist might not need to look far for reasons why the proportions of the sexes should be roughly equal. But acts of suicide and crime were utterly volitional and not obviously explicable in terms of any natural "mechanism."

To Adolphe Quetelet, statistical regularities in willful acts such as crimes placed these acts outside the responsibility of the individual: Quetelet was the first to suggest that crime is "caused" not by wickedness but by society. It should, he said, be attributed "not to the will of individuals, but to the customs of that concrete being that we call the people, and that we regard as endowed with its own will and customs, from which it is difficult to make it depart."[47] In other words, "Society made me do it," or as Quetelet said, crime is a "budget that is paid with frightening regularity."[48] In a mechanistic age, this implied not so much that the conditions of society tended to create a constant proportion of criminals as that there was some "force" that compelled people to break the law until the quota was fulfilled. Within such a philosophy the world is wholly deterministic and free will is nowhere to be found.

Wasn't this just a convenient excuse for criminals? That argument was neatly dispatched with the satirical response that, by the same token, a deterministic force compelled judges to sentence criminals in order to meet the year's quota of prisoners. But there were many, in the 1860s and 1870s, who were deeply disturbed by the fatalism that the statistics of Quetelet and Buckle seemed to invite. William Cyples lamented in the *Cornhill Magazine* that humankind was threatened with "a fate expressive in decimal fractions, falling upon us, not personally, but in averages."[49] In 1860, Prince Albert told the International Statistical Congress in London that some regarded statistics as leading "necessarily to Pantheism and the destruction of true religion, as it deprives, in man's estimation the Almighty of his power of free self-determination, making His word a mere machine."[50]

Fyodor Dostoevsky's alter ego in *Letters from the Underworld* rages against the determinism that statistics threatened:

> As a matter of fact, if ever there shall be discovered a formula which shall exactly express our wills and whims; if there ever shall be discovered a formula which shall make it absolutely clear what those wills depend upon, and what laws they are governed by, and what means of diffusion they possess, and what tendencies they follow under given circumstances; if ever there shall be discovered a formula which shall be mathematical in its precision, well, gentlemen, whenever such a formula shall be found, man will have ceased to have a will of his own—he will have ceased to exist.[51]

If this was possible, as Buckle insisted, then the future was bleak indeed:

> All human acts will then be mathematically computed according to nature's laws, and entered in tables of logarithms which extend to about the 108,000th degree, and can be combined into a calendar . . . In a flash all possible questions will come to an end, for the reason that to all possible questions there will have been compiled a store of all possible answers . . . Man will become, not a human being at all, but an organ-handle, or something of the kind.[52]

To avoid such a rationalistic, mathematical fate, says Dostoevsky's reclusive scribbler, men will always strive to exert their volition, even to the extent of making themselves act irrationally or insanely:

> He might act thus for the shallowest of reasons; for a reason which is not worth mentioning; for the reason that, always, and everywhere, and no matter what his station, man loves to act as he *likes*, and not necessarily as reason and self-interest would have him do . . . [53]

In his more measured way, Leo Tolstoy struggled in *War and Peace* with the questions posed by Buckle's deterministic view of history, with what Tolstoy called the "relation of free will to necessity." He suspected that this "new conception of history" might answer the fundamental question of international affairs: "What is the force that moves nations?"[54] But even conceding the existence of such forces seemed to throw into doubt the notion that we can choose our fate: "A particle of matter cannot tell us that it is unconscious of the laws of attraction and repulsion and that the law is not true; but man, who is the subject of history, says bluntly: I am free, and am therefore not subject to laws."[55] This defiance, Tolstoy concluded, depends on there being a residue of ignorance about the causes of events: "Free will is for history only an expression connoting what we do not know about the laws of human life."[56]

The writer and critic Maurice Evan Hare was more whimsical in 1905:

> There once was a man who said "Damn!
> It is borne in upon me I am
>   An engine that moves
>   In predestinate grooves
> I'm not even a bus, I'm a tram."[57]

Defenders of free will argued (with justification) that because statistical laws are not true laws in the sense of describing cause and effect, as Newton's law of gravitation does, they cannot be applied to individuals and so say nothing about how any one person might act. Conversely, the nineteenth-century faith in the "naturalness" of rationality allowed others to regard statistical regularities as a *demonstration* of free will—for had not Kant argued that free will itself tended to guide men toward orderly behavior? In this view, free will was to be equated not with mere caprice but with its opposite.

From a modern perspective, we can adduce several other considerations. Identifying randomness with maximal unpredictability seems intuitively sound but is not necessarily so, as we shall see in Chapter 10 in particular. Looked at from a distance, randomness becomes total uniformity: the randomly moving particles of a gas push equally in all directions and inflate a balloon into a sphere. Those phenomena that often strike us as the most complex are, in contrast, not random. The random, white-flecked "static" of an untuned television screen presents a chaotic image that in the end looks monotonous and unchanging, while in a movie we can never be sure (this is perhaps a generous assessment) what will happen next.

And there are, as we shall see, many examples of social behavior in which a kind of regularity and order comes not from any predestination in the fates of the participants but from the very limited range of their viable choices. When we walk down a corridor, in principle our options include taking a zigzag path from wall to wall or progressing in Ptolemaic epicycles; but no one whose faculties are intact chooses these over a more direct route.

Moreover, many of the phenomena we shall encounter in this book involve not steady behavior but abrupt changes between certain alternative modes of behavior. In the eighteenth and nineteenth centuries there was a strong belief in the *equilibrium* of society, its maintaining a steady, unchanging state. Here we shall often be more concerned with the making of choices and the sudden shifts this can provoke.

THE WILLFUL DEMON

If social statistics challenged free will at the level of the individual, the Second Law of Thermodynamics as formulated by Clausius posed a deeper kind of determinism that was equally troubling. As William Thomson and Hermann von Helmholtz argued in the 1850s, the inexorable rise of entropy—heat passing forever from hot to cold—implied a cosmic "heat death" in which all of creation was reduced to a vague buzz of faint and useless heat, unable to bring about the organization that life requires. Not with a bang but a whimper, indeed. And the Second Law insists that before this dreary end, all change has a preferred direction, in which entropy increases. Does this not suggest that just as a ball released at the top of a hill has to roll down it, so humans composed of so many dancing atoms have to behave in a certain way? Free will implies that we can do one thing or another; the Second Law seemed to be saying that wherever change is possible, there is only one way it can happen.

Maxwell was as deterministic as Laplace when it came to the operation of Newton's laws of motion; but unlike Laplace, he had a very strong need for the hypothesis of God. He was a devout Christian and could not accept a universe in which God deprived man of free will. How, though, could free will operate without contravening thermodynamics?

Alfred Tennyson sought escape from the fatalism of atoms with the poetic license a scientist could not afford. Maxwell recounted Tennyson's description of a dream in which the poet

> Saw the flaring atom streams
> And running torrents of her myriad universe
> Running along the illimitable inane,
> Fly on to clash together again, and make
> Another and another frame of things,
> For ever.[58]

He noted how Tennyson "attempted to burst the bonds of Fate by making his atoms deviate from their courses at quite uncertain times and places, thus attributing to them a kind of irrational free will."[59]

Maxwell's answer was more sophisticated. He appreciated that the Second Law is a statistical law: a "law of large numbers." In 1867 he saw a way to "pick a hole," as he put it, in this cosmic edict. The statistical inevitability, he said, is just the result of our ignorance: we do not and cannot know the motions of all the atoms even in a tiny scrap of matter. But they were not unknowable *in principle*. Suppose, said Maxwell, a "very observant and neat-figured being"[60] were to exist at a scale that allowed him to watch the atoms in flight, one by one. Such a being could subvert the Second Law by selectively picking out for special treatment atoms moving in a certain direction: by exercising his free will in conjunction with his superhuman knowledge.

Say, for instance, this perspicacious being operated a trapdoor in a wall dividing a gas-filled vessel into two compartments. By opening the trapdoor only to gas particles passing in one direction, he could cause the number of particles, and thus the pressure, in one compartment to increase at the expense of the other. This would contravene the Second Law, as it produces a *less* probable configuration of particles than the one we started with. If we humans were to make a hole in the dividing wall, we would never bring about anything but an equalization of pressures—that is by far the most "probable" outcome.

William Thomson dubbed Maxwell's being a "demon," much to the devout Maxwell's disapproval. Despite the apparent success of his argument, Maxwell feared that his demon alone might not be enough to rescue free will. Through the 1870s he continued to look for other loopholes in physical law that would permit free will to operate without violating the principle of energy conservation (the First Law of Thermodynamics). But this quest was doomed. Decades after thermodynamics had been connected to the theory of information devised by telecommunications engineer Claude Shannon in the 1940s, scientists uncovered the flaw in Maxwell's argument: he had neglected to take into account the thermodynamics involved in the processing of information that the demon must conduct. That is, the demon cannot make a choice about whether or not to open his trapdoor without generating at least as much entropy as is "saved" by letting a particle through. So even Maxwell's demon is not immune to the Second Law.

It is often said that quantum mechanics destroyed the deterministic world of Newtonian mechanics by introducing probability into the very heart of matter. Certainly there is a big difference between uncertainty about the motion of objects *in practice*, which Maxwell and Boltzmann accepted in developing statistical mechanics, and uncertainty about such motions *in principle*, as embodied in Erwin Schrödinger's wave mechanics and, in particular, Werner Heisenberg's Uncertainty Principle in 1927. Quantum mechanics says that there are some things that we not only do not know but *cannot* know.

Yet the path toward the probabilistic physics of quantum mechanics was surely cleared by the introduction of statistics into "classical" physical science in the late nineteenth century. By 1918 the Polish physicist Marian Smoluchowski considered probability to be central to modern physics: "From this trend, only Lorentz's equations, electron theory, the energy law, and the principle of relativity have remained unaffected, but it is quite possible that in the course of time exact laws may even here be replaced by statistical regularities."[61]

The way to statistical science would have been more tortuous if nineteenth-century experience with social statistics had not given scientists the confidence to believe that large-scale order and regularity in nature can arise even when we do not know, or cannot even meaningfully propose, a determining cause for each event. In such situations, we must trust that there are laws within large numbers.

FOUR

THE GRAND AH-WHOOM

WHY SOME THINGS HAPPEN ALL AT ONCE

Nature has her own best mode of doing each thing, and she has somewhere told it plainly, if we will keep our eyes and ears open.   —Ralph Waldo Emerson (1860)

In the statistical method of investigation, we do not follow the system during its motion, but we fix our attention on a particular phase, and ascertain whether the system is in that phase or not, and also when it enters the phase and when it leaves it.
                                                                    —James Clerk Maxwell (1878)

The Attractions of Gravity, Magnetism, and Electricity, reach to very sensible distances, and so have been observed by vulgar Eyes, and there may be others which reach to so small distances as hitherto escape Observation; and perhaps electrical Attraction may reach to such small distances.   —Isaac Newton (1704)

Toward the end of Kurt Vonnegut's book *Cat's Cradle*, something happens that is inevitable from the outset, yet it is no less chilling when it comes: "There was a sound like that of the gentle closing of a portal as big as the sky, the great door of heaven being closed softly. It was a grand AH-WHOOM."[1]

That was the sound of the sea freezing to ice. Not ordinary ice, which one would expect to creep down slowly from the poles if the entire Earth were to drop below freezing point. The sea in Vonnegut's book freezes to *ice-nine*, a hypothetical form of ice that is stable up to a hundred degrees centigrade. When a block of ice-nine falls into the waves, it transforms the water of the oceans instantly into this more stable solid.

This is the nature of freezing: it is abrupt. Either a substance is liquid and mobile (above its melting point), or it is solid and rigid (below this temperature). There is nothing in between—water does not become sluggish and treacly before it turns to ice. There is always a Grand Ah-Whoom.

The same is true when water boils: it is either bubbling in the pan or escaping in wisps of steamy vapor.* There is always a surface at which liquid water changes unambiguously to steam.

Such transformations between solid, liquid, and gas are called *phase transitions*, and they provide the missing link between the kinetic theory of gases and a molecular-scale understanding of the other forms of matter. In ice, water, and steam, the particles are all identical—they are molecules of water, the atomic trio of H_2O.† But they are organized differently: dashing about in isolated, random frenzy in the gas; regimented and immobile in the solid; and jostling each other like a dense, unruly crowd in the liquid.

Maxwell and Boltzmann uncovered the microscopic underpinnings of the behavior of gases. But what could statistical mechanics say about liquids and solids? There was no way to derive the Grand Ah-Whoom from the Maxwell-Boltzmann theory of gases. It was plain enough to every nineteenth-century scientist that the way to make a liquid from a gas is to cool it down. Cooling even further gives you a solid. But if you lower the temperature in the kinetic theory of gases, all that happens is that on average the molecules move more slowly. The bell-shaped curve gets sharper and shifts its peak to lower speeds. Only by cooling to the lowest temperature possible—absolute zero, or minus 273°C—do you end up with particles that have no energy and do not move. But there is no prescription in the theory to make them line up regularly as they do in crystalline solids like ice. And in any event, real solids appear well before absolute zero is reached. And what about the liquid state that precedes them?

This chapter is about phase transitions, which turn out to be one of the central concepts underlying the physics of society. Others have invoked phase transitions, often unknowingly or in a metaphorical way, to explain processes of sudden change in social contexts, for example in René Thom's catastrophe theory or Malcolm Gladwell's concept of a "tipping point" in

---

*The wispy stuff you see above boiling water is, however, liquid—tiny droplets of water that have condensed from the vapor as it rises into the cooler air. Water vapor is as invisible as air itself.
†Strictly speaking this is not quite true, as a few molecules fall apart into electrically charged fragments called ions.

trends, norms, and fashions (see page 321).[2] Postmodernist architect Charles Jencks talks of "the creativity and surprise of a universe that evolves in phase changes—sudden jumps in organization" to justify an architecture based on surprise, "broken symmetry," and apparent instability.[3] Thomas Kuhn's "paradigm shifts" in the evolution of scientific thought[4] have also been touted rather vaguely as an analogue of phase transitions. But we shall see that phase transitions are not merely a convenient allegory for abrupt shifts in modes of behavior or thought. They really do seem to happen in society, and the physical theory developed to understand them is to some extent directly transferable to descriptions of social behavior.

In appreciating the role that phase transitions play in statistical physics, you can get a long way simply by knowing that they exist, that they are abrupt, and that they connect states of matter in which particles are arranged in different ways. But a crucial aspect of the new physics of society requires us to know something of *why* and *how* phase transitions occur. That aspect has to do with the curious abruptness of these changes. For how, if each particle is happily going its own way, do they all suddenly conspire to create a Grand Ah-Whoom?

CONTINUITY PROBLEMS

At face value, it is precisely this abruptness that might seem to make the lack of an explanation of phase transitions in the kinetic theory of gases a less pressing concern. For does it not imply that something special is going on at this point at which steam condenses to water, something that does not apply when steam is content to be steam? The theory's failure to account for the transition meant that it was missing something; but as long as it was used only to explain the particles' behavior above the boiling point of water, wasn't that all right?

Nineteenth-century physicists knew that, sadly, it was not right at all. They knew that it was possible to transform a gas to a liquid *without* going through a phase transition. In the jargon of the time, they knew that there was *continuity* between the two states.

A French nobleman named Charles, Baron Cagniard de la Tour, was the first to show that this was so. In 1822 he found that if ether, alcohol, or water is heated in a sealed tube, it can apparently be interconverted smoothly between liquid and gas, without the transitions of evaporation or condensation. "There can be little doubt," John Herschel concluded in 1830,

> that the solid, liquid and aeriform states of bodies are merely stages in a progress of gradual transition from one extreme to the other, and that, however strongly marked the distinctions between them appear, they will ultimately turn out to be separated by no sudden or violent lines of demarcation, but shade into each other by insensible gradations.[5]

From his careful studies of the "Cagniard de la Tour state" in the 1860s, Thomas Andrews, a chemist at Queen's College in Belfast, concluded that indeed, "the ordinary gaseous and liquid states are, in short, only widely separated forms of the same condition of matter and can be made to pass into one another by a series of gradations so gentle that the passage shall nowhere present any interruption or breach of continuity."[6]

This was possible only above a certain temperature, which varied from one fluid to another. Below this temperature an abrupt phase transition always intervened. We call this changeover point the *critical* temperature or, more loosely, the critical point.*

The Dutch physicist Johannes Diderik van der Waals (1837–1923) accomplished the tremendous feat of explaining all of this while connecting the kinetic theory of gases to the existence of liquids. He did so in his doctoral thesis of 1873, on which, however, he had embarked without this aim in mind at all. Van der Waals had the more modest ambition of explaining certain aspects of the theory of capillarity. Derived largely by Pierre-Simon Laplace in the early nineteenth century, this theory describes how liquids behave close to solid surfaces. But van der Waals dug into ground that by the 1870s was fertile enough to yield a more abundant harvest.

The kinetic theory explains how the pressure of a gas results from the motion of its molecules. Van der Waals knew that a liquid also exerts a pressure on a surface with which it is in contact. It was recognized that this pres-

---

*For water this critical temperature is 374°C. In other words, below 374°C you cannot change water from liquid to gas (or vice versa) without an abrupt phase transition. Above 374°C you can make water progress smoothly from a liquidlike to a gaslike density (by altering the pressure) without any sudden jump.

sure was actually huge, perhaps in the region of several thousand atmospheres, for liquids are much denser than gases and so there are many more molecules bouncing off a given area of surface. But no one knew how to calculate the pressure of a liquid from its known physical characteristics. That is what van der Waals decided to do. The only theory that enabled a pressure to be derived from the microscopic motions of the individual particles responsible for it was the kinetic theory of gases, so this was van der Waals's starting point.

Like gases, liquids are made up of molecules in rapid, disorderly motion. But liquids, unlike gases, have a degree of cohesion: a raindrop holds together as it falls. Laplace supposed in 1806 that this cohesion was due to the tendency of all the particles to attract one another. This attractive force explains why liquids have a surface tension: the mutual tug of molecules at the surface forms a kind of "skin."

The kinetic theory of gases takes no account of such an attractive force. This is not because the molecules in a gas do not attract one another as they do in a liquid. Rather, it is because the attraction has a very short range—typically it becomes negligible once two molecules are separated by a distance of more than their own width. In the gas, the average distance between molecules is much larger than this, so they are simply not close enough to "feel the force." Molecules in a gas do approach one another and collide occasionally, but the attractive force does not cause them to stick together, because they are moving too fast. So it is an effective simplification to ignore attractive forces altogether in a gas. To describe the pressure of a liquid, van der Waals thought of it as a kind of sticky gas, a gas whose behavior is modified by attractive forces.

He included another factor in his theory too. Molecules are very small, but not infinitely so. Yet the standard kinetic theory treats them as infinitesimal "points" which possess mass but no size. Van der Waals supposed that in a liquid, which is much denser than a gas, the space occupied by the molecules is not insignificant. The volume of space in which each molecule can wander is then not the same as the volume of the container in which the molecules are confined, but is less than this by an amount equal to the total volume of all the molecules.

None of this was exactly new. In the eighteenth century Daniel Bernoulli had pointed out the need to account for molecular size, and in 1863 the

French scientist Gustave-Adolphe Hirn had considered how both this size and the existence of attractive forces might modify the way a gas behaves. It was already known that many gases don't behave quite as the gas laws (and the kinetic theory which accounts for them) say they should; and it was generally assumed that factors such as the finite size and mutual attractions of particles probably explain this deviance.

In van der Waals's thesis, "On the continuity of the gaseous and liquid states," however, all these notions came together to unify the understanding of liquids and gases. By making certain assumptions about the character of the attractive forces and of particle-size effects, the Dutchman showed that within a certain range of temperatures and pressures a fluid can exist in either of two different densities. He realized that the denser state must correspond to the liquid and the more rarefied state to the gas. Moreover, he could see that at some point as a gas was compressed or cooled, it would become unstable and switch to the liquid state. He predicted that there would be a phase transition.*

The crucial thing about this switch is its abruptness. You cool a gas, and go on cooling it, and it is still a gas, until suddenly—*whoom!*—there is a liquid. Van der Waals showed that the density of a fluid is not arbitrary: it may take a small value (gaslike) or a fairly large value (liquidlike), but not something in between. These are the only two *stable states* of the collection of particles. That is the key fact we must bear in mind. You might look at the particles and say, "Oh, I'd like to arrange them in *this* way"—but even supposing you could perform such a feat of manipulation, you would probably find that "this way" isn't actually a stable state of the system and the particles quickly revert to one of the two allowed configurations. *When it comes to collective behavior, not all things are possible.*

What about the *continuity* of liquids and gases? After all, this was the topic that van der Waals's thesis announced in its title. His theory predicted that as the temperature gets higher, the difference between the densities of the liquid and gaseous states gets smaller. At a certain temperature this density difference vanishes, and the fluid can then exist in only one state: nei-

---

*Van der Waals could not, however, predict at what point this phase transition would take place as the gas's temperature or pressure was changed. Maxwell, who gave the theory his cautious approval, showed how to do this a few years later.

ther liquid nor gas but something in between. Thus van der Waals predicted the critical point.

The big deal about van der Waals's theory is not immediately obvious, but it is vital to what follows. When the fluid switches from a gas to a liquid or vice versa, the individual particles do not change. They are still behaving like hard little balls, each with its own attractive, short-range force field. But their *collective* state is decidedly (and abruptly) different: one moment a tenuous gas, the next a viscous liquid. No prescription for condensation or evaporation is "built into" the particles—these things just happen, as if all the particles make up their minds to do the same thing at the same time. Except that particles have no minds. So we cannot inspect a single particle and say, "Oh yes, this is obviously a particle that likes to form liquids." Only when a whole mass of particles is considered together can we know what they will do.

Van der Waals identified the ingredients necessary for this behavior. At face value, he seemed to be saying that an attractive interaction between particles was enough to promote liquefaction of a gas. But in fact there is a subtle balance of *attraction* and *repulsion*. Van der Waals thought that molecules experience a kind of repulsion due to heat—due, in other words, to the fact that they are in rapid motion. But the truth was that he had introduced the notion of repulsion as soon as he took account of molecular size, because this implies that once particles are touching one another, they cannot approach any closer. That is to say, there is a repulsion that kicks in once particles come into contact. This seems obvious, but if we neglect size and treat the particles as infinitesimal dots, there is nothing to stop them from coming infinitely close.

So a phase transition arises out of compromises. A balance of attraction and repulsion makes for a stable liquid. If the force of disorder (that is, heat) gets too great, the gas becomes more stable instead. Moreover, the tension between these factors causes not gradual but catastrophic change: a landslide victory for one influence or the other. That landslide is the Grand Ah-Whoom.

A UNIFYING PRINCIPLE

Van der Waals's theory helped bring the liquid state of matter within the province of statistical mechanics. It garnered widespread praise: Maxwell himself averred that "there can be no doubt that the name of van der Waals will soon be among the foremost in molecular science."[7] And so it was, for he was awarded the Nobel Prize in physics in 1910.

But the theory did not seem at first to be saying anything more general about the way matter behaves. It could not describe the phase transitions of freezing and melting that separate a liquid from a solid. Still less did it appear to have any broader relevance for physics.

Yet phase transitions crop up in quite different situations. It has been known for centuries that magnets lose their magnetism if they are heated (and regain it when cooled). For magnetic iron these changes happen at about 770°C, a temperature well within the reach of a medieval blacksmith's forge. In the same year that van der Waals presented his thesis at Leiden, an Englishman named William Barrett proposed that heat-induced "demagnetization" is not a gradual process but happens suddenly at a particular temperature: it is a phase transition.

In 1889 John Hopkinson of King's College, in London, came to the remarkable conclusion that the loss of magnetism, when put in quantitative terms, shares mathematical similarities with the way that a fluid loses its distinct liquid and gas states at the critical point. This discovery is often mistakenly attributed to the Frenchman Pierre Curie, husband of Marie, who came to the same conclusion in 1895. This is because the French physicist Pierre Weiss, who developed a theory of the magnetic transition in 1907, named a magnet's transition temperature the Curie point in memory of his compatriot, who had died in a road accident the year before.*

The atoms in a solid block of iron are arrayed like eggs in an egg carton, immobile and ordered, not at all like the chaos of particles in liquids and gases. And yet Weiss's theory uses some of the same concepts that van der Waals had invoked to explain phase transitions and the critical point in fluids. How could that be?

---

*Hopkinson had equal claim to such a sentimental memorial: he died tragically in a climbing accident in 1898, along with three of his children.

To understand this, we will be helped by considering a model of magnetism devised in 1920, after Weiss's work, by the German physicist Wilhelm Lenz. A lump of iron is magnetic because each of its atoms acts like a tiny magnet. We can think of these atoms as little compasses, with magnetized needles that have a tendency to line up with one another. Physicists call these "atomic needles" *spins* (although this does not mean that they are literally spinning). In iron the orientation adopted by each spin is determined by the magnetic fields created by all the surrounding spins. Thus each atom's spin influences the spins of all its near neighbors in the latticelike array. Typically, these magnetic interactions encourage adjacent spins to align with one another. So the most stable state of the array is that in which all "needles" point in the same direction. In this configuration the little magnetic fields of each atom add up to create one big magnetic field, and the iron is a magnet.

But just as heat's disruptive influence opposes the attractive forces between atoms in a liquid, so too does it tend to undermine the alignment of magnetic spins. The effect of heat is like shaking each of the atomic compasses, subverting their ability to line up with one another. Even though the atoms themselves stay in their serried ranks, sufficient heat can shatter the orderly alignment of spins and leave them pointing in random directions. Then their little magnetic fields cancel one another out on average, and the iron becomes nonmagnetic.

So is the demagnetization transition like the evaporation of a liquid? Not quite. The temperature at which it occurs—the Curie point—is sharp and well defined, but the magnet does not switch abruptly from being a strong magnet to being nonmagnetic. Instead, the strength of the magnetism (called the magnetization) falls steadily toward zero as the Curie point is approached. Thus the Curie point resembles the liquid-gas critical point, at which the distinction between liquid and gas dwindles steadily to nothing.

This is a genuine phase transition: there is a clear-cut change between a nonmagnet above the Curie point and a magnet below it. But it is a different *kind* of phase transition from evaporation, condensation, melting, or freezing. Phase transitions that happen as a system passes through a critical point are called, naturally enough, critical phase transitions (or sometimes, for technical reasons, second-order phase transitions). Ones that involve sudden jumps in some property of the system, such as density, are called first-order phase transitions (Figure 4.1).

Lenz decided to describe all of this with the simplest model he could think of. Instead of regarding iron atoms as compasses with smoothly swinging needles, he assumed that their spins could point in only one of two opposing directions—say "up" or "down." Thus, neighboring spins could either be aligned or opposed, but nothing in between. Some magnetic metals do indeed behave like this (but not iron, as it happens). Lenz also assumed that each atom, which occupies a position on a regular grid, feels the magnetic fields only of its immediate neighbors.

In 1925, at Hamburg, Lenz's student Ernst Ising figured out how this model behaved in the simplest possible case. A real magnet contains a three-dimensional lattice of atoms stacked in ordered layers. This exposes each atom to the influence of many neighbors. A simpler case is a two-dimensional magnet, with atoms sitting on a flat grid like a checkerboard: here the neighbors are fewer. But Ising opted for an even simpler model—a one-dimensional magnet, in which the atoms are lined up in a row, each one flanked by just two others. This of course bears very little resemblance to a lump of iron, so it was perhaps not surprising that Ising obtained a rather disappointing result. His one-dimensional magnet exhibited no phase transition until it was cooled to absolute zero: −273°C. Only then did all the atomic compasses

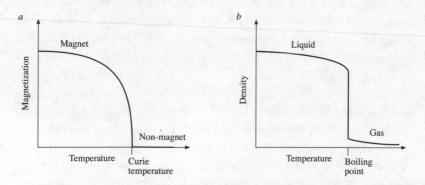

4.1 Two different kinds of phase transition. In a *critical* (or second-order) transition, some property that is characteristic of the entire system (such as the magnetization of a magnet) falls gradually to zero as a "control parameter" (such as temperature) is changed (*a*). In a *first-order* transition (such as freezing or boiling), a property (such as the density of a fluid) changes abruptly at the transition point (*b*). Both types of phase transition are encountered in the physics of society.

become aligned. The slightest heating enabled the forces of disruption to overwhelm those of order, and the row of atoms lost their magnetization.

Ising never achieved anything else of note in science: he became a schoolteacher, fled the anti-Semitism of Hitler's Germany in 1938, and ended up teaching physics in the United States, where he died at the age of ninety-eight. But he has been granted some kind of immortality, because models based on a "lattice" of magnetic atoms are still widely used by statistical physicists and have become known not by Lenz's name but as "Ising models."

In the Ising model, short-range "order promoting" interactions between individual particles compete with the generalized disrupting effect of heat. Yet to extract from it anything vaguely resembling real-world behavior, one has to advance from one to two dimensions. This is the hard part, and it took nearly twenty years more before any progress was made.

The Norwegian physicist Lars Onsager (1903–1976) virtually had to invent a new kind of mathematics before he could deduce how the 2-D Ising model behaves, which he finally accomplished in 1942.* Unlike the 1-D model, this checkerboard magnet (Figure 4.2a) *does* undergo a phase transition from a magnetic to a nonmagnetic state at some temperature above absolute zero, and it is a critical phase transition like the one that happens at the Curie point of real magnets. But if solving the 2-D model was difficult, doing it for the real-world three-dimensional case seems to be impossible. No one has yet succeeded, and there is some evidence that it is an intractable problem.

These days, when scientists are confronted with intractability, rather than throw up their hands in despair they switch on their computers. Although it seems impossible to solve the algebraic equations that describe the 3-D Ising model, it is quite straightforward to simulate on a computer how the model should behave. It is rather like predicting the weather: it is impossible to solve all the equations that describe the patterns of air movement, but we can instead run a computer simulation to see what comes out. Not surprisingly, the 3-D Ising model turns out to have a critical phase transition.

---

*The German-British scientist Rudolf Peierls predicted this behavior in 1936 but was unable to prove it rigorously.

4.2   The Ising model of magnetism (a) assumes that the "magnetic needle," or spin, of each atom can point in only one of two opposed directions. The atoms lie on a regular lattice in one, two (as shown here), or three dimensions. In the magnetic state, all the spins are aligned. Above a critical transition temperature the spins get randomized by heat and there is no overall magnetization. The same model can also be used to describe the transitions between a liquid and gas, and their merging into a single fluid state at the critical point (b). In this case there are two types of grid site in the model: occupied by a particle (corresponding to the dense liquid state) and unoccupied (gas). Below the critical point, the system is either a liquid or a gas. Above the critical point, it is (on average) a single fluid of intermediate density.

I have probably not made a very strong case for why we should expect any analogy between critical phase transitions in liquids and gases and those in magnets. Sure, there are some vague similarities in the atomic-scale picture of interacting particles, but there are some pretty substantial differences too. In 1947 a Japanese scientist named S. Ono suggested that the Ising model could be regarded as a crude model of fluids, because both have a "binary" nature. Instead of an array of cells containing atomic spins that point "up" or "down," one can imagine an array of cells either occupied or unoccupied by a particle, corresponding to (dense) liquidlike and (tenuous) gaslike states (Figure 4.2b). But this too might seem to be stretching a point.

Yet there is in fact an astonishing and deep connection between these different transitions. I shall consider it in more detail in Chapter 10; here I shall sketch only the briefest of supporting arguments. I indicated earlier that the closer you get to the critical point of a liquid and gas, the smaller the density difference between the two states becomes. One can put numbers to this. At 99 percent of the critical temperature the density difference might be, say, a factor of 2 (that is, the liquid is twice as dense as the gas). At 99.5 percent it might fall to 1.5, and so on. There is a characteristic *rate* at which the fluid approaches its critical point.

The same is true of a magnet near its Curie point, where it is the magnetization rather than the density difference that is falling to zero. And the extraordinary thing is that, expressed in terms of percentages, these rates are identical for fluids and for a certain class of magnets. Experimentally, the magnetization of a magnet and the density difference in a fluid fall to zero at precisely the same rate. What is more, a computer simulation of a 3-D Ising model shows that it, too, exhibits the same rate of "critical onset," even though the Ising model is only a very crude model of either a fluid or a magnet. The details differ dramatically in these three cases, and yet their "critical behavior" follows the same trajectory.

Physicists call this *universality*—a term that aims to convey that there are some processes in the world for which the details simply do not matter. It is surprising enough that (as we shall see below) two different fluids, such as carbon dioxide and methane, which have quite different critical temperatures, should approach their critical points at the same rate in relative (that is, percentage) terms. It is baffling that two wholly different *kinds* of system—a fluid and a magnet—also display this universality.

What this suggests is that phase transitions are generic phenomena: they happen in the same way for a wide range of apparently different systems. A physicist can stand up and talk for an hour about a first-order phase transition, and no one will stop him to ask, "Wait, are you speaking here about freezing, or evaporation, or what?" The audience knows that what is being said applies to all such cases. In the same way, a town planner can speak about traffic jams without anyone asking, "Do you mean the snarl-up last Thursday at Parson's Corner?" Every traffic jam involves a different set of vehicles and circumstances, but there are features common to them all.

This still sounds, perhaps, like a very broad generalization to be drawing from some correspondences in the mathematical details of the way magnets and fluids behave. But there is more to phase transitions than this.

NEAR ZERO

Van der Waals's theory described how gases and liquids *in general* are connected by a phase transition below their critical point. What some scientists wanted from such a theory was the ability to predict how and when this change of state would happen for *particular* substances.

All fluids are clearly not the same. Water is liquid under conditions in which carbon dioxide is a gas; and carbon dioxide liquefies at a higher temperature than nitrogen. This is no mystery: different molecules have different sizes and exert different forces of attraction. But van der Waals realized that the critical point is the point of reference that reveals all these specifics.

If you were to survey the heights of all schoolgirls between the ages of twelve and fourteen in a London school and plotted the results on a graph, you would obtain something like the familiar bell-shaped curve of Figure 3.2 (page 61). The number of pupils of each specific height would rise to a maximum and then tail off. If you repeated the survey for all the schools in London and plotted the results on the same graph, the curve would have the same shape but would be much taller—because the absolute numbers (plotted on the vertical axis) are bigger. In a survey of the whole country the peak would be even higher.

But the *proportions* would be the same in each case: the fraction of girls two inches shorter than average would be more or less identical in every survey. So you could collapse all the bell-shaped curves onto a single "master curve" by plotting not absolute numbers but relative proportions: the fraction of the total that are four feet four inches tall, four feet six inches tall, and so on.

If you did the same for boys in the same age range, the shapes of the curves would be the same but they would be slightly shifted on the horizontal axis because the most common height would be slightly greater.* Yet even the two master curves for boys and girls can be superimposed onto a

---

*It is actually only around the age of twelve or thirteen that boys start to attain a greater average height than girls.

single curve, by making the heights relative instead of absolute. That is, you plot the fraction of boys and girls that deviate from the mean by 2 percent, 5 percent, and so on.

Van der Waals did much the same "rescaling" of curves for liquids and gases. He saw that the relationships between pressure, temperature, and density for different substances all fall onto the same "master curve" if expressed in relative terms by reference to the values of these quantities at the critical point. The relative temperature, for instance, is the temperature divided by the critical temperature. This so-called principle of corresponding states shows that all liquids and gases are more or less the same "master fluid" rescaled by some factor related to the critical point. And the rescaling is determined by the properties of the individual particles (molecules) from which the substance is constituted, because the critical temperature, pressure, and density can be calculated, in van der Waals's theory, from these single-particle characteristics: their size and the range and strength of their attractive forces.

The director of the Physical Laboratory at Leiden, where van der Waals was working, was Heike Kamerlingh Onnes (1853–1926). He became interested in the principle of corresponding states for very practical reasons: it allows one to predict how fluids will behave far below the critical point purely from a knowledge of the critical properties (the critical temperature, density, and pressure). By scaling the "master curve" according to these quantities, you can deduce the relationships between pressure, temperature, and density for the entire range at which the liquid state is stable. Kamerlingh Onnes was interested in finding out how cold some gases had to be before they would liquefy.

Helium was especially resistant to liquefaction, and at the turn of the century the existing cryogenic techniques could not produce temperatures low enough to liquefy it. By fitting the experimentally observed behavior of helium at temperatures above its critical point to van der Waals's equations, Kamerlingh Onnes predicted that he would need to reach just 5 or 6°C above absolute zero before he could make it become liquid. This was a reasonable estimate: in fact helium does not condense until 4.2 degrees above absolute zero at atmospheric pressure, a temperature that Kamerlingh Onnes finally achieved in 1908.

Having liquid helium at his disposal, Kamerlingh Onnes was able to use it as a cooling fluid to investigate how other substances behave at such frigid extremes. Physicists at that time suspected that because the heat-induced vibrations of atoms in metals disturb the flow of electrical current through them, the metals might become much better conductors at very low temperatures, approaching "perfect" (resistance-free) conductivity at absolute zero. Yet when Kamerlingh Onnes put this to the test with mercury in 1911, he got a surprise. The electrical resistance of mercury did not decline smoothly as the temperature dropped: it fell abruptly to zero at around the boiling point of helium. At this point mercury becomes a "superconductor," able to carry a current that is totally unimpeded by electrical resistance.

It was soon found that other metals display the same behavior above (but close to) absolute zero. Lead, for example, superconducts at 7.2 degrees above absolute zero. The change to a superconducting state has all the characteristics of a critical phase transition: the resistance plunges swiftly to zero as the superconducting transition temperature is approached, just as the magnetization of iron dips to zero close to its Curie point.

That was not all. In 1937, in Moscow, the Soviet physicist Pyotr Kapitsa discovered that if liquid helium is cooled even further below its boiling point, at a little over 2 degrees above absolute zero it develops extremely bizarre properties. It loses all viscosity, and once it begins to flow, it never stops. This form of liquid helium will even crawl up the side and out of a container. It becomes a so-called superfluid.

A theoretical explanation for superfluidity came in the late 1930s. The superconductivity of metals was harder to understand, and was not satisfactorily explained until 1957. Both of these exotic phenomena happen because at very low temperatures the laws of quantum mechanics take precedence over the laws of "classical" physics, which dominate at higher temperatures. Yet both come about via genuine phase transitions. They are both manifestations of *collective* behavior arising from the interactions between the atomic-scale components of the materials. Thus phase transitions are not restricted to particles behaving in accord with the good old-fashioned Newtonian laws of motion. It is often said that quantum mechanics overturned all of the "classical" physics that went before. But it certainly did not overturn the physics of phase transitions: quantum mechanics merely lends a different flavor to it (one which becomes evident only at low temperatures).

Indeed, phase transitions seem to persist however much one reshapes the fundamentals of physics. To understand the internal structure of sub-atomic particles such as the protons and neutrons that make up the nuclei of atoms, the "traditional" quantum theory of the 1920s is not enough. Physicists in the 1970s had to develop a new theory, called quantum chromodynamics. It uses many of the conceptual tools of statistical mechanics, such as lattice models like the Ising model, and finds abrupt phase transitions taking place among its sub-subatomic constituents: sudden jumps between different stable arrangements of these particles. And many cosmologists believe that a phase transition reshaped the entire universe during an absurdly brief instant a tiny fraction of a second after the Big Bang. They propose that during this "inflationary era" the universe experienced a cosmic phase transition in which it expanded from a size much smaller than a proton to a size comparable to its present gargantuan dimensions. As phase transitions go, this has to be the most dramatic Ah-Whoom ever.

TIME FOR A CHANGE

You might now be imagining that all abrupt change is due to phase transitions. That is not so. If I turn on a light by flicking a switch, there is suddenly light where before the room was dark—but no phase transition has made this happen. I have opened a valve, that's all, and an electrical current has flowed. The key notion about a phase transition is that it happens everywhere throughout a system (that is, *globally*) all at once, and it does so because of a conspiracy of countless constituents.

A phase transition is a sudden global change in behavior arising from the interactions of many constituent particles. Typically these interactions are short-range, *local* ones: each particle takes heed only of its immediate neighbors and neither knows nor cares what is happening further afield. The phase transition happens when some global influence which acts on the particles reaches a certain threshold value. One moment the particles are behaving "normally," as if nothing were amiss; then without warning (or almost so, as we shall see) they switch to some entirely different mode of behavior.

The statistical mechanics of Maxwell and Boltzmann, humbly conceived to explain how gases behave, has now mutated into the discipline called, less archaically, *statistical physics*. Statistical physicists have traditionally been interested in the complexities of how inanimate matter behaves, particularly when it undergoes phase transitions, especially critical transitions. This science is statistical because it generally deals with systems made up of huge numbers of particles or separate components, all interacting with one another so that the average behavior is often all that counts.

Statistical physicists are now starting to glimpse vistas far more enchanting and seductive than gases in pots. To reach this vantage point, they have first had to labor furiously to reshape the tools developed by the great scientists of another century to new uses for which they were never originally designed. Traditional statistical mechanics deals with some extraordinary things, but it struggles to say much about the dynamic world in which we go about our business. Nineteenth-century thermodynamics dealt with *equilibrium* states, to which nothing was ever added or taken away and in which, on average, nothing ever changed. In the following chapter we shall see how statistical physics today is more like life: full of the processes of growth and decay.

FIVE

ON GROWTH AND FORM

THE EMERGENCE OF SHAPE AND ORGANIZATION

The reasonings about the wonderful and intricate operations of Nature are so full of uncertainty, that, as the Wise-man truly observes, hardly do we guess aright at the things that are upon earth, and with labour do we find the things that are before us.
—Stephen Hales (1727)

Are the attributes of a society, considered apart from its living units, in any way like those of a not-living body? or are they in any way like those of a living body? or are they entirely unlike those of both?

—Herbert Spencer (1876)

Order is the handmaiden of all virtues! But what leads to order?
—Georg Christoph Lichtenberg

A phase transition brings magic to high places. Walking in wintry woods in 1856, Henry David Thoreau eulogized the tiny ice crystals that lay sparkling all around him: "How full of the creative genius is the air in which these are generated! I should hardly admire them more if real stars fell and lodged on my coat. Nature is full of genius, full of the divinity; so that not a snowflake escapes its fashioning hand."[1]

Snowflakes are formed in the atmosphere when water vapor freezes into ice: a transformation directly from gas to solid. What can statistical mechanics tell us about these six-pointed crystals, which scientists have studied ever since the seventeenth century?

Until rather recently the answer was, not a lot. If you had asked a statistical mechanician of the 1940s to tell you about snowflakes, he might have shuffled in an embarrassed way and said, "Well, you see, the theory wasn't designed for that." Wasn't designed to tell us about one of the most common phase transitions of one of the most plentiful substances in the world? What kind of theory is this, after all?

Traditional statistical mechanics arguably owes its current form to one man above all: the American scientist Josiah Willard Gibbs (1839–1903). In his book *Elementary Principles in Statistical Mechanics* (1902), this Yale professor brought all the threads together and wove them into a robust and elegant tapestry. Building on the work of Clausius, Maxwell, Boltzmann, and van der Waals, among others, Gibbs made thermodynamics thoroughly coherent and established exactly how its laws arise from the underlying microscopic description of a system.

Gibbs showed that the key consideration in processes of change is *minimization*. Simply put, this is akin to water flowing downhill and gathering in depressions as ponds and lakes. The higher the water is to start with, the more gravitational, or "potential," energy it has. If unimpeded, it will minimize this energy by moving to as low a position as possible. Although it is not immediately obvious, this is a direct corollary of the Second Law of Thermodynamics—the stipulation that entropy increases in all spontaneous change.

When water collects in a lake, it reaches a kind of equilibrium: it stops moving downward and just sits there. Thermodynamics is all about the drive to find equilibrium states like this. It describes processes of change from one equilibrium state to another, even more stable state, like the opening of a sluice gate in a mountain reservoir to let water flow from a higher basin to a lower one. Gibbs stated that for a system to be in an equilibrium state (a stable state with no immediate propensity to change), "it is necessary and sufficient that in all possible variations of the state of the system which do not alter its energy, the variation of its entropy shall either vanish or be negative."[2]

Which is a way of saying that when a system is in equilibrium, you can push, pull, and prod it all you like, and it will simply return to how it was to begin with. This is the paradox of "classical" thermodynamics. It seeks to account for change, but it can't actually say anything about the process of change itself. It can only provide prescriptions for the starting point and the end point, and falls silent on the question of what happens in between.

Now think of the snowflake. It began as water vapor and ended as ice: two equilibrium states, separated by a phase transition. But one snowflake grows into a six-pointed star with branching Christmas-tree arms. Another snowflake has arms ending in cloverlike clusters. A third is made of blocky hexagons. A fourth is like a bulbous six-petaled flower. You can look through the classic snowflake portrait album, *Snow Crystals* (1931) by Wilson A. Bentley and W. J. Humphreys, and see two thousand crystals, none of them identical. Each has a unique history of growth: it clearly matters very much *how* the water vapor became solid.

While snowflakes show nature at its most inventive, they are by no means peculiar in their great variety. Does one tree ever resemble another in exact detail? Is one day's global weather pattern ever like any other? In a river flowing incessantly around a rocky outcrop, is the pattern of water flow at one instant ever exactly like that at another? As the Greek philosopher Heraclitus pointed out in the sixth century B.C., you can never step in the same river twice.

The processes of change that go on around us are very seldom the kind of leaps between Platonic equilibrium states that thermodynamics is designed to handle. Many are ongoing processes—as if that stream of water were forever meandering in the hills, looking for a nice stable basin to fill, while rain forever replenishes its source. They are, in short, processes that are not in equilibrium and never will be, or at least not in our lifetimes.

Even when there do seem to be stable starting and end points to a transition, like vapor and crystal, the *form* that results from the change can be complex and impossible to predict. This is because, in the case of a snowflake, the *growth* process takes place far away from any equilibrium states. It is as though, rather than opening a sluice gate to transfer water smoothly downhill, the entire bank of a reservoir is suddenly destroyed so that the water gushes forth on a course that no one can anticipate.

This is not to imply that thermodynamics is theoretically neat but practically useless. On the contrary, it is immensely useful. It enables us, for instance, to understand and predict the direction of a change, and to know under what conditions the change will occur—to explain, for example, why water freezes at 0°C at sea level. The thermodynamics of equilibrium states can be used to understand energy generation in living cells and in power stations, the formation of minerals in the Earth's crust, and heat dissipation in a computer. But it is a limited theory. If we are dealing with processes taking place far from any stable, unchanging equilibrium, and in particular if we are dealing with processes of growth and form, something else is needed.

HISTORY MATTERS

I have taken my chapter title from one of science's truly classic texts, fit to share the bookshelf with *The Origin of Species*, *Principia Mathematica*, and *Traité élémentaire de chimie*. But although Darwin's book initiated modern biology, Newton's modern physics, and Lavoisier's modern chemistry, the Scotsman D'Arcy Wentworth Thompson's idiosyncratic and eclectic book *On Growth and Form* spawned no definable field of science. Instead, since its publication in 1917, it has awoken generations of scientists to the realization that we live in a world of profound beauty in which there is still much we do not understand.

In this book Thompson (1860–1948) was attempting to challenge the tendency, prevalent in his time (and far from absent today), to answer all questions of biological form by pulling Darwin out of a hat and proclaiming the magic word "Adaptation!" Thompson argued that much in nature could be explained purely on the basis of geometry, mathematics, physics, and engineering, without the need to invoke natural selection.

In many ways *On Growth and Form* was limited by its being ahead of its time, for it laid out many instances of natural form resulting from what we would now identify as *nonequilibrium* growth processes. In 1917 the concept of change away from equilibrium was barely grasped even by the founders of thermodynamics, let alone by a Scottish zoologist, however erudite he was in Greek, Latin, and geometry. When considering ink droplets dispersing in water, cracks spreading in mud, or liquids circulating convectively in a pan, Thompson was forced to be descriptive rather than analytical. He noted that some systems undergoing change reach "a 'steady state,' if not a stable equilibrium,"[3] but could say no more about the nature of these steady states. They were clearly not, however, Gibbs's placid equilibrium states.

Lars Onsager (page 90) knew this. He was one of the first to face up to the shortcomings of standard thermodynamics, and in the 1930s he began to do something about it.

By restricting itself to equilibrium states, thermodynamics might appear

to have painted itself into a corner. How can change happen, except by disrupting that tranquillity? Once you have opened the sluice gates of our
mountain reservoir, the water it contains is no longer in equilibrium. So until all the water has emptied into the lower reservoir and stillness is restored,
nonequilibrium reigns. The same is true of water freezing to ice. It cannot do
so infinitely fast, and while it is freezing, thermodynamics can no longer describe what is going on.

Classical thermodynamics gets around this with ingenious legerdemain.
It treats the process as if it were happening very, very slowly—strictly
speaking, infinitely so. That makes it possible to look at the system at any
moment and say, "Why, nothing is happening: it is still in equilibrium." It is
like drilling the tiniest of tiny holes into the bank of the reservoir so that the
trickle it releases is barely perceptible. True, the lower reservoir then fills up
drip by drip; but at any one moment both the higher and lower reservoirs appear to be in stable equilibrium states.

This might seem to be at odds with my earlier suggestion that phase transitions are abrupt changes. But by "abrupt" I do not necessarily mean "instantaneous," although I must confess that invoking Kurt Vonnegut's Grand
Ah-Whoom rather implies that interpretation. I mean that next to no change
in the prevailing conditions is required to induce the transition. To cross the
boundary from liquid water to ice, you need only alter the temperature by an
infinitesimally small amount. At one hundredth of a degree above 0°C, the
equilibrium state is that in which all the water is liquid. At one hundredth of
a degree below zero, all is ice.

Many transformations in the real world, unlike the kind of pseudo-
equilibrium processes that thermodynamics can handle, occur in circumstances when the "driving force" is much greater than the minimum
necessary—for example, when the temperature of water falls suddenly far
below zero. Snowflake formation is such a case: it happens at temperatures
several degrees below water's normal freezing point.

Onsager considered that the best starting point for developing a thermodynamics of nonequilibrium states was one in which the states depart only
slightly from equilibrium, impelled by relatively small driving forces. We don't
open the sluice gates but simply make the hole rather larger than a pinprick.

Not even nonequilibrium processes escape the Second Law; they, too,
proceed by increasing the total entropy of the universe. But whereas equi-

librium thermodynamics tells us only that the entropy will be greater *after* than *before*, an unfolding, dynamic nonequilibrium process prompts us to consider how entropy is changing over time. Onsager began wondering about the *rate* at which such processes generate entropy.

For equilibrium states the criterion for determining how a system will behave is the Gibbs condition: the system will configure its components so as to minimize the energy.* Onsager looked for a rule equivalent to the Gibbs condition that would determine the character of a nonequilibrium "steady state." Even away from equilibrium, systems are capable of adopting states that in some sense remain the same. A river is not at equilibrium, since its waters are constantly flowing from high ground to low, but it generally stays in a steady state in which the waters are confined between its banks at a roughly constant level. Living cells are also in such a "dynamic" steady state, maintaining their integrity and their function while constantly burning up energy and churning out wastes. Nonequilibrium steady states are ubiquitous: a whirlpool's vortex, a cruising automobile, the ebb and flow of the tides.

Onsager did not, in truth, get very far with elucidating the criteria that make specific nonequilibrium steady states preferred over other possible states of a system. He showed that close to equilibrium, there are general rules relating the driving force to the rate of entropy production in a system, which was in itself a significant achievement in virgin territory, deemed worthy of a Nobel Prize in 1968. But he found no universal Gibbsian principle for nonequilibrium thermodynamics.

There is a good reason for this: it is almost certain that no such principle exists. Ilya Prigogine, a chemist born in Russia and working in Brussels, believed he had found the magic formula in the 1940s. He claimed that the most favorable nonequilibrium steady state, at least for only small departures from equilibrium, is that which minimizes the rate of entropy production. One still sees this criterion quoted as fact, but sadly it is not universally true.

---

*By a "system," I mean any collection of components that we choose to define, such as the water molecules that make up a snowflake. We could equally well consider as a system a mountain lake and the surrounding slopes and valleys, or the planets orbiting the sun. What matters is that we be able to define, to a sufficient degree of accuracy, the relevant components of the system and how they interact with one another. A further point is that "energy" is actually too vague a term here; Gibbs gave a more precise definition in terms of a quantity called "free energy," which is a well-defined portion of the system's total energy.

Does this mean, then, that away from equilibrium, anything goes—that all things are possible? Clearly not. Nonequilibrium processes do seem to make choices that, while not always predictable in advance, are nevertheless consistent and reproducible. And what is truly extraordinary is that many of these choices correspond to states that are not chaotic and messy, but instead display a high degree of orderliness.

One of the classic examples of an ordered, nonequilibrium steady state was discovered by the French scientist Henri Bénard in 1900.* He warmed a thin layer of liquid in a copper dish, which set up convection currents: warmer, less dense liquid at the bottom rises to displace cooler, denser liquid at the top. This is a nonequilibrium system because of the differences in temperature in different parts of the liquid; at equilibrium, the temperature is equalized everywhere and there is no convective flow. The system is driven away from equilibrium by the heat coming from below and will be prevented from equilibrating as long as heat is supplied (since the top surface, where heat escapes, will always be cooler).

If the heating is very gentle, however, no convection takes place; heat is simply redistributed by conduction through the fluid. Only when the temperature difference between top and bottom reaches a certain threshold do convection currents start to circulate from bottom to top and back again. Bénard saw that the currents organize themselves into roughly hexagonal cells in which the fluid rises in the center and sinks around the edges (Figure 5.1a).

Under different experimental conditions, a variety of other convection patterns can be seen (Figure 5.1b, c). D'Arcy Thompson noted this diversity in convection patterns in smoke and commented on their similarity to cloud patterns, "as of a dappled or mackerel sky"[4]—for indeed clouds can be sculpted by orderly convection currents in the atmosphere. But he could not say much about *why* they are formed. In 1916 the physicist Lord Rayleigh was able to explain the roll-like convection state shown in Figure 5.1b using the theory of fluid mechanics; a better theoretical understanding of the other convection patterns emerged in the 1960s. Yet there is still no theory that can reliably predict which pattern will appear in any given experiment un-

---

*Although these patterned states are associated with Bénard's name, they were actually first observed thirty years earlier by the German physicist Georg Hermann Quincke.

a   b

c

5.1 Henri Bénard found that a convecting fluid can become patterned into circulation cells defined by rising and sinking streams of fluid. In *a* these cells, visualized by adding metallic flakes to a liquid, form an orderly hexagonal arrangement. In *b* and *c*, for a different fluid and a different visualization method, the cells are roll-like.

der a given set of conditions. There is no Gibbs-like criterion for nonequilibrium pattern formation.

Searching for a priori predictability is in fact futile because Rayleigh-Bénard convection patterns (as they are now called) can differ even under seemingly identical conditions, if their means of preparation are different. That is to say, if you take two different experimental routes to reach the same end point—heating at different rates, say, or stirring and not stirring initially—then the resulting patterns may differ. These nonequilibrium steady states depend on their *histories*.

Rayleigh-Bénard convection patterns are examples of *dissipative structures*: organized arrangements in nonequilibrium systems that are dissipating energy (the convection patterns are maintained by the constant supply of heat, for example) and thereby generating entropy. In the 1950s and 1960s, Prigogine and his co-workers suggested that dissipative structures are reached when a nonequilibrium system is driven to a crisis point, called a *bifurcation*. Close to equilibrium, the system might not do anything very remarkable. The fluid in a Bénard dish, for example, just sits there conducting heat, apparently at rest. At the bifurcation point, however, the system is abruptly compelled to change its state dramatically.

As the term implies, there are generally two choices on offer. In the roll-cell Rayleigh-Bénard state (Figure 5.1*b*), adjacent rolls rotate in opposite directions like the rollers of a mangle. But any one roll could rotate in either of two ways—clockwise or counterclockwise—provided the directions of all the others were reversed too. So there are two possible "global" roll-cell states, equivalent in every respect except the direction of rotation. What determines the choice? Pure chance—or more properly, *fluctuations*, which physicists often call *noise*.

Noise is everywhere. At any temperature above absolute zero, atoms are shaking with thermal energy. This sets up a random background "buzz" that pervades all matter. The buzz gets "louder" as the temperature rises: the forces of disorder exert themselves relentlessly. Because of this haphazard aspect of atomic motions, all processes incur small random variations, or fluctuations. If one could measure the pressure on a very small area of a balloon's rubber wall with tremendous sensitivity, one would find tiny variations caused by the moment-to-moment differences in the number of gas molecules striking it. Indeed, scientists now routinely make high-precision measurements of quantities such as pressure and temperature at very small scales, and invariably find that these quantities fluctuate around some average value.

Normally the effect of these fluctuations is negligible, in keeping with their tiny magnitude. But at a bifurcation point, a nonequilibrium system is poised on a razor's edge. Will it take this path, or that one? The smallest chance fluctuation can tip the balance and irrevocably determine the future fate of the system. Prigogine explained that "in the neighborhood of the bifurcation points fluctuations play an essential role and determine the 'branch' that the system will follow."[5]

Increasing the driving force of a nonequilibrium process even further beyond a bifurcation point can produce a switch to another steady state: a different pattern, attained by a second bifurcation. In general, says Prigogine, there is a cascade of bifurcations through which a nonequilibrium process might progress as it is driven further from equilibrium (Figure 5.2).

At each branching point the options are well defined, but the choice is arbitrary. So two systems that are wholly identical at the outset might end up on quite different branches while experiencing the same driving force, sim-

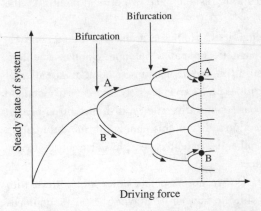

5.2   Ilya Prigogine predicted that nonequilibrium steady states are reached via a cascade of bifurcations, at each of which the system must choose between two alternatives. Because this choice is determined by random fluctuations, two initially equivalent systems (here denoted A and B) driven out of equilibrium can end up on different "branches"—in different steady states—while experiencing the same driving force, because they have a different history of choices.

ply because they happened to take different paths at each junction. "Time forks perpetually towards innumerable futures," as Jorge Luis Borges says in his story "The Garden of Forking Paths."[6] But whereas the people in the book written by Borges's character Ts'ui Pên may venture simultaneously down both of the routes open to them, real systems can take only one of the possible ways ahead. So, of course, can real people, and a long and countless succession of such choices makes the world what it is and not some other place. "In this way," says Prigogine, "the bifurcation introduces *history* into physics and chemistry, an element that formerly seemed to be reserved for sciences dealing with biological, social, and cultural phenomena."[7]

Away from equilibrium, then, Gibbs's determinism gives way to historical contingency. Ironically, perhaps, it is this that stymies Prigogine's attempt to find a minimization principle for nonequilibrium thermodynamics—for in *any* steady state out of equilibrium, what matters is not just the prevailing conditions influencing the system's components but how these conditions were arrived at.

Nevertheless, there is a striking and important similarity between nonequilibrium bifurcations and equilibrium phase transitions. A bifurcation is a sudden global change to a new steady state. That sounds familiar. Indeed, the bifurcation point bears a strong resemblance to a critical point, like the Curie point of a magnet.

As a metal is cooled through its Curie point, it changes from a nonmagnet to a magnet.* In the nonmagnetic state all the atomic "compass needles" (spins) are randomly oriented, while in the magnetic state they are aligned in an orderly fashion. This critical phase transition therefore corresponds to the onset of an *ordering* process. Similarly, when a dish of fluid is heated above the threshold for convection to begin, this nonequilibrium bifurcation orders the fluid into ranks of roll cells. In both cases, physicists say that *symmetry breaking* has taken place.

Why symmetry *breaking*? Do we not associate order and pattern with symmetry, and randomness with the lack of it? Maybe so, but randomness has its own kind of symmetry. A system in which all the components are moving at random is, on average, *more* highly symmetrical than one in which

---

*Note that not all metals are magnetic, and not all magnetic metals align all their spins at a Curie point.

they are moving in concert in a certain direction. In the random state, any direction is indistinguishable from any other. In a convecting roll-cell state, the direction parallel to the rolls is clearly different from the direction perpendicular to them: the rolls "point out" a special direction in space. So when a uniform fluid is transformed to one that is circulating in roll cells, some symmetry is lost, or broken. The same is true of a magnet: the aligned spins point out a special direction in space whereas in the nonmagnet there was none.

A bifurcation offers two equivalent choices of steady state, and so does the transition to a magnet. Think of Ising's model, in which each spin may point in just one of two directions: "up" or "down." The spins can line up in either of these two directions, with equal but opposite magnetization (see Figure 4.2, page 91). And what of the liquid-gas critical point, which we compared with the Curie point? The fluid can adopt either a gaslike or a liquidlike configuration below the critical point.

Moreover, these choices at a critical phase transition are, like those at a bifurcation point, at the mercy of fluctuations. There is no reason for the magnet to prefer one direction of alignment over the other: but a chance excess of one type of spin in some part of the system can tip the balance throughout. So systems undergoing critical phase transitions become hypersensitive to fluctuations. We shall see later that this gives rise to very special and remarkable behavior at critical points.

These correspondences do not imply that there is after all some nonequilibrium statistical mechanics that parallels that of equilibrium systems. No, the truth is both simpler and deeper. The two kinds of transformation—phase transitions and nonequilibrium bifurcations—have features in common because they are both fundamentally of the same ilk: they are both *collective* modes of behavior arising from the mutual, local interactions of many individual components. There are conditions both in equilibrium and away from it for which these interactions can make one part of a system almost miraculously sensitive to what is happening far away. Every particle is suddenly in touch with all the others via intricate networks of mutual nudges—and all at once, a new steady state emerges.

THE SHAPE OF CULTURES

To develop a physics of society, we must now take a bold step that some
might regard as a leap of faith and others as a preposterous idealization. You
may have guessed it already: particles will become people. To make that
bold step a little easier, I shall introduce a stepping-stone that will bring life
into the picture before we have to worry about such things as free will.

It would be hard to make the case that bacteria, lacking anything resembling a brain or a nervous system, possess free will. Yet they are certainly
alive. Possessing only primitive means of communication, bacterial cells
can display rich and varied patterns of collective behavior in nonequilibrium growth processes.

The Japanese physicist Mitsugu Matsushita of Chuo University discovered in the 1980s that these patterns derive from a kind of bacterial physics.
Louis Pasteur once said that fortune favors the prepared mind, and Matsushita was prepared for his foray into the life sciences by his background in
the statistical physics of nonequilibrium growth. When he saw the complex
branching pattern of a growing colony of *Bacillus subtilis* bacteria (Figure
5.3), he knew what he was looking at.

It was a *fractal*: a structure that reiterates at successively smaller scales

5.3 Under certain condi-
tions, the bacterium *Bacillus
subtilis* grows in colonies that
have a complex *fractal* branching shape, resembling the pattern seen in many growth
processes in nonliving systems.

what it displays at larger ones. Matsushita recognized the kind of fractal branching pattern that is characteristic of a process called diffusion-limited aggregation (DLA), which occurs, for example, when metal collects at a negatively charged electrode dipped into a solution of a salt: the phenomenon of electrodeposition.

What captivated physicists in the early 1980s, when they first began to study fractal shapes like this produced in nonbiological processes like electrodeposition, was that these patterns resemble the "organic" forms of nature: branched fractal mineral deposits in rocks, for example, are sometimes mistaken for fossilized ferns. Now here in Matsushita's laboratory was a genuinely biological fractal.

The fractal shape of a DLA deposit looks fairly chaotic, but it is not arbitrary. No two deposits are ever identical, but they all have a feature in common, namely the efficiency with which the branches fill up the space available to them. No matter how big the structure grows, there will always be "fjords" of empty space penetrating deep into the ramified form. There is a mathematical measure of this efficiency of space filling, called the *fractal dimension*.* All DLA growth patterns have the same fractal dimension. It is a badge of identity which can distinguish the branching growth from some other superficially similar structure.

In 1981, U.S. physicists Tom Witten and Len Sander proposed a theoretical model of the DLA process, motivated by their attempts to describe how dust particles clump together in air. The model invokes particles undergoing a Brownian random walk (that is, diffusing; see page 42) in air and sticking together when they come into contact. They found that this process creates unstable branch tips that cannot help splitting in two at every opportunity. In 1984, Matsushita showed that the clusters that emerge from this DLA model are precisely like those formed in electrodeposition in a flat dish.

---

*The fractal dimension is not a whole number, implying that fractal objects incompletely occupy the space they inhabit. A block of stone is a three-dimensional object: it fills up a specific three-dimensional volume of space. Likewise, a square drawn on the flat, two-dimensional plane of a piece of paper occupies a specific surface area: it is a two-dimensional object. And a simple line is one-dimensional. But a fractal shape like the one in Figure 5.3 is somewhere between one-dimensional and two-dimensional. This particular shape is around 1.7-dimensional: it has a fractal dimension of 1.7. This means that it "fills up" more of the two-dimensional space than does a simple line, but not as much as say, a disk.

Matsushita's branching colonies of *B. subtilis* did not just resemble DLA clusters; they had the same fractal dimension. This suggested (although it did not prove) that the formation processes in both cases shared the same essential features, most notably constant tip splitting induced by random fluctuations in the advancing edge of the cluster.

DLA is a nonequilibrium process, and fractal patterns are one of the motifs of nonequilibrium growth. The immediate and irreversible sticking of particles wherever they strike the growing cluster leaves them no opportunity to find their most stable equilibrium configuration. The DLA cluster is a map of frozen accidents of history.

Matsushita and his co-workers wondered what might happen if they modified the driving force pushing their fractal bacterial colonies away from equilibrium. So they altered two factors that they presumed would affect the growth process. The colonies were cultured on a layer of transparent gel called agar, suffused with the nutrients needed for the cells to grow and multiply. As the amount of water in the gel diminishes and the material becomes more rigid, the cells stick ever more firmly to the surface. So by varying the agar-to-water ratio in the gel, the researchers could control the cells' mobility. And by changing the nutrient content they could alter the "health" of the colony—its ability to generate new cells.

When they altered these two parameters, Matsushita's team found that the colonies developed growth shapes very different from the branching DLA structure. At high nutrient levels the colony was dense, with fat, fingerlike branches only at the perimeter, rather like a blob of lichen on a rock face. This shape resembles that produced in a theoretical growth model devised by the mathematician M. Eden in 1961 to describe the development of cancer tumors.

Both the DLA-like and Eden-like colonies appear on hard agar surfaces, on which the cells cannot move. The colony advances as new cells are generated at the edge. But if the gel is softer and the cells can move, other patterns appear: thin, radiating branches at low nutrient levels, and Eden-like concentric rings if nutrient is abundant. If the bacteria are both fully mobile and abundantly fed, they advance rapidly in a roughly circular front, leaving behind a colony so sparse as to be barely visible.

Thus the "space" mapped out by the latitude and longitude of gel hardness and nutrient content (the "control parameters") is divided into distinct regions, each with its own characteristic growth pattern (Figure 5.4). The changes from one pattern to another happen relatively abruptly—that is, as a result of relatively small changes in one or another of the control parameters. This reminded Matsushita and colleagues of the "phase boundaries" that separate the gas, liquid, and solid states of a substance in a space mapped out in temperature and pressure. Crossing a phase boundary means undergoing a phase transition, and the map of the boundaries is called a phase diagram. The Japanese researchers considered that they had discovered a kind of phase diagram for bacterial growth.

5.4   Bacterial colonies grown in a petri dish of agar gel develop patterns that depend on two factors: the amount of nutrient available and the hardness of the gel. These patterns can be organized in a *morphology diagram*, as shown here. The switches between distinctly different types of pattern can be fairly abrupt. The boundaries between morphologies are shown here in gray, while the dashed line indicates the change from immobile (left) to mobile (right) cells as the gel medium becomes softer.

The changes in colony shape cannot be *rigorously* compared with phase transitions, however. For one thing, the system is not at equilibrium. It is probably better to compare these changes with those in a dissipative system such as Rayleigh-Bénard convection as it switches from one pattern of convection cells to another. Researchers studying bacterial pattern formation often speak of a "morphology diagram," an analogue of a phase diagram that classifies the shapes of colonies and the conditions under which they occur (*morphology* here means nothing more than "shape").

Some other kinds of bacteria show different, complex growth patterns, and also undergo abrupt changes from one to another as the growth conditions are varied. We shall see in the next chapter that some of these patterns can now be understood by making simple assumptions about what controls the cells' movements, in the same way that the complex DLA cluster can be understood from the model of particle motions and interactions developed by Witten and Sander. From simple "rules" dictating how individual cells behave, we can deduce the global patterns into which they can form.

ICE FLOWERS

The concept of a morphology diagram for nonequilibrium growth processes has a history, for in the 1930s the Japanese scientist Ukichiro Nakaya, working at Hokkaido University, drew a similar picture for snowflakes. Bentley and Humphreys's book is filled with wondrous six-pointed stars, but in the final pages one finds something quite different. Here are snowflakes that look more like architectural designs: columns topped with tablelike plates, rectangular blocks, and even what appear to be ice sundials (Figure 5.5). These are snowflakes formed under unusual atmospheric conditions, such as extremely low temperatures. Nakaya and his co-workers conducted experiments that mimicked the formation of natural snowflakes. They grew flakes attached to a single strand of rabbit's hair placed within a stream of moist air in a very cold room.

The researchers found that as the air temperature changes, the shape of the snowflakes alters at certain thresholds. Hexagonal columns of ice grow below about −25°C, for example. The air humidity also affects the shape. Between about −5°C and −22°C the flakes are flat plates if they are formed in

5.5 Not all snowflakes are branched. Some have less familiar shapes, such as columns and plates.

relatively dry air, whereas the familiar star shapes appear if the moisture content is higher. So temperature and humidity are the control parameters that map out a "morphology space" for snowflakes, just as gel hardness and nutrient level do for bacteria.

Even though no two snowflakes, nor any two bacterial colonies, are identical, they can nevertheless be divided into distinct *classes* of growth pattern, separated by what we might call shape transitions. There is, in other words, a kind of order underlying the apparent profusion of shapes: each individual growth pattern may be uniquely embellished, but for a given set of growth conditions there is an inevitability in what we might regard as its Platonic form. It is in this sense that an experiment in snowflake growth is repeatable: the details may differ, but the form stays the same.

Yet where does this individuality come from? It illustrates the contingency of nonequilibrium growth: a snowflake arm branches *here*, and not *there*, because of some fluctuation during the growth process which sparked off the budding of a new arm at that stage. Snowflakes experience so-called growth instabilities which amplify small random bulges into new branches. The same happens in the DLA process, resulting in the ramified fractal patterns of Matsushita's bacterial colonies. But snowflakes are not so random: their alluring beauty lies in their six-pointed symmetry, which was first clearly identified by the German astronomer and mathematician Johannes Kepler in 1610. This hexagonal symmetry is imposed by the regular arrangement of water molecules in ice: a crystal lattice that picks out six "special" directions in space. New branches of crystalline ice sprout preferentially in these directions. In effect, the growth of the crystals is constrained by an underlying hexagonal grid, such that the geometric orderliness at the molecular scale becomes manifest as a regularity evident at the much larger scale of the whole snowflake. This interplay of chance and regularity in snowflake growth is a subtle affair whose details came to be understood only in the 1980s.*

The influence of an underlying symmetry on an otherwise random growth process has been demonstrated strikingly by growing fractal bacterial colonies on a layer of gel stamped with a hexagonal grid of grooves. The grid channels the branching process so that the resulting colony looks rather like a snowflake (Figure 5.6).

5.6 "Bacterial snowflakes" can be grown by letting a colony develop on a sheet of gel stamped with a hexagonal grid of grooves. This biases the direction in which new branches sprout and extend.

---

* I discuss the modern understanding of this process in my book *The Self-Made Tapestry*.

Nonequilibrium growth and the patterns it produces are an active area of research in physics. But while a great deal is now understood, there is no overarching theory comparable to the statistical mechanics that describes the states and phase transitions of systems at equilibrium. The structures and patterns that arise in systems away from equilibrium are often complex and rather subtle. Nevertheless, as we have seen, many of the tools and ideas developed for equilibrium statistical physics can be applied or adapted to nonequilibrium situations, which are by no means immune to predictability and regularities of behavior. Since most of the processes that the new social physics seeks to understand are nonequilibrium phenomena, this is a reassuring message to carry forward with us.

SIX

THE MARCH OF REASON

CHANCE AND NECESSITY IN COLLECTIVE MOTION

It is not nearly as bad to explain a phenomenon with a little bit of mechanics and a strong dose of the incomprehensible as to try to explain it by mechanics alone.
—Georg Christoph Lichtenberg

A condition of order at the junction of crowded city thoroughfares implies primarily an absence of collisions between men or vehicles that interfere with one another.
—Edward A. Ross (1901)

He smiled and said that he would go so far as to assert that, if a craftsman were willing to construct a mechanical figure to the specifications he had in mind, he could use it to perform a dance that neither he nor any other accomplished dancer of his time would be capable of imitating.   —Heinrich von Kleist (1810)

Thomas Hobbes was an elderly and venerable scholar when Antony van Leeuwenhoek first saw bacteria through his homemade microscope in the 1670s. A Delft linen merchant by trade, Leeuwenhoek ground magnifying lenses which he used to inspect cloth, and he gradually became immersed in the microworld. Nothing was too dangerous or too distasteful that he would shrink from examining the wonders it held beyond the resolution of the human eye. Among his microscopic specimens were explosive gunpowder and his own excrement.

There was no theory, no natural philosophy, that could have prepared him for the multitude of "little animals" that seemed to swarm in just about any material derived from or touched by living matter. Marsh water was full of them, a miniature menagerie. "And the motion of most of these animalcules in the water," he said in his report to London's Royal Society in 1674, "was so swift, and so various upwards, downwards, and round about, that 'twas wonderful to see."[1]

Here, then, was Leeuwenhoek, peering with a godlike eye into the Commonwealth of the Bacteria and seeing them going avidly about whatever business such creatures might have. Who could guess their purposes?

When Israeli physicist Eshel Ben-Jacob became fascinated by Mitsugu Matsushita's bacterial patterns in the 1990s, those purposes were the key to understanding what was going on in these multicellular colonies. As someone who had helped decode the mystery of the snowflake's branching shape, Ben-Jacob, like Matsushita before him, recognized in the bacteria's delicate traceries a fingerprint of processes familiar from the inorganic world. Here were living cells behaving like blind and insensate gas particles. Could the rules that governed their motion really be so simple?

Ben-Jacob and his co-workers set out to explore the factors that dictated the growth patterns of *Bacillus* cells and soon discovered marvelous new, uncharted regions on the map of shapes called the morphology diagram. Some colonies expanded by sending out baroque curling tendrils (Figure 6.1a), while others formed whorls whose spiraling arms ended in blobs (Figure 6.1b). When they inspected these blobs under a microscope, the researchers

b

a

6.1   "Chiral" growth in a colony of *Bacillus* bacteria (a). The "vortex" growth mode (b); in the dot-like blobs at the ends of each branch, the cells swarm in circles.

saw something that surely never passed before the eyes of Leeuwenhoek. The blobs were vortices. Tens of thousands of long, thin cells swarmed in loops like vehicles in an immense traffic circle. These vortex-forming bacteria, which the researchers called the vortex morphotype, were mutants of the original strain in which the propensity for turning in circles seemed to be genetically hardwired. If a few cells were extracted and used to seed a new colony, this too developed vortical blobs.

This remarkable behavior was not unprecedented: something similar had been reported in the 1940s in the movements of a species of *Bacillus* bacteria christened *B. circulans.* What particularly struck Ben-Jacob and his colleagues was that the cell movements were coordinated, as though the cells had all agreed to move in the same manner. The branching patterns seen by Matsushita could be interpreted by assuming that the cells followed a random walk across the agar gel, like gas particles executing their erratic paths in space. But in both the curling and the vortex growth patterns there was a kind of organized *collective* motion.

Humans organize themselves to move together in all kinds of elaborate ways. But there is no mystery about the coherence and precision of a military parade or a display at an Olympic opening ceremony. These patterns are the product of hours of training and drilling; they are brought about by prior mutual consent and depend for their execution on each participant paying careful attention to detail. Someone has determined where the marchers will go and what, at each stage, they will do next: there is a guiding intelligence whose instructions are being followed by every individual. Social scientists have tended to assume that any kind of complex behavioral pattern like these requires complex motives and planning.

But bacteria do not have complex motives, for they do not have brains. They cannot conspire to decide what to do next. Neither is there some Master Bacterium dictating their movements. And yet the coordination of the spiraling motion is astonishing. If, as this indicates, such patterns of motion can arise without volition, is it conceivable that people might sometimes fall into similar patterns unguided by any plan or intention?

THE CHOREOGRAPHY OF SWARMS

Bacillus bacteria are not freaks of the microbial world which happen to have an unusual penchant for group activities. Collective behavior is a common phenomenon in nature, and these feats of cooperation sometimes go well beyond swimming in circles. A microbial Hobbes would have no difficulty pointing to his Leviathan, his "Multitude united in one Person"—he need only cite the case of the slime mold *Dictyostelium discoideum*. These single-celled organisms go their individual ways when times are good, when food and water are plentiful and the weather is warm. But come drought, famine, or bitter cold and the slime cells seek mutual support.

They cluster into aggregates of cells, each heading for a population center like starving peasants streaming into a city. Once clustered in groups of tens to hundreds of thousands,* they behave like a single multicelled organism: a "slug." The slug moves as a coherent mass. Eventually it sets down roots and changes shape. It begins to resemble a bizarre plant, with a narrow stalk and a bloblike head called the fruiting body. This body contains cells that have become spores, which can survive hardship virtually without nourishment, waiting for better times to come along. It is just as Hobbes pictured the "Essence of the Commonwealth": "One Person, of whose Acts a great Multitude, by mutuall Covenants one with another, have made themselves every one the Author, to the end he may use the strength and means of them all, as he shall think expedient, for their Peace and Common Defence."[2]

How are these "mutuall Covenants" agreed on by cells that can neither speak nor hear nor see? For *Dictyostelium*, as for all single-celled organisms that display collective behavior, communication is effected by a kind of sense of smell. In times of stress, some cells exude a chemical substance that acts as an "attractant" to other cells, rather as higher animals emit pheromones to attract mates. Certain *Dictyostelium* cells become "pacemakers," emitting pulses of the attractant, and nearby cells then follow the trail toward the source of these waves. This sort of chemically induced cell motion is called chemotaxis.

---

*The clustering process involves some of the most fantastic spontaneous patterning seen in nature, and is again discussed in my book *The Self-Made Tapestry*.

At the stage at which *Dictyostelium* cells are converging into their aggregate citadels, they usually organize themselves into circulating vortices, rather like the vortex morphotype of *Bacillus* (Figure 6.2*a*). This kind of movement is familiar to biologists of a quite different stamp: for it is common in schools of fish (Figure 6.2*b*). Yet fish do not communicate by chemotaxis; they have eyes with which to see one another. Moreover they have brains, albeit small ones, which allow them a far greater and more sophisticated range of responses to their environment. Is it mere coincidence, then, that we find the same collective mode of behavior in slime molds and in fish?

Many kinds of animals aggregate into co-moving groups, often for specific and readily identifiable reasons. Groups can protect vulnerable young, and there is safety in numbers in the face of predators. A throng of bees regulates the temperature of a hive using body heat; ants forage more effectively en masse. But as biologists Julia Parrish and Leah Edelstein-Keshet have pointed out, "It is hard to argue the case that all animal aggregations have a

a

b

6.2   Vortex motion in cells of the slime mold *Dictyostelium discoideum* (*a*). Some fish swim in the same kind of vortex pattern (*b*).

functional purpose . . . Pattern and structure can arise . . . through nonlinear interactions whether or not the units are alive."[3] In other words, seeking a "biological" (which is to say, an adaptive) explanation for all patterning and collectivity in animal populations runs the risk of invoking a contingent explanation for something that is in fact an immediate consequence of the "physics" of the situation.

Regardless of questions about purpose, there is still the matter of *how* animals coordinate their behavior and motion. A striking example of collective motion in the animal world is to be seen at dusk above the treetops, as flocks of starlings swoop and dive this way and that in near-miraculous close formation. Again, no individual is calling the shots—there is no leader. Yet every bird seems to make the same decision at the same time. Bees do something similar when they swarm. This is fertile ground for those who wish to believe that science is overlooking some deep and mysterious aspect of nature.

Until the late 1980s there was no satisfactory explanation for the maneuvering of flocks of birds. One researcher was even forced in the 1930s to propose that birds use a kind of thought transference. In 1984, Wayne Potts of Utah State University suggested that maneuvers are instigated by single individuals and then spread like waves through the other members of the flock. Potts claimed to identify this behavior in film of dunlin flocks swooping over Puget Sound. He asserted that each bird watched the approaching wave and timed its maneuver to coincide with the wave's arrival, just as a chorus line synchronizes its high kicks.

But it was not really a convincing explanation, for it required that each bird be aware of what was happening a long distance from itself, and the waves Potts reported moved too fast for each bird simply to be reacting to its nearest neighbors. The chorus-line theory asks rather too much of avian brains. So something else was needed. That extra ingredient eventually came not from biologists but from physicists.

NEWTON'S PUPPETS

Scientists tend to get drawn into the tropes of their fields. Zoologists look for explanations of animal behavior in the characteristics of individual animals; molecular biologists demand a genetic explanation. Physicists, on the other hand, are more accustomed to thinking in terms of interactions *between* things, whether those things are atoms, electrons, quarks, or gas molecules. But one thing is clear: in a phenomenon such as flocking, the precise path for each day's flying cannot be genetically preprogrammed into every bird.

Craig Reynolds is not exactly a physicist. But as a software engineer with the Symbolics computer company in California in 1987, he was more familiar with the physicist's rule-based, axiomatic mode of thinking than with the biologist's phenomenological approach. So when he used to sit and watch blackbirds flock in the local cemetery, he began to wonder what rules they were following. "All evidence," he later wrote, "indicates that flock motion must be merely the aggregate result of the actions of individual animals, each acting solely on the basis of its local perception of the world."[4] This, he felt, was the key: *local perception*. A bird cannot possibly anticipate or keep track of what all its flockmates are doing, but it can respond very quickly to what its immediate neighbors do.

So Reynolds wrote a computer program in which particles were commanded to obey a set of three simple rules, each based only on information about the particle's nearby neighbors. It was hardly fair to call such responsive, "sentient" agents "particles," so Reynolds called them "boids": robotlike virtual birds, a fusion of bird and sci-fi droid. Each boid reacted to all the others within a certain distance of itself—let's call this the local sphere. The rules of motion were these: each boid would try to match its speed to the average speed of others within its local sphere; it would move toward the center of mass of this group;* and it would avoid collisions with others. The stipulations on speed and direction are cohesive rules: they encourage boids near one another to stick together as they move. But there is nothing in this prescription that dictates the overall behavior of the flock. Nothing allows boids on one side of the group to directly influence the movements of boids on the other side. The rules contain no built-in tendency to form a coherent flock.

Yet that is just what the boids do. Their motion, simulated with computer graphics, is startlingly reminiscent of the movement of real birds. Reynolds was able to include additional influences on the motion, such as permanent obstacles or a tendency to move toward a certain position (as if the birds

---

*The center of mass is like the center of balance of a pirouetting dancer, typically located in that case somewhere in the belly.

were seeking a new roost or a food source). In this way he could mimic a wide range of the flocking movements seen in the real world. So convincing was his software that it was adapted by Hollywood moviemakers for animated special effects such as a swarm of bats in *Batman Returns* and a herd of stampeding wildebeest in *The Lion King*.

A boid is a form of *automaton*: a robotlike creature programmed to follow a set of rules that dictate its response to its immediate environment. The behavior of an automaton is entirely deterministic: it takes a look at its surroundings, performs some "mental" computation, and applies to the result a set of criteria that prescribe its next act or move.

The concept of particle-like automata stems from the work of Hungarian mathematician John von Neumann (1903–1957) in the 1930s, when he was formulating early ideas about computers. He was interested in the possibility of thinking machines that could reproduce and increase their complexity. The Polish mathematician Stanislaw Ulam suggested to him a tractable, simple model of this process: a checkerboard universe with an automaton in every square (or cell). Each of these cellular automata can exist in one of several different states, and which of them it chooses is determined by the states of the automata in neighboring cells. In effect, one can think of each automaton as a kind of memory cell holding information, just as each memory element in a computer can exist in one of two binary states—1 and 0—like a switch that is "on" or "off." Von Neumann and Ulam looked at how certain patterns of information might be able to duplicate themselves on the grid.

One can devise all kinds of games to play with cellular automata, each with different rules specifying how a cell responds to its neighbors. Yet it was difficult to investigate the dynamic world of cellular automata until real digital computers were invented. In the late 1960s, John Horton Conway, a mathematician at Cambridge University, devised a kind of checkerboard cellular automaton game which he provocatively called the Game of Life. It was supposed to be a crude model of how living cells or organisms proliferate. Alone they die; in a community they thrive and multiply. But if the community gets too overcrowded, they expire through lack of food and resources. In the Game of Life, each cell can be alive or dead. A living cell stays alive if it has two or three living neighbors adjoining it; if there are less or more, it dies. A dead cell is revived (one should really think of it as an empty grid space that is reoccupied) if it has exactly three living neighbors.

Conway's Game of Life is the prototype for research on "artificial life." The sheer diversity in the shape and behavior of the living cell clusters the Game of Life generates is legendary; there are several Web groups devoted to exploring them. Some clusters propagate across the grid, winding like snakes or gliding like birds. Some clusters eat others; some spit out a stream of new clusters. It is a strange world, full of richness and surprise. And it all arises from a few simple rules about local interactions between cells.

Craig Reynolds's boids are automata that can move under their own steam and are not confined to a grid, but they are automata all the same, rigidly bound by rules. They are arguably the first of such artificial-life games to reproduce the kind of complex phenomena seen in the living world. Boids are one of the star exhibits in the catalogue of research commonly grouped now under the banner of "complexity theory," whose key concept is *emergence*. Flocking emerges spontaneously from the behavior of the boids. It is not programmed in; the rules specify only the motion of individuals. Yet something about the way those individuals interact produces a kind of coherent group behavior. Emergent properties show that the whole can be more than the sum of its parts.

In this context *complexity* has become a word that can mean pretty much whatever you want it to. Often it has little to do with physics as such. Most of the work conducted under this umbrella is a kind of empirical computer science; it is about devising games for automata and playing them to see what happens. Much of this, fascinating though it is, comes up in the end against a brick wall, for it has no underlying theory. What it yields is a description, or rather a prescription, not a deeper theoretical understanding. As Harvard biologist E. O. Wilson puts it, "By itself, emergence can be no explanation at all if you don't have any insight into the mechanics of the system."[5]

That is why boids and all of the other marvelous computer models of collective animal behavior, from ant colonies to herd animals on the prairie, would have remained little more than high-tech parlor games if physicists had not begun to see that they are really a form of nonequilibrium statistical physics.

THE PHYSICS OF GROUP MOTION

Tamás Vicsek had never heard of boids when he and Eshel Ben-Jacob teamed up in the early 1990s to figure out why mutant *Bacillus* bacteria were going around in circles. Like Ben-Jacob, Vicsek had established himself as an expert on the growth and form of nonequilibrium systems, such as the branched clusters formed by diffusion-limited aggregation (see page 111). He recognized that whatever rules the vortex bacteria were following, they must be simple ones.

But bacteria are not like gas particles. Most important in this context, they have their own form of propulsion. Bacteria burn up nutrients and use the energy released to whirl their propeller-like flagella, strands of protein that drive the cells along. When gas particles collide, they obey Newton's laws and conserve momentum (more or less); bacteria can flagrantly disregard such strictures, stopping or speeding up at will. This self-propulsion is what makes a bacterial colony a nonequilibrium system: as long as the cells are burning up fuel, they are driven away from an equilibrium state.

In 1994, Vicsek and his student András Czirók devised a model to describe bacterial motion. They treated each bacterial cell as a "self-propelled particle"—like one of Maxwell's dancing gas particles but with its own source of propulsive power, and with a simple "program" dictating its motion. This program specified only that all cells maintain the same speed and that each one travels in the average direction of motion of all the others that lie within a certain distance. These rules are similar but not identical to those governing Reynolds's boids.

Vicsek and Czirók also assumed that there would be an element of randomness in the bacterial motions. The bacteria would not all obey the rules perfectly, for real life is never so neat and tidy. This randomness is a kind of background noise, like the white-noise hiss of a poor sound recording which interferes with the signal. If the noise is too great, the signal is swamped—which in our case means that the random element of cell motion would overwhelm the cells' tendency to coordinate their movements.

The researchers simulated the movements of their bacterial automata on a computer and found something they recognized at once. If the noise level was low, the cells displayed collective behavior: they all moved in the same direction. Remember that each cell is instructed to pay heed only to its neighbors within its local sphere, so there is no guarantee in the underlying rules that the whole group will move as one. As the noise was increased, the degree of coordination decreased. At some critical noise level all semblance of coherence was lost: each of the self-propelled particles did its own thing (Figure 6.3). In this case, the average velocity of the group fell to zero, since for any cell moving in one direction there was likely to be another moving in the opposite direction.

Vicsek and Czirók were reminded of the behavior of a magnet as its temperature is altered. At low temperature all the magnetic spins point in the same direction, and the atomic magnetic fields add up to create an overall magnetization. At high temperature the needles point in random directions and the fields cancel out. Between the two states—magnet and nonmagnet—there is a phase transition at a critical temperature.

6.3   A phase transition in self-propelled particles. If the noise is low, the particles align their motions (a). For a higher noise level, the motion breaks up into smaller, co-moving clusters that tend to circulate like vortices (b). If the noise exceeds some critical value, the motions become random (c). The direction of a particle's motion is shown here by an arrowhead, attached to a short line that shows its most recent trajectory.

The self-propelled particles were undergoing an analogous phase transition from an aligned to a nonaligned state, the average velocity playing the role of the magnetization and the noise playing the role of temperature. Just as the liquid-gas critical point can be "mapped" onto the phase transition of a magnet, so were the researchers able to map their model of bacterial motion onto this phenomenon.

It's more than a mere analogy. Despite some important differences connected to the fact that the ordering of self-propelled particles is a *nonequilibrium* transition, one can demonstrate a formal mathematical equivalence between the two processes. They both share "universal" features.

MOB RULES

The cooperative behavior of "social animals" was not lost on Thomas Hobbes, and he anticipated that it might provoke objections to his bleak prediction about humanity's savage State of Nature:

> It is true, that certain living creatures, as Bees, and Ants, live sociably one with another, (which are therefore by *Aristotle* numbered amongst Politicall creatures;) and yet have no other direction, than their particular judgements and appetites . . . therefore some man may perhaps desire to know, why Man-kind cannot do the same.[6]

Hobbes's answer was that people are different: their thoughts, desires, and sensations are more complex, and provoke them into conflict. For example, "Men are continually in competition for Honour and Dignity, which these creatures are not . . . [and] these creatures, having not (as man) the use of reason, do not see, nor think they see any fault, in the administration of their common business."[7]

This answer will ring true for many social scientists, who have long assumed that people are just too complicated to yield to any mathematical model of behavior. We are moved by a thousand impulses in a blend unique to each of us. So what is the point of making idealizations of human activity?

The objective of any physical model of the kind I have been discussing so far is to get out more than you put in. If I were to devise a model of a crowd

in which every person had a preprogrammed and complicated trajectory, replete with sudden stops to check direction signs or peruse a shop window,
that would not be much of a model; any information I got out would be pretty
much what I had put in.

In 1971, L. F. Henderson of the University of Sydney saw that beyond
this miasma of individuality, there might lie some quantifiable *statistical*
characteristics of what humans do in moving groups. In a sense this is obvious, but it is worth stating. Football crowds move into stadiums before a
game, spread themselves among the seats, and flow out again after the game
is over. The spectators are not so lost in their individual internal worlds that
these collective movements are rendered impossible. Shoppers in a busy
street move along the sidewalk mostly in one direction or its opposite. People do not travel on foot at twenty miles per hour; they do not walk with their
eyes closed, colliding with all in their path. There are some general rules,
some constraints, some trends and averages.

Henderson's decision about how to search for these statistical properties
is telling: he wondered whether they might fit the Maxwell-Boltzmann kinetic theory of gases. That is to say, he suspected that the distribution of velocities of people walking along a sidewalk would trace out the bell-shaped
curve used by Maxwell and justified by Boltzmann. He tested this idea by
observing various moving crowds: students walking along a footpath on the
University of Sydney campus, people using a pedestrian crossing in a busy
street, children moving in all directions in a playground. In all cases, the velocities fitted the Maxwell-Boltzmann curve quite closely—but with one curious difference. Each of the crowd curves had two distinct peaks, giving the
appearance of two intermingled populations of Maxwell-Boltzmann particles
with slightly different average velocities. These, Henderson considered, were
males and females, who seem to move in the same general manner but at different rates.

Henderson suspected that perturbations to the crowd motion might bring
about changes in the "global" state of the throng:

> At one end of the passage, the crowd may be forced to slow down, probably by the
> need to surrender a ticket at a barrier, when it may be expected that people will
> bunch into a densely packed crowd which moves along with a slow elbow-to-elbow
> shuffle. This transition will be called a phase transformation—the loosely packed
> phase is a crowd gas and the densely packed phase is a crowd liquid.[8]

In other words, he foresaw phase transitions in crowds, brought about in a manner analogous to liquefying a gas by compressing it. Yet he did not in fact report any observations of transitions of this sort.

Although it was valuable in establishing the potential connection to statistical physics, Henderson's study did not really contain anything very surprising. We'd *expect* people in a crowd to have a more or less bell-shaped spread of velocities; it is hard to imagine anything else. Not all bell-shaped curves are mathematically identical to a Maxwell-Boltzmann curve, but Henderson's data on just several hundred moving individuals were unlikely to reveal subtle distinctions between them. And in any case, hadn't Adolphe Quetelet long ago established the bell-shaped error curve as the characteristic fingerprint of social statistics?

When physicist Dirk Helbing began to think about models of crowd motion in the late 1980s, when he was at the Georg-August University at Göttingen in Germany, he thought Henderson's "gas-kinetic" description would be a good starting point. But he realized that it said nothing about what was actually motivating the walkers: what their intentions were and how they responded to their surroundings. Only by including such factors could a model hope to capture the complex patterns of movements that develop in human crowds. Henderson had gone on to propose that pedestrian motion might be treated like the flow of a fluid. But if so, it was a fluid with a mind of its own, a fluid that could defy Newton's laws by stopping dead or breaking into a run.

Helbing considered that the motion of any individual was conditioned by two influences: internal, or "personal aims and interests," and external, or "perception of the situation and environment." Often these will conflict. I am forced to slow my habitual pace because there is a slow-moving group in front of me. I want to cross the sidewalk diagonally to enter a shop, but there is a stream of people in the way. In a particularly dense crowd, I am prevented from coming to a stop by the crush of people behind me.

External influences on behavior are caused by interactions with others. In 1945 the psychologist Karen Horney identified three ways in which people interact: "moving towards people," "moving away from people," and "moving against people." Motion here is to be understood largely metaphorically, but it may also be interpreted literally. People walking with friends or partners, or as members of a group, tend to stick together; famous people at cocktail parties attract a coterie of admirers and sycophants through their own personal (or manufactured) magnetism. But in most crowds people do not know one another, and so have no propensity to cluster together as they move. Rather, they try to keep their distance from strangers. Such interactions resemble forces of attraction and repulsion. (Horney's "moving against" is a rather special case: she had in mind deliberate obstruction, motivated by aggression and conflict. It happens in some social dynamics but is rarely to be seen in simple walking motions in crowds.)

The American social psychologist Kurt Lewin argued that the kinds of attractive and repulsive interactions sketched by Horney could be broadly applied in the social sciences. He postulated in the 1950s that there might be an analogy between the forces of electromagnetism that act on particles with electrical charge and the social pressures that determine the behavior of people. According to Lewin, individuals could be considered to "move" in an abstract field of ideas, beliefs, habits, and notions. The field is conditioned for each person by the behavior he or she sees in others, and it pushes and pulls the person toward certain dispositions.

This is very much the modern equivalent of Hobbes's mechanistic vision of humanity. Motivated by Lewin's concept of "social forces," Helbing and his colleague Péter Molnár at the University of Stuttgart developed a physics-based mathematical model of pedestrian movement. Of course, it is one thing to speculate about "forces" between people, but quite another to describe them by mathematical equations like those one can write for electromagnetism or gravity. Yet that is what Helbing and Molnár did.

They assumed that walkers simply want to move in a particular direction at a certain preferred speed. This impulse is tempered by external factors, foremost of which is the wish to avoid collisions—indeed, to avoid coming close enough to others to compromise their own "personal space." Thus, people in crowds behave as if there were a force of repulsion between them that increases in strength as the distance between two people gets smaller. Modern theories of liquids invoke a similar "soft" repulsion between particles, which starts small but rises steeply as the particles approach one another. Recall that van der Waals's theory of fluids implicitly included a "hard" repulsive force that was felt only when two particles collide.

Van der Waals's theory also took account of attractive forces between particles. Helbing and Molnár's model could incorporate attractive forces between people if the researchers wanted to investigate certain special situations, such as group cohesiveness, but otherwise they simply gave their walkers a shared aversion to closeness. These walkers are not like boids or Vicsek's self-propelled bacteria, because they do not possess any tendency to align their motions and thus to form coherent flocks.* Instead, each individual—which, borrowing from David Bowie's "Diamond Dogs," we might call a peoploid—ploughs his or her own furrow, subject to the constraints that the other pedestrians impose.

In this sense, the model is one of self-centered people with no real social graces or courtesies. And yet when Helbing and Molnár ran their computer simulations, they found that certain types of group dynamic did emerge spontaneously. Some of these even looked like good behavior.

For example, peoploids walking in opposite directions down a corridor tended to organize themselves into counterflowing streams, which reduced the need for collision-avoiding maneuvers (Figure 6.4). This happens frequently in real life. Barriers placed along the center of a thoroughfare, such as pillars and trees, can encourage this stream formation, even when it is not

6.4   Pedestrians moving along a corridor in the Helbing-Molnár model become organized into counterflowing streams. The arrows show the direction of motion; their length is proportional to the person's speed. Black and gray particles are moving in opposite directions down the corridor. Such simulations can be seen at www.helbing.org/Pedestrians/Corridor.html.

---

*Of course—and as we shall see—people sometimes do exhibit herdlike collective behavior. One can see this manifested at soccer matches in the form of the Mexican wave, or La Ola, which became a fad at the 1986 World Cup in Mexico. Spectators would leap to their feet with their arms in the air and then sit down again, in a wave that passed swiftly round the crowd. Helbing, Vicsek, and their colleague Illés Farkas have modeled these waves by assuming that spectators switch between three states: excitable (those ready to leap up as the wave approaches), active (those who are leaping), and passive (those who have done their bit and sat down again). These simulations can be seen at the Web site http://angel.elte.hu/wave. The waves generated in this model are closely analogous to the concerted ripples of cardiac tissue that make the heart beat.

specified on which side people should pass. The lanes become organized spontaneously, but it is a matter of pure chance to which direction each of them becomes designated.*

When two groups of peoploids try to pass in opposite directions through a single doorway, alternating bursts pass first in one direction and then in the other (Figure 6.5). Each group temporarily "captures" the doorway—a trailblazer passes through and several others follow in its wake. The other group almost appears to stand back to let this stream through, though this apparent courtesy is the result of nothing more than the wish to avoid close contact.

This model can be used to improve pedestrian walkways so as to reduce discomfort, congestion, and inconvenience. Dividing a corridor by installing pillars is one way to improve the flow. To ease the bottlenecks caused by doorways, one might imagine that making the door wider would help. But it is not that easy; with a wider door, the alternations in direction of flow sim-

6.5 People pass through a doorway in alternating bursts that allow passage first in one direction (*a*) and then in the other (*b*). The simulations can be seen at www.helbing.org/ Pedestrians/Door.html.

---

*This may not be quite true in practice—people generally have a preferred passing direction around obstacles (including other walkers). This preference is culturally determined; in the United States and Germany, for example, it is to the right. The bias seems to be learned, since children (and tourists) tend to lack it. The learning is subconscious but is beneficial to all individuals since it reduces the risk of collisions from two approaching walkers who choose to swerve the "same" way: one to the left, the other to the right. Among adults, lane directions in a divided corridor will also no doubt be influenced by instincts acquired from road driving.

Helbing and his co-worker Kai Bolay have been able to simulate this learning process and have found that if walkers are allowed to adapt their behavior to improve collision avoidance, they develop a preference for passing on a particular (but arbitrarily determined) side.

ply become more frequent. A better solution is to introduce two doors. Even if it is not specified which is to be used in each direction, a crowd will automatically organize itself into two counterflowing streams that pass through one door or the other (Figure 6.6). Two doors are thus more efficient than a single door of their combined width.

Intersections are tricky. In the simulations, the flow at intersections never settles into a stable state in which collisions and congestion are minimized. But temporary solutions often emerge in which the pedestrian traffic circulates at the intersection in one direction or the other (Figure 6.7a). Good planning can help these efficient flow modes to appear. An obstacle at the

6.6 If two doorways are available, each will find itself assigned to pedestrians moving in a particular direction—even though this is not specified at the outset. Here I show not instantaneous snapshots, as in Figures 6.4 and 6.5, but the trajectories of black and gray walkers over several consecutive time steps.

6.7 At intersections, the motions of pedestrians often settle temporarily into a circulating mode that reduces the chances of collision and obstruction (a). Here the arrows are shaded differently for different approach directions; each person is trying to cross the intersection to the opposite side. Intersections can be designed to promote the efficient circulating mode by using barriers and signs (marked "!") that draw walkers toward them (b).

center of the intersection, for example, blocks off routes that impede circulation. Slanted barriers give further guidance, stabilizing circulation in a given direction—especially if helped by posters or displays that draw walkers to one side of the corridor (Figure 6.7b).

In 1997, Helbing and Molnár, together with computer scientist Joachim Keltsch in Tübingen, used this pedestrian model to deduce how trails evolve organically in open spaces over which people walk. If we are crossing a grassy meadow or park, we have a tendency to follow the route trampled flat by innumerable feet before us—even if the route is not the most direct one. Why? Perhaps it is easier walking on flat, smooth ground than on grass. Maybe an authoritative voice from childhood tells us not to wander off the path (even if the path is the arbitrary construct of other walkers). The reasons do not really matter; the fact is that we do it.

How do the trails get there in the first place? Before people started crossing it, the space was uniformly covered in grass. No doubt the first pioneers had a variety of destinations. A trail system that Helbing and colleagues observed on the campus of Stuttgart University revealed several common entry and exit points, defined by the various university buildings around the lawn. But the trails did not follow the shortest path between any of these points (Figure 6.8).

6.8   Trail patterns formed spontaneously on an open space at Stuttgart University. Notice the lack of sharp intersections. None of the paths represent the most direct route between any two points of entry or exit.

The trails build up from the passing footprints of every walker who crosses the space. The researchers emulated this in their model by assuming that each peoploid crossing a space wears away the grass to a small degree. But the grass regrows at a steady rate, so that trails that remain unused for a long time will disappear. The more worn a trail becomes, the more visible and tempting it is to subsequent walkers, who, in the model, are attracted toward a particular trail segment with a pull that depends on its distance from them and its visibility.

Helbing and his colleagues let loose a steady stream of peoploids across an open space, all coming from and going to a few destinations at the periphery. They found that the trails that evolved depended on how strongly peoploids were attracted to existing paths. At first, peoploids simply trod down fairly direct routes between the various destinations (Figure 6.9a). If the attraction is small, these direct routes persist and become well-trodden trails. But if the attraction is appreciable, the direct routes evolve into something else: a trail system that represents a compromise between directness and the tendency to follow existing paths (Figure 6.9b). In this case, the diagonal routes between opposite corners and "edge" routes between adjacent corners merge into gently curved trails that take all walkers slightly out of their way. No one passes through the center: the trails intersect to leave a central island, as they tend to do in real life (Figure 6.8).

6.9   The model developed by Helbing and his colleagues simulates the way trails develop. In the early stages (a), walkers take fairly direct routes between destinations at the four corners. The final state (b) represents a compromise between the various "most direct" trails.

Park and garden planners often incline to linear thinking: they create paths that are straight lines, intersecting at sharp angles. In contrast, trails that evolve organically—that, in the jargon of the field, are *self-organized*—curve and merge smoothly. Because of this difference, planned paths in open green spaces are often augmented and subverted by new trails worn into the grass where people have followed their own instincts. In one Stuttgart park, Péter Molnár noted the futile attempts of park officials to obliterate such trails and reinstate the original ones that the planners had specified. Yet without any concerted intention to undermine this design, pedestrians soon reestablished their preferred routes.

How much more effective it would be if planners could anticipate the walkers' wishes and build accordingly. A model like this one might help them to do just that, producing routes that are natural and comfortable. It is a simple matter, says Helbing, to gear the model to constraints such as a budget limit on the total trail length, so as to find the best compromise between economy and efficiency, or to figure out how best to extend existing walkways. In this way, planning can be fitted to human nature.

THE LANGUAGE OF SPACE

In open park spaces, pedestrians have the freedom to ignore the official paths and collectively define their own trails. In the city or a building, that is seldom possible—you can't walk through a wall if the fancy takes you. Can urban and architectural spaces also be planned to conform with human needs and impulses, rather than the other way around?

Geographer Michael Batty and his colleagues at University College in London have simulated the way visitors move around the Tate Gallery (now called Tate Britain), an art museum in London. They compared the results with observations of people moving through the real gallery on a single day in August 1995. Visitors tended to circulate among the rooms containing the classical and British collections, to the left of the gallery's central axis, in preference to the modern collection on the right. Was this because people prefer classical and homegrown art to modern works? Not at all, said the researchers. Their simulations, in which every room in the gallery exerted an equal "attractive force" on the visitors, showed the same left-right asymmetry, implying that the apparent preferences were simply being dictated by the way the rooms were laid out: the room structure in the modern section was more convoluted. In other words, what people view at the gallery depends not only on what they like, but also on how the gallery is arranged. The ground plan of the Tate was dictating the experience of visitors in ways that were not foreseen or intended by the gallery designers.

Bill Hillier, a specialist in urban planning at University College, believes that there is a kind of logic—a "syntax"—in the way people relate to and navigate space. He and his co-workers have used computer modeling to look at how people use and navigate a built environment at scales ranging from individual room arrangements to entire city networks. They have applied this kind of modeling to help plan shop interiors and malls, galleries and museums, commercial office buildings, hospitals, schools, airports, and railway stations. These models can provide diagnoses of problems that arise in existing street networks and building layouts.

But the bigger challenge is to figure out the *rules*—Hillier's space syntax—that pedestrians are using in these places. Hillier believes that lines of sight play a significant role in this visual language of space. If the language can truly be decoded, it could be programmed into the motivations of Helbing's peoploids to allow them still more realistic ways of moving in space, making possible more powerful predictions about how people would like their environment to be laid out.

Walkways in new city spaces have in the past often been planned through a mixture of guesswork and modernist architectural thinking that imposes an arbitrary aesthetic on places in which people have to live. Hillier and his colleague Julienne Hanson say that whereas there are distinct patterns of pedestrian movement in traditional urban environments, people in some newer developments such as housing projects seem to use the space randomly: they look, and most probably feel, lost in it. Few urban dwellers will have trouble identifying such places as this, in which the design feels at odds with the way we instinctively navigate and use open spaces. At the Elephant and Castle in South London, near where I live, there is a sprawling development of concrete tower blocks built in the 1950s. At that time it was considered to be a model project, a shining example of how huge numbers of people could be efficiently and comfortably housed. Shops and community centers were incorporated into the buildings; walkways took people across the roads out of the reach of the traffic below.

Now many of those shops have been closed for years, and the walkways have been demolished because they were convenient escape routes for muggers. The problems of this project run far deeper, of course, than a failure of design. But one does not need to spend long in this environment to know that it is not a design for living, or for making one feel part of a community.

That is an old and familiar lament, and some of the salutary lessons of postwar urban design have now been learned. Yet Hillier and Hanson argue that there is more to all this than poor planning. Urban design, they say, is political: "The nineteenth-century dreams of a social order, in which the benefits of capitalism are retained through the creation of a quiescent working class, are dreamed in a strongly spatial form."[9] In other words, urban communities were redesigned in the Victorian age to reproduce and reinforce social hierarchies. By developing mathematical and computer models of the spatial patterns of communities ranging from traditional villages to modern towns, Hillier and Hanson have shown that urbanization tended to increase and diversify people's interactions up until the time when new templates for planning were introduced during the Industrial Revolution. Whether consciously or not, these templates reduced social encounters and fragmented communities, discouraging collective activity and keeping people passive under an imposed authority. High-rise blocks, for example, pack living spaces together densely while reducing the frequency of encounters that generate a sense of social solidarity. "It is wrong to say that high-rise estates are unsuccessful," Hillier and Hanson argue. "For their unmanifest purposes of community reduction they are extremely successful."[10] The "soft" solution of creating garden cities, in which urban spaces are separated into small, relatively isolated zones, looks and feels more benign but has the same result.

If Hillier and Hanson are right, the barriers to congenial urban design are created by more than simple ignorance of how people use space. Nevertheless, the more we can understand and predict the ways in which we instinctively want to move around our environment, and the better we are able to model the diverse human movements that these spaces must accommodate, the more likely we are to create places where people feel relaxed, comfortable, and considerately housed.

QUICK EXIT

On November 28, 1942, the Coconut Grove nightclub in Boston was jumping. American servicemen were making the most of their leave, beer and jazz helping them briefly forget the war that was spreading across the world. Then the fire started.

There was only one exit, and everyone fled to it. But the first people to arrive at the doors discovered that they opened inward. Before they could be opened, the pressure of the crowd at the back pushed the first-comers against the closed exit, and there was then no way out. In the panic and smoke, no message could be relayed to those at the back who were feverishly trying to move forward. Four hundred and ninety-two people died in the blaze.

On April 15, 1989, soccer fans in Sheffield, England, filled the streets around the Hillsborough stadium, desperate to get inside and claim their places for the FA Cup semifinal between Liverpool and Nottingham Forest. The steady trickle of fans through the turnstiles did little to ease the crowd, so ten minutes before the game started the police decided to open an exit gate. The consequent rush into the stadium pushed fans at the front up against riot-control fences, where ninety-six were crushed to death.

At face value, predictability seems to vanish when panic strikes a crowd. Safety measures planned on the assumption that people will use them rationally can rapidly become useless when they must serve a horde of terrified people. What hope is there of anticipating human behavior in these situations?

Yet irrational does not mean unpredictable. Quite the contrary: a panicking crowd has only one objective in mind—to escape as quickly as possible. Fear, though, is not the only cause of stampedes. Injuries and deaths have also resulted from the rush for seats at concerts, for example when eleven people were killed at a concert given by The Who in 1979 in Cincinnati.

In 1999, Helbing went to Budapest to work with Tamás Vicsek. They were intending to use the pedestrian model to study the formation of animal traits, but Vicsek felt there was still more to learn from it about human crowds. The researchers found that if people's motions became too erratic (too "noisy"), this could lead to a crowd jam in a corridor. Viewed as a problem in physics, this is a counterintuitive result. Making the movements of peoploids more erratic is like raising the temperature of a group of particles, making them jiggle more frantically. Yet the consequence is that the crowd "freezes." In other words, the "people fluid" could be frozen by "heating" it, whereas a normal fluid, like water, is frozen by cooling.

Vicsek noted that this jamming of an overexcited crowd looks something like the effects of panic. So he, Helbing, and Vicsek's colleague Illés Farkas, of Eötvös University in Budapest, began to use the model of pedestrian motion to study what happens when crowds get out of control. They reasoned that one thing in particular makes panic different from normal motion: people lose their inhibitions and come into contact. So much so, in fact, that the pressure can become life-threatening. The force of a surging crowd has been known to bend steel barriers and collapse brick walls.

This does not mean that people's natural aversion to physical contact ceases, just that it no longer dominates the motion. When people are touching, their movement is restricted. In a densely packed crowd it can become impossible to slip past others or even to turn around. There is a kind of friction between people that hinders movement.

So the researchers gave their peoploids frictional properties, as if they were billiard balls coated with sandpaper. They also included a further grim ingredient. If the pressure on a person became too great, it was assumed that that unfortunate individual would become injured and unable to move. This only makes matters worse, of course, as the injured person then becomes an obstacle to the movement of others. By estimating the level of pressure likely to result in injury, the researchers hoped to derive some indication of when and how casualties arise in dense crowds.

There are some experiments that should be done only on a computer, and lighting a fire in a crowded room is clearly one of them.* Helbing and his colleagues put a crowd of peoploids in a room with just one exit and gave them a good reason to head for the door: a fire encroaching steadily from the opposite wall. If the peoploids are able to contain their fear and move at a calm speed—less than about four feet per second—then they are able to

---

*Researchers at the University of the Philippines have run real-world tests of models of human escape-panic by watching mice swim to safety through a doorway in a flooded chamber. They found modes of behavior similar to those seen in their computer models, which were closely related to the model developed by Helbing, Farkas, and Vicsek and described here. In particular, clogging at the doorway could make the mice's escape inefficient and sporadic.

evacuate the room in an orderly manner. Clustered around the door, they give one another enough space to allow a steady stream to pass through.

If the peoploids try to move at speeds greater than this, the outcome is chillingly different. Converging on the doorway, the crowd presses up against one another, and friction takes hold. The peoploids become locked shoulder to shoulder, unable to pass through the door easily even if it stands open in front of them. The crowd panics and jams (Figure 6.10a).

This clogging effect is familiar enough: salt grains can get stuck in a shaker even though no grain is bigger than the hole. Because of their mutual friction, the grains form arches above the hole, held together under their own weight. In a crush of peoploids, these arches can eventually become loosened because each individual is able to keep moving; but their repeated formation and collapse means that peoploids no longer pass steadily through the doorway. Instead they come out in sporadic bursts, making evacuation very inefficient.

So, although each peoploid is "programmed" to try to move faster, the room empties more slowly. This panic state appears in a fairly abrupt transition once the peoploids try to move farther than about a meter and a half per second. At slower speeds, the time taken to clear the room becomes shorter as the speed increases; for faster speeds, the time rises steadily the more quickly the peoploids want to go (Figure 6.10b). Faster becomes slower. It

6.10   If people try to leave a crowded room too quickly, they push up against one another and become jammed around the doorway. This is a panic state (a). The time it takes for people to empty a room initially decreases if they move faster. But above a certain threshold speed, corresponding to the appearance of a panic state, the emptying time increases as people try to go faster (b). See "Pedestrian simulations" at angel.elte.hu/~panic/.

looks as though there is some form of nonequilibrium phase transition between a relaxed and a panicked state.

Once the desired velocity exceeds about five meters per second—about eleven miles per hour, a running pace—the pressures in the clogged region become so great that peoploids are injured.* The number of casualties rises steadily with the speed. Faster, then, means not only slower but more dangerous.

Many nightclubs are dark enough to begin with, and once smoke starts to fill a room it can become impossible to see more than a few feet ahead. Under these conditions, people do not even know in which direction to run. Helbing and colleagues wondered what might happen then.

One's instinct might be to rush around madly and randomly looking for a way out. But if you see several people rushing past in a certain direction, you might conclude that they know something you don't. There is a natural tendency to follow the crowd in a herdlike way. The researchers assumed that people would search for obscured escape routes with a mixture of individualistic (random) and herding (collective) behavior.

A little herding, they found, is a good thing. It reinforces success: when someone finds an exit, others are more likely to follow them to it. This effect is self-reinforcing: the more people move to one place, the more inclined others are to follow them. But there is a drawback. Herding can persuade nearly everyone to head for a single exit and forget about finding other ways out even if they know that others exist. This is what Helbing and his colleagues found when they included a herding element in the simulations. Herding was added in much the same way that Vicsek allowed for flocking in his self-propelled particles: the peoploids were given a tendency to match their direction to the average direction of the group immediately around them.

As the herding instinct increases (which might be expected to happen as panic rises), peoploids in a room with several invisible exits at first begin to leave more efficiently, because they are able to capitalize on one individual's success in finding a way out. But beyond a certain herding level, the reverse effect kicks in rapidly. A single exit then becomes clogged with individuals who have followed the majority, while other exits remain unfound or barely

---

*The onset of injury depends on the size of the crowd: it happens at slower speeds if the crowd is larger.

used. The most efficient use of exits, therefore, happens for a certain optimal level of herding. Too much or too little, and exits are wasted.

For safety engineers, these are telling results. Often, says Helbing, safety measures are planned on the assumption that doors are used uniformly. Engineers note the rate at which people file out of a single door, and use that to calculate how many doors are needed in a room with a certain capacity. The crowd is treated rather like water flowing through a sieve: all the holes are used equally. But if the doors are used very unevenly by a panicking crowd, this calculation might seriously underestimate how many exits are needed. A simulation of panicking peoploids might be a far better way to deduce how long it will take to clear the room.

Michael Batty has shown how pedestrian modeling might be used preemptively to avoid the kind of uncontrolled and dangerous crowd dynamics seen in these simulations of panic. He and his colleagues have simulated the distribution and movements of crowds during London's Notting Hill Carnival, an annual two-day event that now draws up to a million people into a small (about one-square-mile) neighborhood in northwest London. The carnival poses immense safety problems: in 2001 there were over five hundred accidents, a hundred of which required hospital treatment. Getting ambulances in and out of the densely packed crowd is a big challenge (Figure 6.11).

Batty's model incorporated a tendency for the pedestrians to flock and

6.11   An ambulance attempts to maneuver through the crowd at the Notting Hill Carnival in London in 2001.

swarm, creating mass movements between the various attractions around the carnival route. There was a total of thirty-eight widely used different entry points into the carnival area, of which the five subway stations were the most important. The researchers searched for crowd-control measures, such as barriers, street closures, and capacity limits imposed on entry points, that would reduce the risk of overcrowding, which tends to become focused at particular places along the circular route of the carnival procession.* London police, politicians, and a body called the Carnival Review Group, run by the Greater London Authority, are making an ongoing assessment of the various options for changing this route to alleviate such dangers. Pedestrian modeling provides a means of testing these alternatives without having to use each annual festival as a huge and hazardous one-off experiment.

Batty and his colleagues point out that this kind of modeling has the potential to reorient the way we think about controlling urban systems. "In the past," they say, "there has been an unwitting assumption that cities can be modeled as quasi-natural systems in which control is something that is applied after the fact."[11] Of course, an event like the Notting Hill Carnival is already highly constrained by the route plan, the entry points, and so forth, before any additional crowd-control measures are proposed. Yet none of these constraints is a "given": routes can be altered, entry points can be closed. With an ability to predict crowd behavior, the entire planning process becomes much more malleable: prediction, say the researchers, becomes blended with prescription. Planning, rather than being imposed from the top, becomes interwoven with its outcome in an iterative and interactive manner. There is nothing in these models of pedestrian motion that dictates what people *ought* to do. Rather, the aim is to find out what people *will* do, using some simple assumptions about what motivates them and allowing for the constraints they encounter. That is in the true spirit of the modern physics of society.

These are still early days in our attempts to understand human motion, but already we can see that the seeming complexity of human behavior does not extinguish our ability to understand and predict at least some aspects of it. Equally important is the demonstration that abrupt changes in group be-

---

*The results of the model simulations can be seen at http://www.casa.ucl.ac.uk/research/urbanstudies/index.html.

havior do not necessarily require concerted changes in everyone's intentions. These switches of collective motion can instead emerge spontaneously even as individual predispositions are altered only incrementally.

There are potential pitfalls too. Parrish and Edelstein-Keshet warn about the hazards of physical modeling of animal group dynamics: "Many sets of rules can lead to lifelike group behavior, so that the results, though visually appealing, may be uninformative—it is not always possible to deduce individual behavior from emergent properties."[12]

In other words, just because you get what looks like the right kind of collective motion, that doesn't mean you have the right rules, which in turn means that you won't necessarily make the right prediction under different circumstances. The "social force" model of pedestrian dynamics does seem versatile enough to cope with situations other than those for which it was originally designed; it pays to remember, however, that such models of collective motion may not furnish unique explanations for a given phenomenon, nor are they necessarily suited to every situation. They are just a beginning.

CITY LIMITS

Once there were three hills, named Tothill, Penton Hill, and the White Mound, or Tower Hill. Tracks wound their ways between them, and the tracks became lanes and roads, and settlements grew up around them, and the place was called London.

Or so say some legends, which are probably fanciful. But one way or another, dwellings multiply and byways grow until hamlets become villages and towns and great cities. Not all of them grow as organically as London, of course, with its dense web of streets like a warren practically begging to be populated by Dickensian ruffians. It would be fascinating to know whether pedestrian models have anything to tell us about the evolution of urban geography, but that question has not yet really been explored.

Models of city growth *have*, nonetheless, emerged from physics. In the nineteenth century, a group of sociologists who have become known as the boosters argued that cities exert an attractive force on people and on trade which is entirely analogous to gravitational attraction. The boosters envisaged a science of city growth every bit as mechanical and deterministic as

Newton's laws describing bodies moving in gravitational fields. But modern physics-inspired theories recognize the complex "organic" character of urban expansion and take their cue from studies of nonequilibrium growth processes of the kind we encountered in the previous chapter, not least those displayed by bacterial colonies.

And why not? The modern city is, after all, alive. It is what makes the city so thrilling, so terrifying, so liable to swallow its inhabitants. London, Tokyo, Delhi, and Los Angeles pulsate; they groan and sigh and spread their many tendrils. "Whether we consider London as a young man refreshed and risen from sleep, therefore," writes the city's biographer Peter Ackroyd, "or whether we lament its condition as a deformed giant, we must regard it as a human shape with its own laws of life and growth."[13]

To see this shape as human requires a novelist's imagination; rather, the mass that straddles the Thames is amorphous. But "laws of growth" are there for sure; and no one, it seems, can control them. Disturbed by the grotesque overcrowding within and around the city walls, Elizabeth I prohibited "any new buildings of any house or tenement within three miles from any of the gates of the said city of London."[14] Did this stem the city's outward sprawl, a process of accretion that had long since burst the bounds of the encircling stone walls? You might as soon try to contain a wildfire.

In 1787, Henry Kett compared the expansion of London to an epidemic, and complained,

> Mansions daily arise upon the marshes of Lambeth, the roads of Kensington, and the hills of Hampstead . . . The chain of buildings so closely unites the country with the town that the distinction is lost between Cheapside and St George's Fields. This idea struck the mind of a child, who lived at Clapham, with so much force, that he observed, "If they go on building at such a rate, London will soon be next door to us."[15]

Today all these districts are in inner London; some people consider them appealingly central.

Every major city has similar tales to tell. And while the consequences of urban sprawl are perhaps not quite as grim as they were in Elizabethan England, still they give cause for deep concern both in developed and developing nations. Faced with soaring urban air pollution, noise, traffic congestion, and lack of open space and the tensions of overcrowding, many U.S. citizens identify urban sprawl as their most significant local anxiety. Containing urban expansion has become a hot topic on the political agenda. Donald Chen, director of Smart Growth America, an organization that seeks to improve urban conditions, argues that urban sprawl is "undermining America's environment, economy and social fabric."[16]

If urban planners are to mitigate sprawl, they must first understand its causes. But as Chen points out, "the theories explaining why sprawl occurs are as numerous as they are politically controversial."[17] For example, the accusation that sprawl is encouraged by government spending on infrastructure such as highway construction and subsidies for power and water supplies is somewhat undermined by the fact that the decrease in U.S. public subsidies in recent years has not diminished the growth of American cities. Those amoebas go right on growing.

The organic, formless shape of big cities was evident to Lewis Mumford in the 1930s:

> Circle over London, Berlin, New York, or Chicago in an airplane, or view the cities schematically by means of an urban map and block plan. What is the shape of the city and how does it define itself? As the eye stretches toward the hazy periphery one can pick out no definite shape, except that formed by nature: a banked river or a lakefront: one beholds rather a shapeless mass, here bulging or ridged with buildings, there broken by a patch of green or the separate geometric shapes of a gas tank or a series of freight sheds. The growth of a great city is amoeboid: failing to divide its social chromosomes and split up into new cells, the big city continues to grow by breaking through the edges and accepting its sprawl and shapelessness as an inevitable by-product of its physical immensity.[18]

The comparison with the spread of a microbial colony is more than just an emotive metaphor. In the 1990s, Michael Batty recognized in the clumpy, irregular outlines of cities something akin to the shapes of clusters of particles formed by diffusion-limited aggregation—like those Mitsugu Matsushita saw in bacterial growth.

This was a challenging image. City planners strive to see the *regularities* in urban form, naturally enough, since those are the things that they and their predecessors have put in place. The grid structure of American cities—an urban schema first developed by imperial Rome—is the first thing you notice from city maps or in the view from a Manhattan penthouse. But the city boundaries become anything but neat and formal as they reach out into the map's emptier expanses. Envisaged as diffusion-limited aggregation (DLA) clusters (see previous chapter), cities shed any pretense of rational planning and begin to look like truly organic things, growing with a life of their own—a "crystallization of chaos,"[19] in Mumford's memorably apt phrase.

Batty and his colleague Paul Longley adapted a theory of DLA-like growth, called the dielectric breakdown model (DBM), to describe the shapes of cities. Whereas DLA clusters grow by the accumulation of particles at their wispy edges, the DBM embodies a process in which the ragged tips advance by pushing their way outward into some surrounding medium—arguably a more realistic description of urban sprawl. Batty and Longley "tuned" the model to adjust the density of the clusters: crudely speaking, whether the branches were thick or thin. They found that it could mimic, in an approximate way, the manner in which a city expands while constrained by natural obstacles such as rivers and coastlines (Figure 6.12). In general, as a city grows it becomes increasingly dense, filling up more of the available space: its fractal dimension (see page 112) increases. Batty and Longley calculate that between 1820 and 1962 the fractal dimension of London increased from 1.322 to 1.791.

In the universality of these fractal growth models we can find an echo of

a   *b*

6.12   A fractal growth model developed by Michael Batty and Paul Longley can give a reasonable imitation of the shape of the Welsh city of Cardiff, confined by rivers and sea: *a* shows the real city, and *b* is the computer simulation "grown" from the model.

Herbert Spencer's words from 1876: "When we say that growth is common to social aggregates and organic aggregates, we do not thus entirely exclude community with inorganic aggregates: some of these, as crystals, grow in a visible manner."[20]

But the sad truth is that DLA-type clusters don't look much like real cities. For one thing, the former consist of one continuous mass, dense in the center and getting progressively more diffuse as one moves outward. But real cities are more clumpy: they tend to grow by linking to and then engulfing outlying settlements, just as London has swallowed Kensington, Clapham, and Hampstead. Moreover, the presence of the city promotes the formation of such satellite communities in the first place: the suburbs, whose residents enjoy their proximity to the bright lights without actually having to live there. Businesses in these small communities feed off the trade provided by the city as well as that from locals.

This more complex picture of city growth has been simulated in a model of nonequilibrium growth developed by physicists Hernán Makse, Shlomo Havlin, and Gene Stanley at Boston University in 1995. To Makse, the irregular shapes of cities looked like the forms of fluids percolating into porous rock. Some simple models of this phenomenon (which is important in oil mining) treat the progress of the fluid through the tortuous pore network as a random process, so that each advancing tendril is independent of the others. But Makse found that a better description emerged if he assumed that one region of the fluid "feels" the effect of others over relatively long distances. In the terminology of physics, this means that they are *correlated*.

Makse and his colleagues applied the model of "correlated percolation" to the growth of cities. They reasoned that clusters of development are correlated, too: new housing and businesses are more likely to spring up in locations where there is already housing and commerce nearby. They regarded city growth as a process of the same type as DLA: it happens by the accretion of new particles ("development units"), generally around the periphery. But whereas in DLA particles stick at random, in correlated percolation they are apt to go where other particles are. Furthermore, the model allowed for new centers of development to appear that are not physically connected to the main cluster (while still feeling its influence through the correlations).

The shapes that result depend on how strong these correlations are. Within a certain range of correlation strength, the growing clusters start to look very much like real cities (Figure 6.13). One way to make a more precise quantitative comparison is to consider how many small towns there are around the main city. (In the model, towns correspond to outlying clusters of connected particles.) Clearly there are more small towns than large ones, and the data for London and Berlin show that there is a precise mathematical relationship between the size of a town and the total number of towns of that size. This relationship (it is in fact the *probability distribution* of town sizes—see page 45) is accurately mimicked by the model.

6.13   The shape of a city like Berlin (*a*), here shown as it was in 1945, is mimicked by the clusters of particles that develop in a model of "correlated percolation" (*b*). The detailed features are quite different, but the two structures have a similar appearance. In fact, they share rather precise mathematical properties in terms of the distributions of "particles."

The striking thing about this relationship is that it is the same for Berlin in 1920 as for Berlin in 1945 and even for London in 1981, even though these two cities have very different sizes and have no doubt been subject to very different planning regulations. There seems to be an inevitable "shape" to big cities, which the correlated-percolation model can reproduce and which does not fundamentally change as the city grows. Planning seems to have had no effect on this law of growth: looked at in these terms, the "shape" of London in the 1980s bears no imprint of the Green Belt policy implemented in the 1960s in an attempt to contain its sprawl within bands of protected parkland. "You could say," argues Gene Stanley, "that the lawmakers do what they want to do, but people will live where they want to live."[21] This collective process creates its own physical laws of shape and form.

But planning is not always invisible in the overall structure of a city. Rui Carvalho and Alan Penn at University College, London, say that a slightly different way of measuring a city's "shape" reveals a division into two classes, which show the influence of different planning regimes. They use a mapping scheme introduced by Bill Hillier and Julienne Hanson, in which open spaces in the urban environment are represented as straight ("axial") lines that are extended until they cross another open space. This creates a so-called axial map (Figure 6.14). Carvalho and Penn looked at the probability distributions of the lengths of the axial lines for thirty-six cities in fourteen different countries, and found that twenty-eight of these fitted onto two different "master curves": that is, there seem to be two distinct types of city structure. One class corresponds to relatively "open" structures, with many axial lines that span the whole urban space—this group includes Bangkok, Eindhoven, Seattle, and Barcelona. The other group has a denser web dominated by short lines, and includes London, Hong Kong, Athens, and Dhaka. The researchers think that the growth of the first group of cities has been governed by "global" planning of the large-scale structure, while the second group has been guided only by "local" planning, so that there are fewer "city scale" features such as long avenues. A handful of cities, such as New Orleans and The Hague, fall in between these two classes, indicating a mixed influence of local and global planning.

The U.S. economist Herbert Simon points out that an absence of central planning does not necessarily mean that all cities are poorly "designed." On the contrary, they are (or at any rate, they once were) often remarkably effective in arranging for goods to be transported, for land to be apportioned

6.14   Another way to depict the underlying spatial structure of cities is to represent them as "axial maps" in which straight lines, designating open spaces, are extended until they cross another such line. Here we can see the rather different structures of Tokyo, Bangkok, Athens, and Las Vegas. The probability distributions of these axial maps for many cities fall into two distinct classes of city structure, corresponding to different planning regimes. A few cities do not fit into either class.

between residential, business, and manufacturing districts, and for a lot of activity to be fitted into a small area:

> I retain vivid memories of the astonishment and disbelief expressed by the architecture students to whom I taught urban land economics many years ago when I pointed to medieval cities as marvelously patterned systems that had mostly just "grown" in response to myriads of individual human decisions. To my students a pattern implied a planner in whose mind it had been conceived and by whose hand it had been implemented. The idea that a city could acquire its pattern as naturally as a snowflake was foreign to them. They reacted to it as many Christian fundamentalists responded to Darwin: no design without a Designer![22]

Sadly, the laws that dictate city growth are today swelling many cities to proportions that few would perceive as "marvelous." That this process is beyond the reach of planners may seem a pessimistic conclusion for those who worry about urban sprawl.

Maybe so; but maybe not. If cities are going to grow come what may, perhaps it is better to focus attention not on how to fence them in but on how to make them more attractive places in which to live. Good public transport and services, low-emission vehicles, protected green spaces, varied local shops, attractive architecture—all seem perfectly feasible if the will exists, and all seem more appealing than grand plans for imposing futile designs on places that, as one nineteenth-century traveler said of London, already seem to have "no beginning and no end."[23]

SEVEN

ON THE ROAD

THE INEXORABLE DYNAMICS OF TRAFFIC

A social planner can usefully contemplate traffic signals. They remind us that, though planning is often associated with control, the crucial element is often coordination. People need to do the right things at the right time in relation to what others are doing.   —Thomas Schelling (1978)

One of the principal objects of theoretical research in any department of knowledge is to find the point of view from which the subject appears in its greatest simplicity.
                                                    —J. Willard Gibbs

> Down to Gehenna or up to the Throne,
> He travels the fastest who travels alone.
> —Rudyard Kipling (1890)

How would you feel about spending a whole working week in every year sitting in a box the size of a large cupboard with nothing to do, unable to get up, stretch, sleep, read, or watch TV? Well, if you live in Washington, D.C., Boston, or Denver, you may be doing that already. And in Los Angeles you are likely to be experiencing a full week and a half of it.

I'm talking about sitting in traffic jams. As cities sprawl and commuters have farther to travel, as many public transport systems are eroded and cycling becomes a life-threatening activity, sitting in an automobile in stationary traffic has become a regular part of urban life. The time wasted in this largely solitary and stressful manner has more than tripled, on average, in large American cities over the past two decades. In Los Angeles you can anticipate fifty-six hours of jam time every year: one and a half times the length of a normal working week.

And it is sure to get worse. Londoners never tire of telling one another (usually when stuck in a traffic jam) how the average vehicle speed through the city today is as slow as it was a hundred years ago, when horsepower meant exactly that. The city has recently introduced "congestion" tolls in a desperate attempt to ease the flow in the gridlocked center.

In holiday seasons it is not unusual for jams in Europe to grow more than a hundred kilometers long. The hidden cost of motor transport caused by traffic jams is phenomenal, and would surely give us pause if it were reflected in the price of a vehicle or a tankful of fuel. The economic cost of all the wasted time is estimated at around $100 billion a year in Germany, and that's without considering the environmental impact. The price of the delay and wasted fuel in, say, Houston (where fuel is cheap!) adds up to around $850 per person per year.

Emissions from road vehicles are one of the biggest sources of pollution. In a densely urbanized country such as Germany, 60 percent of the noxious carbon monoxide and nitrogen oxides pumped into the atmosphere comes from road traffic. Pollution from road vehicles is the most likely cause of the recent increase in childhood asthma in cities. The greenhouse effect of carbon dioxide emissions from car exhaust is probably helping to break up the polar ice caps. Engines idling in a traffic jam add to these problems while fulfilling no useful function. In Seattle more than eighty gallons of gas are burned per person each year by cars that are motionless or barely moving.

One option is to build more roads. But this simply attracts more traffic, and you are soon back where you started. "Building more roads to ease traffic," says Richard Moe, head of the U.S. National Trust for Historic Preservation, "is kind of like trying to cure obesity by loosening the belt."[1] The ideal solution is to encourage greater use of public transport, more cycling, and more walking. All these things are possible, but our addiction to the internal combustion engine is going to be hard to kick, and it seems to be getting worse all over the globe. If reducing the volume of traffic is a daunting task, might we at least in the meantime direct it more efficiently?

Despite the use of computers to optimize the design of traffic control measures and street networks, every city has its black spots where traffic planners and road engineers have clearly not got it right. Perhaps they are simply faced with an impossible task, trying to make a decades-old infrastructure cope with twenty-first-century traffic volume. And how well we all know the perversity of traffic flow: the way, for instance, "phantom" jams on a highway can form for no apparent reason.

Statistical physics can now help us make sense of how traffic moves, and why sometimes it doesn't. Researchers worldwide are devising conceptual models derived from the physics of liquids and gases that can predict when and where congestion will occur and what form it will take. Traffic has been found to have its own peculiar laws of motion, which are often surprising and occasionally infuriating.

The physics of traffic will certainly not solve all our transport problems, but it can guide us to better safety measures, improved highway design, and more accurate forecasting of traffic jams. It will render traffic control less of an empirical activity and more of an exact science. And not a moment too soon, for in the world of transport planning, laissez-faire is no longer an option.

KEEPING TRACK

Navigating around busy road systems is like trying to play the stock market: you are always making choices on the basis of incomplete information. Traffic reports on local radio and in-car computerized route planners give only a limited picture, and not necessarily as quickly as you need it. But the German city of Duisburg shows what we might expect in years to come. On the Internet, you can call up a map of the instantaneous traffic flow throughout the city, updated every minute.

Traffic flow can be measured by wiring up roads with monitors that register passing vehicles. A single monitor such as an "induction loop," a pressure-sensitive wire pinned across the road, can clock the number of cars passing per minute. The most useful measure of traffic flow, however, is not numbers but *density*. Ten cars could pass an induction loop in a minute either nose-to-tail in a slowly moving queue, or more widely spaced at a higher speed. In the former case the traffic density is high; in the latter it is low. To deduce the density, one needs to know not only the traffic flux in cars per minute but their speed. And to measure that requires two induction loops, closely spaced, at each monitoring point.

In principle a city's traffic density can be fully mapped by staking out every road with pairs of induction loops. But this is completely impractical. In Duisburg, the measurements are made only at a small number of key points; most of the streets are not monitored. The gaps are filled using a computer simulation of traffic flow: a model of moving particles much like those we encountered in the previous chapter. The real data are used to keep the simulations honest: the calculations are continually tweaked to make sure that they give the right values for traffic density at the places where this is known. The assumption is that the bits in between will then not be far from the true picture.

The model was developed in the early 1990s by Michael Schreckenberg, a physicist at the University of Duisburg, in collaboration with Kai Nagel, then at the University of Cologne. Nagel has worked on a similar project in the United States to create real-time traffic maps of Dallas and other American cities.*

WAVES AND PARTICLES

The Nagel-Schreckenberg (NaSch) model was by no means the first attempt to use physics to understand traffic. James Lighthill (1924–1998), one of the great twentieth-century experts on the physics of fluid flow, proposed in the 1950s that the behavior of traffic moving on a road was rather like that of a liquid flowing down a pipe. In collaboration with Gerald Whitham at Manchester University he fashioned this insight into a rough-and-ready theory of traffic flow.

In the Lighthill-Whitham model, the individuality of drivers is entirely submerged beneath average driving behavior, just as the theory of fluid motion ignores the vagaries of individual molecules. This is ironic, for Lighthill himself was anything but average in his driving habits. He was a persistent speed offender, but would explain in court that as Lucasian Professor of Mathematics at Cambridge (the chair once occupied by Isaac Newton) he was fully aware both of the laws of mechanics and of his social duty not to waste energy. As a result, he told the hapless judges, he felt obliged to desist

---

*See http://www.traffic.uni-duisburg.de/OLSIM/ and http://www-transims.tsasa.lanl.gov/.

from braking when going downhill. It seems that this defense was occasionally successful.

Perhaps the monotonous, beetlelike procession of vehicles on a highway might make us less nervous about the apparent psychological naïveté of constructing a physics of traffic as opposed to a physics of pedestrians. But behind every wheel is an individual in command of a machine, and that can do strange things to human behavior. If pedestrians have their quirks, how much less predictable are those of a car driver, trying to cope with anything from exhaustion to squabbling children, from the influence of narcotics to surging hormones?

Again, however, we can consider the matter in terms of averages and fluctuations around them. Most drivers manage to drive sensibly and predictably, without suddenly changing speed or colliding with everything in their path. Most variations from the average are minor: some people like to drive at ten miles per hour below the speed limit on a highway, others at the same speed above the limit. Few make the journey from Birmingham to Southampton entirely in third gear, although I'm told my grandfather used to try.

Various researchers have tried to improve on Lighthill's work by allowing for the responses of real drivers. A group at General Motors in Warren, Michigan, in the 1950s devised one of the first so-called car-following models. This model treated vehicles as discrete objects, rather than regarding traffic as some quasi-continuous fluid, and it assumed that each driver modified his or her speed in response to what the car ahead did. The driver would accelerate or brake (decelerate) depending on two factors: the distance to the car ahead and the relative speed of the two cars. (Drivers tend to brake more sharply if they are both traveling at 60 mph rather than 20 mph.) In 1974, Rainer Wiedemann at Karlsruhe University went further, devising a model in which each driver was governed by a complex suite of "psychological" rules. But the more complex the model, the harder it becomes to know what outcomes are in any sense "fundamental" aspects of traffic flow and which follow from the details of the rules.

The model that Kai Nagel developed in collaboration with Schreckenberg certainly had the virtue of simplicity. It was basically a kind of cellular automaton. The highway is divided into a series of cells, each of which can be either empty or occupied by a vehicle. The vehicles move from cell to cell in each time step, like pieces on a Parcheesi game board advancing on each round. So the flow evolves in a succession of freeze-frames. As in Dirk Helbing's pedestrian model, each driver wants to reach a certain preferred speed. Every car will, on a clear road, accelerate until it reaches this speed.

The drivers also aim to avoid collisions. This entails maintaining a safe distance from the car in front, which gets larger as the relative velocity of the two vehicles increases. The third ingredient of the model is an element of randomness, or "noise." No driver estimates accelerations and decelerations perfectly—in particular, a driver might overcompensate when slower vehicles come into view ahead by braking more than is necessary. Fluctuations in speed are also commonly caused by roadside distractions: haven't you ever braked to read a billboard? These were the rules of the road. How did the traffic move?

Nagel and Schreckenberg found that there were two distinct types of flow, distinguished by how the flow rate varies as the traffic density increases. The flow rate is a measure of how many vehicles pass a certain point each hour (or minute, or whatever time interval you choose). The traffic density is the number of vehicles on each kilometer (or mile or whatever unit) of highway.

As the traffic gets denser while remaining light enough for drivers to drive as they please, the flow rate increases as the density increases: you pack more cars into each mile of road without their needing to slow down, so more vehicles pass any given monitoring point each hour. But at a particular "critical" density this tidy pattern ("free flow") breaks up. The cars start reacting to one another's presence by slowing their speed. Then the increased throughput owing to greater traffic density is compensated by a decreased throughput owing to slower motion. At the critical density, the flow rate begins suddenly to decrease rather than increase as the traffic density rises (Figure 7.1). In the parlance of traffic studies, we have gone from "free" to "congested" traffic.

RISK AND CONTINGENCY

In a version of this model explored by Nagel and his colleague Maya Paczuski in 1995, each driver's preferred speed is maintained rigidly whenever possible, mimicking the kind of cruise-control facility available in some vehicles. In this scenario the transition between free and congested flow can be postponed: instead of undergoing a transition to congested flow at the critical density, the traffic continues to flow freely, increasing its flow rate as the density increases (see Figure 7.1). It is as if the drivers have collectively decided to take the *risk* of maintaining their speed as the traffic gets heavier.

In this cruise-control model, the critical density thus represents a bifurcation point at which two options become available. One is the safe bet: all slow down. The other is the gambler's choice: all stay fast. As long as no one loses his or her nerve or concentration, this dense, fast-moving traffic can survive without collision. But it is a perilous and precarious state. If one person brakes, the car behind must do so too, and the one behind that . . . and all of a sudden the high-flow state collapses and we are on the congested, slow-moving branch. Of course, these simulated drivers do not *know* they are taking a risk if they keep moving fast, just as molecules do not know that they are in a gas or liquid state. The point is simply that the fast-moving state is a feasible collective state, even above the critical density.

But this state is so fragile that it falls apart at the slightest provocation.

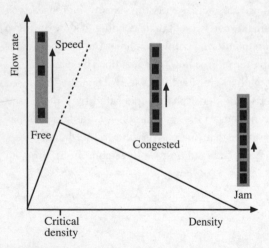

7.1 The transition from free to congested traffic can be seen on a graph of flow rate plotted against traffic density. Above a certain density, the flow rate begins to decrease as density increases, because the cars are forced to go slower. But there is an alternative, metastable "free" state above the critical density (indicated by the dashed line).

Any random fluctuation converts it, in a Grand Ah-Whoom, to a congested state. It is not, in other words, stable. Physicists know all about states like this: they call them *metastable*, which means "next to stable." Metastable is not the same as unstable—the fast-moving state can persist above the critical traffic density as long as no one rocks the boat.

Gases, liquids, and solids can be metastable too. Each can exist under conditions where another state is actually more stable. A liquid can be cooled below its freezing point without its seizing into rigidity; it may pass by the phase transition as though there were nothing unusual about that point at all. Such a liquid is said to be supercooled.

A metastable liquid survives because freezing has to start somewhere. Water does not freeze everywhere at once; it begins with a few tiny ice crystals, which then grow steadily through the liquid. Typically these "seed" crystals are germinated by irregularities in the liquid, such as particles of dust or scratches on the walls of the vessel. At such places it is slightly easier for water molecules to come together in an icelike configuration. If the water is kept scrupulously free from impurities and other sites of ice "nucleation," the only way ice can begin to form is for the randomly moving water molecules somewhere to come together by chance in an icelike arrangement. That is to say, some random fluctuation must trigger the transition.

In principle this becomes possible the moment water falls below its freezing point. But in practice it may not happen for a very long time—although the chances that it will do so rapidly increase the further the temperature falls below freezing. So far, the record cooling for supercooled water stands at around minus 39°C. Below this, it is virtually impossible to prevent water from freezing.

We might think of the nucleation phenomenon as rather like the spread of group chants through a football crowd. One sometimes gets the impression that an incoherently yelling crowd is metastable with respect to one that is chanting in unison. But the chant has to start somewhere; the spectators won't all pick it up at the same instant. Many small groups of fans might try to start a chant or song, only to give up when it doesn't catch on with their neighbors. Once the chant spreads to a group of some critical size, however, it seems to take on a life of its own, and soon a whole section is voicing it as one.

Incidentally, the freezing of the oceans triggered by a shard of ice-nine in Kurt Vonnegut's *Cat's Cradle* (see Chapter 4) implies that liquid water is metastable relative to this imaginary form of ice—that, given the chance, water would freeze to "warm ice." In this picture the oceans are just waiting to freeze solid, except that a seed of ice-nine is too difficult to form by relying on random fluctuations alone. Only when a preexisting seed is added does the transformation become possible. In the late 1960s a group of Russian scientists thought they had discovered a new, "gummy" form of water that was more stable than liquid water under everyday conditions of temperature and pressure. Some feared that this so-called polywater might indeed congeal the oceans if it came into contact with them. Fortunately, polywater was just a figment of a few experimenters' imaginations.

Here I must make a slightly technical point about metastability that is important for what follows. I distinguished earlier between first-order and critical phase transitions. Freezing, boiling, and so forth are first-order transitions. The onset of magnetism at the Curie temperature, and the cooling of a fluid below its critical temperature so that it separates into liquid and gas, are critical transitions. It turns out that only first-order transitions can support metastable states, which temporarily "ignore" the transition. Critical phase transitions, on the other hand, cannot be avoided, for something special happens at a critical point to ensure it is not overlooked—as we shall see in Chapter 10.

The switch between free and congested traffic in the NaSch model is thus a kind of first-order phase transition. The existence of the metastable "free flow" branch has a further implication. Suppose the density of traffic gradually increases as rush hour approaches on a city street. The critical density for congestion may be exceeded without a congested state appearing, so that the traffic persists in a metastable free-flow state. But before long a fluctuation caused by some nervous driver triggers the collapse to congestion. The flow rate plummets to nearly zero and the traffic density surges.

Then the traffic eases as rush hour passes, and the density begins to fall. But the traffic cannot make the transition back to the free-flow branch until the density falls below the critical density, because until this point is reached congested flow is always more stable (Figure 7.2). In other words, metastability is a one-way affair. We can get to a metastable free-flow state by increasing the traffic density from low values, but not by decreasing it from high values. In the same way, one can make supercooled water by cooling the liquid but not by warming up ice close to but below the melting point.

Flow rate

Initial and
final state

Jam

Critical
density

Density

7.2   If free flow persists into the metastable regime, it can be triggered into switching to congested flow at any moment by a random fluctuation. As the traffic subsequently lightens, free flow cannot be regained until the critical density is reached. So increasing the traffic density has different consequences from decreasing it: there is a "loop" in the flow history that can be traversed only in one direction.

Thus the state of the traffic depends not only on its density but on its *history*—on whether it was previously denser or less dense. As the traffic density rises and then falls, the flow rate follows a loop like that shown in Figure 7.2, which can be traversed only in one direction. Physicists call this one-way behavior *hysteresis*.

The NaSch model gives an inkling of how traffic jams can form without any apparent cause. Suppose that while a column of traffic is moving in the metastable free-flow state, one driver decides for some reason to brake suddenly. Perhaps there is a genuine cause: a dog runs out in front of the car. Or maybe the driver slows instinctively as he reaches for his ringing mobile phone. This situation can be idealized in a simulation by programming a vehicle to suddenly change its speed from that of the surrounding traffic to some low value, before immediately accelerating again to the same speed as the others. It is the briefest of disturbances, but look what it does to the flow (Figure 7.3).

Here each line of dots rising steeply from the lower left to the upper right is a plot of how a single car's position changes over time in such a simulation. A straight, sloping line corresponds to a car moving with constant

7.3   Jams triggered by a single spontaneous fluctuation in metastable flow. The figure shows how the distance x for vehicles along a highway varies with time t in a computer model of traffic flow. A straight line of dots sloping steeply toward the top and the right denotes a vehicle moving at constant speed. Each vehicle entering the highway (at distance zero, along the bottom edge of the plot) begins a new line. The dark bands sloping down from the upper left correspond to jams. Here a vehicle's smooth progress is disrupted, creating a kink in its "time line." A single disturbance at the top left, caused by the sudden braking of one vehicle, triggers a whole set of congestion points as time progresses.

speed. A car brakes at the top left, and then quickly resumes its previous upward "climb." But the cars following behind—the lines to the right—respond by slowing their speed to avoid collision. The wave of kinks in the sloping lines corresponds to a wave of cars that abruptly slow their speed. You can see that many cars are affected, including those that enter this stretch of highway long after the braking car has vanished from the scene. The thicker the web of black lines, the denser the jam.

And there is more. If the jam stayed at the point where it was initiated, the wave of disturbance would trace out a horizontal line in this diagram. But instead the wave slopes downward as time increases to the right, indicating that the jam moves upstream—in the opposite direction to the traffic. In other words, a jam initiated at one place can move spontaneously to another place in a stream of traffic. And the initial disturbance splits into several branches, so that a car arriving later (toward the right of the diagram) at this stretch of road will encounter not one jam but a succession of them. These jams show no sign of dying away—if anything, they proliferate as time goes by. So a single, small, brief fluctuation creates several moving waves of congestion.

These events probably sound all too plausible, and indeed they do arise in real traffic. But the NaSch model is not ideal for describing them. It turns out to be too simplistic and excessively sensitive to small disturbances. Before looking at more accurate ways of modeling traffic, however, let's see what observations tell us about real-world behavior on the road.

THREE FORMS OF TRAFFIC

In 1965 a team of researchers at Ohio State University tracked a number of cars by means of aerial photographs as they moved along a highway. They saw precisely the kind of "jam without a cause" that the NaSch model can produce, moving upstream in a persistent wave (Figure 7.4). This type of jam

7.4   A spontaneous jam in real traffic data. This is the same type of graph as Figure 7.3, with each line representing the time line of a single vehicle. The kink sloping to the lower right is a moving jam.

is a consequence of driver overreaction: people braking harder than they really need to.

The general picture described by Nagel and Schreckenberg received more substantial support in 1996, when the German researchers Boris Kerner and Hubert Rehborn, of the Daimler-Benz research laboratories (now DaimlerChrysler) in Stuttgart, made detailed observations of traffic flow on a section of the A5 South autobahn connecting Giessen, in Germany, to Basel, in Switzerland. The highway is particularly busy where it passes close to Frankfurt, and this stretch was fitted with double induction-loop detectors to measure the velocity of every passing vehicle.

Kerner and Rehborn saw precisely the behavior forecast by the NaSch model. The flow rate increased as the traffic density increased, up to a point at which the stream could suddenly collapse into a jam. The experimental data (Figure 7.5) staked out a graph very much like the bifurcating plot of the model (Figures 7.1 and 7.2).

What exactly does this graph imply for an individual motorist? In Figure 7.5 a series of data points are sequentially numbered. These points represent consecutive measurements of the flow every three minutes at one location on the highway. At a time corresponding to point 1 the traffic is freely moving but at a density greater than the critical density of around 20 vehi-

7.5   Free, congested, and metastable free flow show up in observations of real traffic made on a German highway in 1996. The numbered data points represent 1-minute averages over a 12-minute flow sequence at one point on the highway (see text).

cles per kilometer. Subsequently it begins to slow down, carrying the flow rate off the metastable branch (points 2 and 3). Then suddenly a jam coalesces, rendering the vehicles almost immobile. Things stay that way for the next ten minutes or so—points 4, 5, and 6—before the traffic gradually starts moving again and the flow rejoins the free-flow branch. Notice that free flow is regained at around the critical traffic density (point 12). In other words, the traffic emerges from the jam more gradually and at a lower density than it enters: the hysteresis effect. Again, this "time history" is a one-way phenomenon: we never see the time sequence from 1 to 12 occurring in reverse order.

All of this looks like good news for the NaSch model. But life on the road is not quite that simple. Figure 7.6 shows a different set of traffic flow data, which reveals a new facet of the motion. The general form of the results is similar: a free-flow branch with a metastable region, and a congested branch splitting off from it. But the congested branch is a complete mess! Vehicles do not simply enter a jam that congeals out of free flow, slow to a near standstill, and then accelerate away at the end of it. Instead, their speeds (and thus the overall flow rates) vary considerably once free flow has turned into congestion.

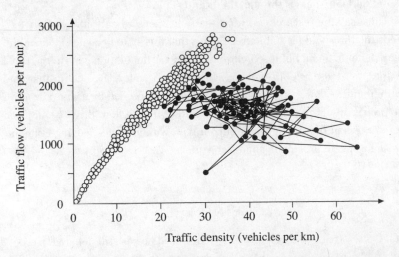

7.6   Vehicles diverging from the free-flow branch can sometimes follow a much more erratic time line, experiencing a wide and unpredictable range of different traffic states.

Kerner and Rehborn proposed that *two* distinct flow states can be identified in congested traffic. In jams movement is minimal, the traffic is very dense, and the flow rate falls close to zero. This corresponds to the right-hand extremity of the "congested" branch: points 4, 5, and 6 in Figure 7.5. But congested traffic can keep moving at a respectable flow rate even at high density if all the vehicles (in all lanes) move with more or less the same speed. In this case, they said, the traffic has become *synchronized*.

Thus, said Kerner and Rehborn, there are not two but *three* basic flow states: free flow, synchronized flow, and jams. In the transition from free to synchronized flow the vehicles keep moving and the flow rate stays quite high, but the density increases sharply. In a transition either from free or synchronized flow to a jam, the vehicle speeds drop sharply to near zero, and the density is about as high as it can be, with vehicles nose to tail.

Does this sound familiar? In the phase transition from gas to liquid, the particles remain mobile but the density jumps to a higher value. But when gases or liquids freeze to solids, the particles become very densely packed and immobile. The three states of traffic look remarkably like these three states of matter. Moreover, Kerner and Rehborn say that the transition from free flow to jam rarely happens directly; synchronized flow normally appears as an intermediate state, in the same way that matter changing from a gas to a solid normally passes through the liquid state.

Thus, say the researchers, what really happens when traffic density exceeds the threshold for stable free flow is that this state becomes metastable relative not to a jam but to synchronized flow. A fluctuation can tip the traffic into the dense, slower-moving synchronized state. They claim that the jump from free flow to synchronized flow is accompanied by a sharp drop in the probability of passing in multilane traffic. In free flow, drivers can pass more or less at liberty. In synchronized flow, with all lanes moving at about the same speed, passing is almost absent.

DISTURBING THE FLOW

Exactly how free flow gives way to synchronized flow is still debated. Kerner and Rehborn are convinced that synchronized flow is a fundamental traffic state, just as liquids are a fundamental state of matter. Others, such as Dirk

Helbing, suspect that its appearance in traffic may be contingent on external influences: perturbations to the free-flow state. That is to say, the synchronized state does not congeal spontaneously out of nothing but has to be triggered by disturbances such as bends, hills, bottlenecks, and entry and exit lanes. These could act like the particles of dust that nucleate a solid from a super-cooled liquid, or, more aptly in this case, a liquid from a supercooled gas.

Kerner and Rehborn agree that synchronized flow does seem to be commonly triggered by "impurities" on the highway, such as entry and exit points at junctions where vehicles enter and leave the road. They suggest that when synchronized flow appears spontaneously it is a transient state that persists for typically only half an hour or less, whereas it can last for several hours upstream of on-ramps.

In some models synchronized flow is itself metastable and apt to congeal into a jam if perturbed. But Michael Schreckenberg and his co-workers at Duisburg maintain that in real traffic synchronized flow is actually quite ro-bust and results from a factor not accounted for in many traffic models: the drivers' desire for smooth and comfortable driving.

Most models assume only two things of drivers: that they aim to reach a certain preferred speed and that they avoid collisions by slowing down. In principle this could lead to a very jerky ride: drivers speed up whenever they have enough room, only to brake when another vehicle comes into view ahead. Schreckenberg and his colleagues argue that few people actually drive this way; instead they prefer to avoid sharp acceleration and deceleration. When this rule is added to the cellular automaton model, synchronized flow emerges as a fundamental and stable state, which persists even in the face of a serious perturbation such as an upstream-traveling jam moving through it.

This sort of refinement promises to enhance the predictive power of traf-fic models, but it also shows us something deeper: these collective flow modes are an irreducible aspect of traffic flow. Making the models more psy-chologically complex may alter the precise conditions under which the flow modes appear, but it will not alter the fact that traffic, like atomic matter, really does seem to have fundamental states.

Are there truly just three such states? Well, yes and no. It appears that any one vehicle will indeed find itself at any moment either in light and un-correlated traffic, dense and synchronized moving traffic, or dense and near-motionless jams. But Dirk Helbing and his collaborators have found that there are many ways in which these flow modes can be mixed and permuted over the course of time.

In collaboration with Martin Treiber at Stuttgart, Helbing devised a model in 1998 that shared features with both the cellular automaton of Nagel and Schreckenberg and the fluid-flow models originally conceived by Lighthill and Whitham. Helbing and Treiber dispensed with individual cars and treated the traffic as a smooth fluid. But it was a most peculiar fluid. In the traditional theory of fluid motion, called hydrodynamics, each little "parcel" of fluid affects those around it through viscous drag: it exerts a frictional force that slows down the movement of the surrounding fluid. In Helbing and Treiber's model, the interactions between parcels of "traffic fluid" are more complex, capturing the same kind of driver responses as the particles in the NaSch model. Again, drivers are assumed to speed up and slow down as they react to what is happening ahead, aiming to reach a particular velocity and to avoid collisions. It is a fluid with a mind of its own—indeed, with a multitude of minds.

At low densities this model generates free flow. As the density increases, free flow becomes metastable: small fluctuations can be absorbed and dispersed, but larger ones will create localized jams that propagate upstream. As the density increases further, these jams can evolve into cascades, like those seen in Figure 7.3 (page 166): congested waves separated by free flow. This "stop and go" motion is familiar to anyone who has sat behind a driving wheel. At even higher densities, the traffic becomes uniformly congested in a slow-moving jam.

Helbing, Treiber, and Ansgar Hennecke at Stuttgart looked at how their traffic fluid behaved in the presence of an irregularity such as an on-ramp. They instigated a localized wave of congestion in the traffic ahead of the ramp and watched what happened when this wave, moving upstream, drew level with the ramp. For low traffic flow, the congestion simply dispersed. In heavier traffic it could trigger all kinds of behavior: waves of localized jams separated by free flow, waves of congestion with almost no respite in between ("oscillatory congested traffic"), congestion pinned to the position of the on-ramp, uniformly congested traffic behind the ramp, and so forth.

The researchers illustrated this diverse behavior on a kind of phase diagram that shows the conditions under which each state forms (Figure 7.7). This can be regarded as analogous to the "morphology diagram" drawn up

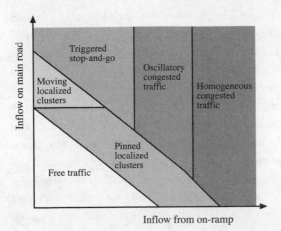

7.7 Traffic states in the "fluid" model developed by Helbing and his colleagues. Homogeneous congested traffic is a jam that grows steadily in length. In oscillatory congested traffic the jam also lengthens over time, but comes in waves so that individual vehicles in the jammed region alternately slow to a halt and then move again. Moving localized clusters are knots of dense traffic that move steadily upstream, in the opposite direction to the traffic flow. Triggered stop-and-go is a similar state in which such knots occasionally throw out a smaller dense cluster that moves downstream before triggering a new upstream-traveling knot. Pinned localized clusters are knots that stay fixed at one point on the road as vehicles move into and out of them.

for bacterial growth patterns (page 113); after all, traffic flow, like bacterial growth, is a nonequilibrium process. The jumps between different flow states occur quite suddenly as the "control parameters"—the flow of traffic on the main highway and the influx at the ramp—are altered in the model. Traffic, it seems, is at the mercy of a series of nonequilibrium phase transitions.

JAMS TOMORROW?

Are these various models truly capturing the behavior of real traffic, or are they little more than fancy computer games? Only if the model effectively mimics reality can it be trusted to provide accurate traffic forecasts. With that in mind, Helbing, Treiber, and Hennecke have compared their predictions with real traffic data taken from various German and Dutch highways fitted with induction loops. They were heartened to find that all the various states they predicted could be identified in these traffic flows. What's more, they found that by feeding into their model the basic features of the real-life traffic (such as the ratio of cars to trucks), they could predict with uncanny accuracy how it would develop as the hours passed—even when the flow pattern was by no means simple or regular (Figure 7.8).

This gave the researchers some confidence that what comes out of this kind of traffic model does not depend too delicately on what goes into it. Provided the drivers are assumed to share certain general characteristics— accelerating to a preferred speed on an open road, braking to maintain a velocity-dependent distance from the vehicle in front, and having imperfect responses that make overreaction likely—it does not matter exactly what one assumes about details such as the reaction time, the shape of the road, or the number of lanes. The same general flow modes will emerge in any case.

The observations have convinced Helbing and his colleagues that most jams are caused by anomalies: bottlenecks, on-ramps, hills, or perhaps just an aberrant maneuver by a single driver. In more than one hundred real jams that they studied, all but four (at most) were caused by some kind of stationary bottleneck.

7.8   The traffic-flow model of Helbing and his colleagues can predict how various traffic states will evolve over the course of several hours. The model is fed the traffic density and speed measured at some upstream location, and from this it calculates the likely speed and density at several points farther down the highway. Shown here is the traffic flow (speed and density) at two different points on the A5 South highway near Frankfurt, over the course of about two and a half hours. Congestion was caused by a lane closure at some point along this section. The solid line is the real flow measured by induction loops; the dashed line shows the model predictions for these locations. The model misses much of the fine detail but correctly predicts the basic features of the flow, such as the time and place of jams and their alleviation.

Traffic in cities does not behave like traffic on highways; in particular, its flow is dominated by intersections. Ofer Biham and colleagues at the Hebrew University of Jerusalem have studied a cellular automaton model in which vehicles moved from south to north and from west to east on a square grid: a fair approximation of any American city. The motions of vehicles were timed to mimic the alternation of traffic lights at intersections. Ominously, the researchers found that as the traffic density increased there was an abrupt phase transition from free flow to a stationary jam—to total gridlock, in other words. (It has been claimed that "gridlock" is an oxymoron, since grids provide, almost by definition, points of free entry and exit. Studies like this warn us not to trust such simple intuition.)

Traffic physics is already proving its worth. Nagel's Transportation Analysis and Simulation System, based on his cellular automaton model and developed at Los Alamos National Laboratory in New Mexico, has been used to plan road networks in Dallas. Knowing the likely effects of intersections, crossings, and lane narrowings should help planners place them wisely. Schreckenberg has plans to expand his own system in Duisburg to encompass the whole autobahn network of North Rhine–Westphalia, generating hourly traffic forecasts.

Modeling studies of freeway traffic can also provide useful insights into driving regulations. A model developed by Helbing in collaboration with Bernardo Huberman of Xerox's research laboratories in Palo Alto, California, shows that a mixture of cars and trucks in multilane traffic can produce what they call "coherent flow," in which all the vehicles move at the same speed without making any lane changes. This is reminiscent of, but not identical to, synchronized flow. Coherent flow is a robust state, quite stable in the face of fluctuations, and even though it prevents all drivers from reaching their preferred velocities it is nevertheless very efficient: the high traffic density and moderate speeds mean that the flow rate is almost as high as it could possibly be.

Moreover, coherent flow is a safe state because it reduces two of the main causes of highway accidents: differences in speed and lane changes. So it is worth looking for ways to promote the transition to coherent flow when traffic becomes heavy—for example, by imposing appropriate speed and lane-changing restrictions.

Helbing and Huberman say that their model also reveals how American lane rules, which allow drivers to travel at any speed in any lane, can make roads more efficient than European rules, under which the lanes are graded from slow to fast. Because trucks stay largely in the slow lane, car drivers tend to avoid that lane even when there are no trucks immediately ahead, reducing the capacity of the highway by up to 25 percent. Helbing and Treiber, meanwhile, have shown how real-life congestion and delays on a German autobahn could have been eliminated by introducing speed limits that alter in response to changes in the traffic density. A limit imposed only during the rush hour could prevent jams and reduce average travel times for all vehicles.

One of the major triggers of jams in dense traffic is fluctuations. Drivers lose concentration, get too close to the vehicle in front, and then slow abruptly. If random perturbations like this could be smoothed out, the threat of jams would be much reduced. One way would be to design traffic regulations which have the effect of forcing drivers to pay more attention in areas containing potentially jam-forming obstructions such as bottlenecks. Another possibility would be to implement traffic-control measures at on-ramps that adapt to changes in traffic density.

A more farsighted option would replace the human driver with one that never tires or miscalculates. Nearly all traffic models are collision-free, since the rules are designed explicitly to avoid them. The "virtual drivers" do not need to be granted any particularly sophisticated responses or capabilities in order to avoid crashes with other vehicles; they need to know little more than how far away the car in front is and how fast it is approaching. With this information, cars could apparently be piloted safely by robotic autopilots—real automata!

That possibility is already being explored by some automobile manufacturers, who are developing "driver assistance" systems that use radar and other sensors to track what is going on around the car and to respond accordingly. One of the main aims is to avoid crashes, or collisions with pedestrians or cyclists, by allowing automated braking or maneuvering to kick in, which has a faster response time than any human can achieve. Once such systems are installed in vehicles, they could help with the more mundane driving tasks too, such as lane changing and cruise control.

As we've seen, jam-inducing fluctuations can be caused by drivers overreacting in heavy traffic. This could be eliminated by driver-assistance systems. Traffic simulations by Treiber and Helbing have shown that some jams in heavy traffic can be smoothed away completely if just 20 percent of the cars are equipped with automated systems that enable them to respond optimally to changes in traffic flow.

Think of linking these autopilots to satellite navigation systems, digital road maps, and an up-to-the-minute traffic forecast provided by simulations pegged to real data, and you begin to have a prescription for making your journey efficient and safe without ever having to lay your hands on the wheel. Automobile associations, which are fond of lobbying for greater "driver freedom," will need to accept that it is sometimes this very freedom that makes the roads hell. Robotic drivers could be not only more considerate, law-abiding, and skillful but also better informed and more foresighted—in short, far better placed to use the roads wisely.

EIGHT

RHYTHMS OF THE MARKETPLACE

THE SHAKY HIDDEN HAND OF ECONOMICS

There is nothing which requires more to be illustrated by philosophy than does trade.
—Samuel Johnson

If all economists were laid end to end, they would not reach a conclusion.
—George Bernard Shaw

Nobody should be rich but those who understand it.
—Johann Wolfgang von Goethe

When the Scotsman Adam Smith (1723–1790) invented political economy as a field of philosophy, he had no giants' shoulders on which to stand. No one had previously pondered the question of how a market economy sustains itself, for trade was considered a worldly affair and hardly the stuff of philosophy. But Hobbes's *Leviathan* was essential reading for Smith, since its blueprint for constructing a stable Commonwealth could not afford to neglect matters of how trade was to be managed, or wealth and land were to be distributed.*

Yet the world in which Smith wrote *An Inquiry into the Nature and Causes of the Wealth of Nations* (1776) was utterly different from that of Hobbes a century earlier. There was a new social order, and its sovereign was Mammon. For Hobbes, wealth and production meant land and agricul-

---

*All the same, Hobbes has been criticized for giving insufficient attention to economics as the real basis of power. In *Oceana* (1656), Hobbes's rival James Harrington argued that the unrest of the English Civil War stemmed largely from the changes in land and property ownership that had taken place in the early seventeenth century.

ture; for Adam Smith it meant industry. While Hobbes allowed that trade could be an asset to a sovereign, for him the true measure of a nation's wealth was gold, gathered by whatever means necessary: plunder, conquest, royal marriage. Smith, on the other hand, knew that the rise of the mercantile system had made commerce the determinant of a nation's treasury.

Smith had no rosier a view of human nature than did Hobbes. Men, he concluded, are basically acquisitive: out for what they can get. So he found in Hobbes's political model the foundations for a theory of how the market system works. While Hobbes's individuals seek to accumulate power (primarily by buying it from their fellows), Smith's merchants and traders aim to amass wealth.

For Hobbes, the lust for power can be curbed only by surrendering all authority to a supreme sovereign; the alternative is a life nasty, brutish, and short. But no one ruled the marketplace: it was a free-for-all. Here was the nub of the conundrum that Adam Smith sought to explain. How, in a society awash with greed and lacking any centralized means of market regulation, do the common people nevertheless generally manage to obtain goods at prices that do not bankrupt them? What is to stop traders from charging whatever they will for their wares, if not an all-powerful authority?

The answer, in a word, is competition. Traders who inflate their prices will lose their customers to those who maintain prices at an affordable level, and they will go out of business. The pressure of competition in a free-market economy, argued Smith, should ensure that goods will always be sold at their true value or "natural price"—an honest measure of the quality and quantity of materials and labor that went into their making. Anyone charging more can always be undercut by a competitor. Smith argued that this self-regulating market will ensure that all the things people need will be supplied to them. There is no need for a central authority to assign people to jobs so as to guarantee an adequate number of tailors, cobblers, bakers, dairymen, and so forth. Wherever there is a demand, there is profit to be made, and someone will step in to make it.

Smith's classic treatise was the first real analysis of the workings of the new market economy of the Age of Enlightenment, and it contained an idea that not only pervades all of economic theory today but also resonates profoundly with the physics I have described in the preceding chapters. Not only are there indeed laws of the markets, but these laws *emerge* from the push and pull of trade, from the interaction of myriad agents. They are not imposed from outside. Two opposing forces—self-interest and competition—suffice to generate a self-regulating steady state, a kind of equilibrium. This is the "hidden hand," as Smith said, that keeps order in the market. As the economist Robert Heilbroner says, "One may appeal the ruling of a planning board . . . but there is no appeal, no dispensation, from the anonymous pressures of the market mechanism."[1]

In Smith's view, "good" (meaning a fair and comprehensive market) can emerge from selfishness. The message of *The Wealth of Nations* is one of laissez-faire: leave the market alone and it will take care of itself. It is precisely the opposite of the dictatorial straitjacket that Hobbes proposed to curb humankind's lust for power and was a welcome message to industrial capitalists seeking to expand their empires without interference from governments.

Adam Smith's world contained the seeds of ours, but it was not the same. His capitalists were all small fry: shop owners, merchants, and small-scale industrialists. In the 1770s a "manufactory" with a dozen workers was a large one. The guild regulations that had governed business practice in the Middle Ages were dissolving, and few external rules and laws had been introduced to replace them. There were no labor organizations or trade unions, and no large corporations, let alone multinationals. It was a highly fragmented market, some would say an "atomistic" one. But even on its own terms Smith's economic theory was too simplistic to cover the whole story. It was just a beginning, still lacking the kind of certitude that Newtonian mechanics had given physics.

All the same, we can see already that this primitive economic theory betrays an implicit aspiration to be a science of the same stature. Enlightenment intellectuals had long suspected that this was possible. "Nothing," said Daniel Defoe in 1706, "obeys the Course of Nature more exactly than Trade, Causes and Consequences follow as directly as Day and Night."[2] Adam Smith never wrote of market forces as such, but the concept of causative forces analogous to Newton's gravity is evidently there in his great work. His contemporary Josiah Tucker, the dean of Gloucester, was more explicit: "The Circulation of Commerce may be conceived to proceed from the Impulse of two distinct Principles of Action in Society, analogous to the Centrifugal and centripetal Powers in the Planetary System."[3] By the early nineteenth century the idea that there were fundamental and immutable laws of economics, just as there were in physics, had taken firm root. That belief has, by and large, remained in place ever since.

Smith's hidden hand became a part of the lore of the business world, and to many it was a force that society hindered at its peril. Here is Ralph Waldo Emerson in 1860: "Wealth brings with it its own checks and balances. The basis of political economy is non-interference. The only safe rule is found in the self-adjusting meter of demand and supply. Do not legislate. Meddle, and you snap the sinews with your sumptuary laws."[4]

Smith's economic principles had begun to take on the aura of an immutable physics of the marketplace, and they seemed to promise the enticing prospect of a calm and stable economy. If the market bucked, that was because someone had been tampering with it. Emerson again:

> The laws of nature play through trade, as a toy-battery exhibits the effects of electricity. The level of the sea is not more surely kept, than is the equilibrium of value in society by the demand and supply: and artifice or legislation punishes itself, by reactions, gluts, and bankruptcies. The sublime laws play indifferently through atoms and galaxies.[5]

The trouble is, even now no one has discovered what these laws are. In spite of the elegance of Emerson's words, considerations of supply and demand evidently do not suffice to explain the sometimes wild behavior of the market. Economic models have been augmented, refined, garlanded, and decorated with baroque accoutrements. Some of these models now rival those constructed by physicists in their mathematical sophistication. Yet they still lack their "Newtonian" first principles: basic laws on which everyone agrees.

This is not because economists are incompetent. On the contrary, the finest economic minds have equaled any in natural science. But it does mean that economic models are persistently incompetent, in the sense that they repeatedly fail to make the accurate predictions of which scientific theories are routinely capable. It is not only scientists who point this out—some within the economics community admit as much. John Kay of the London Business School says, "Economic forecasters . . . all say more or less the same thing at the same time; the degree of agreement is astounding. [But] what they say is almost always wrong."[6]

And it was ever thus. One week before the Great Crash of October

1929—the largest in history—Irving Fisher of Yale University, perhaps the most distinguished U.S. economist of his time, claimed that the American economy had attained a "permanently high plateau."[7] Three years later the national income had fallen by more than 50 percent. No one, not a single economist, had seen it coming.

Over the past decade, statistical physicists have begun to suggest that economists might want to rethink some of the basic assumptions upon which they construct their models. By importing ideas from physics, say the physicists, economists can start to make sense of the erratic and so far unpredictable behavior of the world's markets. This endeavor has even coined its own neologism: *econophysics*.

How have economists reacted to this encroachment on their territory? Some econophysicists feel they have been met with the deafening silence of indifference, punctuated only by bursts of irritation. Some economists, meanwhile, regard the apparent isolation of econophysicists as self-inflicted. They point out that economics has always been an eclectic, catholic discipline, little concerned with professional affiliations. What matters, they say, is not academic credentials but whether new ideas are *useful*. They wonder how much of the physicists' contribution is genuinely useful to them.

It is true that the physicists have not always made their task easier. As a group they are not renowned for their tact and humility in pointing out the shortcomings, as they perceive them, of other areas of human inquiry. And the things that interest physicists are not always those that interest economists. Moreover, physicists are used to systems whose fundamental laws remain the same in all places at all times. Economic laws are almost certainly not like this, which is one reason economics is harder than pure physics. (The economist Paul Krugman consoles himself with the thought that "luckily it is not quite as hard as sociology."[8])

Yet it seems inevitable that in time these tensions will ease, the misunderstandings will be resolved, and each side will come out of its trench to negotiate pacts and truces. Several leading economists are already listening respectfully, if cautiously, to what the physicists are saying. And most econophysicists are only too eager to engage in discussion, to be put right on their shaky grasp of economic principles and concepts, to find common ground.

The first instinct of physical scientists is one not always shared by economists: to learn from experiment. First they want to know what the fundamental economic phenomena are that one should seek to explain. In economics there is really only one experiment against which to compare theories, and it is vast and ongoing. What happens in the real marketplace?

THE IRON LAW

Smith's model of the market was purely descriptive. But by the nineteenth century confidence in science, and specifically in a mechanistic description of the world, was so buoyant that it seemed possible to bring everything within its embrace. And so economists sought rigor in mathematics.

Jeremy Bentham (1748–1832) was one of the first to propose that the ebb and flow of trade and production could be couched in terms of pure mathematics. His utilitarian philosophy, which he called the Felicific Calculus, took a more beneficent view of humanity than did Smith or Hobbes, proposing that people aim to maximize not their power or profit but their "pleasure," albeit by a kind of cold-blooded profit-and-loss reckoning.

Building on this image of man as a "pleasure machine," the Irish economist Francis Edgeworth (1845–1926) brought the mathematical approach to an early pinnacle in 1881. Despite its now exotic-sounding title, his book *Mathematical Psychics* was an attempt to develop a precise theory of political, social, and economic behavior replete with abstruse differential calculus and a good dose of statistical analysis. It was the precursor to a welter of theories that, drifting ever further from any foundation in the real world, tried to subsume human behavior beneath pages of complex calculations. According to Paul Krugman, this tradition is alive and well in modern economics: "It is cynical but true to say that in the academic world the theories that are most likely to attract a devoted following are those that best allow a clever but not very original young man to demonstrate his cleverness."[9]

But the most influential of "scientific" economic theories in the nineteenth century, at least politically speaking, was that developed by an intemperate and almost permanently poverty-stricken German Jew named Karl Marx (1818–1883).

In Adam Smith's worldview, labor was simply another commodity on the market. The laborer was a trader, and he traded in his own time and sweat. This commodification of labor was a product of the Industrial Revolution, from which there emerged a class of men and women who, in contrast to the trained craftspeople of the Middle Ages, were just so many pairs of hands. These hands were ready to operate machinery, shovel coal, or do the increasing number of low-skilled but specialized tasks demanded by the division of labor in a factory system. In return, these "proletariat" workers gained their wages.

Smith proposed that wages reach their "fair" value just like any other market price: the cheapest laborer captures the market. This meant that the "true" or "fundamental" value of labor was a subsistence wage: a worker could not afford to work for less than this, whereas those who asked for more would be undercut. So as economist David Ricardo (1772–1823) pointed out in the early nineteenth century, the industrial capitalist could always manage to employ a workforce for no more than the cost of keeping it alive: this was Ricardo's grim Iron Law of Wages. Generally, employers did just that; in fact sometimes not even subsistence was guaranteed to the proletariat.*

Marx considered the future prospects for this capitalist system. His conclusions spelled hope for the oppressed workers and despair for their employers. Capitalism, he said, was doomed to exhaust itself and to be overthrown by proletarian revolution. This was not mere wishful thinking or the desired outcome of any moral imperative; he believed he had proved his forecast with scientific rigor. Marx's approach seemed the epitome of scientific model building: he idealized, he simplified, he removed irrelevancies. His economic landscape held a two-tiered society comprising only workers (who sell their labor in return for a wage) and capitalist factory owners (who buy labor and sell goods).

The capitalists seek profit. But as Smith had pointed out, profits are always being eroded: by rising wages, and by the force of competition, which constantly pulls the market price of goods down toward the cost of produc-

---

*Adam Smith pointed out that "subsistence" wages are not merely equal to the cost of keeping a worker alive: "they must upon most occasions be somewhat more; otherwise it would be impossible for him to bring up a family, and the race of such workmen could not last beyond the first generation." Smith does not endorse the grinding poverty implicit in his words, but that was the reality of his world—a world where "one-half the children born . . . die before the age of manhood" (A. Smith, *An Inquiry into the Nature and Causes of the Wealth of Nations* [abridged], ed. L. Dickey [Hackett: Indianapolis, 1993], p. 33). The industrial world was not relentlessly ruthless, however: real wages more or less doubled during the Victorian era.

ing them. The only way a company can maintain a profit is by expanding. And so (said Marx) it must constantly bid for additional labor. This increase in demand for labor means that workers can ask for higher wages, which bites back into the capitalists' profit margins. To escape this cycle, Marx assumed the factory owners would do what he saw them doing all about him: introduce labor-saving machinery. Marx's economic model thus embraces technological change. This, as we shall see, has come to be regarded as a crucial element in modern economic theory, and rightly so, for new technology continues to change the market demand for workers at a pace at least equal to that in Marx's time.

But here's the catch. In a working day (which in the mid–nineteenth century could be sixteen hours long), workers can produce goods worth more than a subsistence wage. This "surplus labor" is the source of the capitalists' profits. Machines, however, do not offer surplus labor—in a competitive market, industrialists have to buy them at a cost equal to the value of the goods they can generate. So mechanization renders workers unemployed but does not rescue the capitalists' profits.

The result is an economic recession: wages are low and unemployment is rife. By the mid–nineteenth century it had become clear that the market was prone to recessions, such as that caused by the notorious South Sea Bubble of 1720. But they were dismissed as the consequence of some external factor perturbing "business as usual." The belief was that left to itself, the market attained equilibrium in the manner suggested by Smith. Marx, in contrast, implied that recessions are an inevitable part of the way the market works.

They are, however, transient. As wages fall, the profit margin rises once more. It becomes more viable for capitalists to start employing workers again and to expand their businesses, and the economy is revitalized. Recession provides its own cure. By the same token, an economic boom is self-limiting because it boosts wages and reduces profitability. This is again, then, a self-regulating market. But it is not the steady, equilibrium market of Adam Smith. Instead it is plagued by cycles of boom and bust.

A physicist or engineer will see at once that Marx had identified a mechanism of *negative feedback*, whereby a change moderates its own cause. In general, negative feedback promotes stability, since it returns the system to a stable state if fluctuations drive it away. So a scientist will rightly object that Marx's scheme does not in fact guarantee boom-bust cycles, as he supposed. This will happen only if the negative feedback *overcompensates* for the change, swinging the balance back in the opposite direction. There is nothing in Marx's formulation that makes this inevitable.

Yet Marx's economic vision contained the crucial concept of a market that was potentially unsteady and apt to oscillate between boom and bust. And in Marxist theory these fluctuations are an intrinsic, irreducible element of the system, not something imposed from outside. Why, though, should a fluctuating market be destined for collapse, rather than simply for perpetual fluctuation? Here one cannot avoid the impression that Marx's political predispositions overtook his analytical thinking. Each recession, he said, will be worse than the last, for each one bankrupts ever larger firms. In the end, the misery caused by lack of production, unemployment, and low wages during a deep recession will prompt a proletarian revolution. "What the bourgeoisie therefore produces, above all," said Marx, "are its own gravediggers."[10]

Come the revolution, private ownership will cease and the means of production will pass into communal control. *The Communist Manifesto*, which Marx wrote with Friedrich Engels, sought to hasten this end with a call to arms: "Let the ruling classes tremble at a Communist revolution. The proletarians have nothing to lose but their chains. They have a world to win."[11]

In fact, the proletariat had rather more to lose, as they discovered in France, Austria, and Germany, where the attempts at revolution in 1848 were violently suppressed. In Russia, of course, it was to be a different story.

CAN THE MARKET BE STEADY?

If Marx's historical prognosis was flawed, he had nevertheless found the pulse of the capitalist economy. It was a stuttering pulse.

The wild fluctuations of the market seem to taunt economists trying to construct models of how trade and commerce work. These chaotic leaps between boom and bust reveal, more than any other measure, the disparity between economic theory and practice. A great deal of the theory developed since Marx can be bluntly characterized as a futile attempt to tame the untamable.

Nothing is more revealing of the desperation this engenders than the insistence on speaking of "the business cycle." To a physicist this is almost offensive. Take a look at the changes in the U.S. economy over the past century or so, as measured by the annual percentage change in national output per capita. Using this (rather than the absolute change) as a measure of the "size" or "health" of the economy has the advantage that it removes the effects of inflation and business expansion, which create an overall upward trend in output. In a stable economy, the relative change would stay at zero. In reality, it fluctuates by up to 20 percent or so (Figure 8.1a).

When economists speak of the business cycle, they are referring to the fact that booms—peaks in the graph—are followed by recessions, or valleys.* But a cycle is surely something that happens repeatedly with the same interval—that is, periodically, like the change of seasons or the swing of a pendulum. There are plenty of peaks followed by valleys in this graph; but can you persuade yourself that you are seeing something rhythmic here? The graph looks more like the random static of a detuned radio signal, traced out on an oscilloscope. Like noise, in other words.

Actually it is worse than that. The data in this graph are annual—there is only one datum point for each year. Figure 8.1b shows the quarterly growth rate of real U.S. gross national product, a similar measure of the size of the U.S. economy. Now the data points come every three months, and we see even finer-scaled randomness in the static. Business cycles? Where?

Yet the business cycle is a part of standard economic dogma, and the very term betrays a desire to impose order on the chaos. Some economists have claimed to perceive periodicity—regularity—in the recurrence of boom and bust. By doing so, they hoped to be able to forecast when the next crash was coming, and so avoid the kind of catastrophe that wiped out investments and ruined lives in 1929. The Austrian Joseph Schumpeter claimed in the 1940s that there were at least three business cycles: one short-term, one of seven to eleven years, and one with a period of about fifty years. The apparent randomness of the fluctuations was, he said, a result of

---

*To some degree, the word *cycle* is used simply to indicate that the whole economic system is affected in the same way: growth or decline applies across different economic sectors, even if there is no direct relation between them. That does not make the choice of this word any less misleading, however.

8.1   The national output per capita in the United States fluctuates annually by up to 20 percent or so, in a manner that looks utterly unpredictable (*a*). (Notice the difference between fluctuations before, during, and after the Second World War, however, which shows that external events can have a big impact on the economy.) The quarterly growth rate of real U.S. gross national product shows how economic measures fluctuate on a finer timescale (*b*).

the out-of-step interactions between these three cycles. The Great Depression was caused by the coincidence of troughs in all three cycles.*

The ingenuity of economists in conjuring cyclical behavior out of the market's paroxysms has to be applauded, if not lauded. In the 1930s, for instance, a retired U.S. accountant named Ralph Elliot claimed that markets wax and wane in an eight-step pattern of five "advancing" waves and three "declining" waves.† This pattern, he said, could be discerned over several time frames, ranging from day-to-day fluctuations to "supercycles" lasting over 200 years. Elliot waves thus divide each cycle into stages whose structure echoes the famous Fibonacci sequence, in which each number is the

---

*Paul Krugman calls Schumpeter's two-volume tome *Business Cycles* "turgid, almost meaningless" (P. Krugman, *Peddling Prosperity* [W. W. Norton, New York], p. 26). Indeed, he comments that attempts to account for the business cycle have led many economists to produce their worst work. Clearly, they are as perplexed as they are troubled by it.

†Elliot had good cause to want to understand market fluctuations, having previously lost his job and a good part of his savings in the 1929 Wall Street crash.

sum of the previous two: one cycle can be divided into two parts (declining, or "corrective," and advancing, or "impulsive"—the "bull" and "bear" phases of the economy), composed respectively of three and five waves, giving the sequence 1, 1, 2, 3, 5, 8. Smaller-scale subdivisions of these waves yield higher numbers in the Fibonacci sequence. This piece of numerology, ostensibly constructed from the theory of bull and bear markets devised by the Wall Street analyst Charles Dow, leaves market speculators even today juggling with Fibonacci numbers and the "golden ratio" that the series encodes in a quasi-mystical belief in hidden patterns.

Schumpeter listed a whole slew of purportedly periodic economic cycles: Kitchin cycles, Juglar cycles, Kuznets and Kondratieff cycles. The last of these are particularly notorious. The Russian Nikolai Kondratieff asserted in 1926 that capitalist economies go through boom-bust cycles lasting fifty to sixty years. Whether or not this identified (as hoped) a fatal flaw in the capitalist system, it did not protect Kondratieff from the Stalinist regime when he proposed the "reactionary" idea that land expropriated by the Party should be returned to the peasants. He was banished to a gulag, where, driven to madness, he died in 1938. Some economists insist that Kondratieff waves appear in the long-term history of Western markets: they claim to identify one with the Industrial Revolution, from 1787 to around 1842; another with the "bourgeois" period, from 1843 to 1897; a third with the growth in electrical power and car manufacture, in the first half of the twentieth century; and a fourth, covering the postwar period until the present. But there is simply not a long enough historical record to permit the identification of such a long-period cycle with any statistical conviction. On such timescales, one must suspect that a cyclical model becomes more a way of systematizing history than a way of understanding the economy.

The truth is that dips and peaks in the economy resolutely refuse to recur in any predictable manner, making attempts to construct cyclical theories of economics look increasingly like Ptolemy's elaborate schemes for predicting the motions of the planets while retaining an earth-centered universe (except that at least there *is* an underlying rhythmicity to the planets). Irving Fisher admitted in 1925 that the "business cycle" might be inherently unpredictable, an idea that he seems to have forgotten four years later.

Having failed to tame these untidy irregularities with notions of periodicity, economists then tried to banish them from conventional economic theory. Although Marx's pioneering analysis of boom and bust placed the causative factors *within* the workings of the economic system (it was, in the jargon, an *endogenous* theory), the aim soon became to seek external forces that drive a potentially stable market into these paroxysms. This is still largely true of conventional economic theory today, for apparently the idea of an inherently unstable market remains too unsettling for many to contemplate.

Schumpeter, for example, proposed that the fifty-year Kondratieff cycles were associated with episodes of major technological advance, which transformed methods of production and thus altered the economic system that they fed. This idea of technology as an external perturbation on the economic system resurfaces in one of today's most orthodox economic theories, discussed in Chapter 9. The British economist Stanley Jevons, meanwhile, attempted in the 1870s to explain the business cycle as a consequence of the sunspot cycle—not quite as exotic an idea as it sounds, although it is wrong. Sunspots become more numerous every eleven years or so, a symptom of periodic changes in the output of solar radiation. Jevons suggested that these changes affect harvest yields and thus the price of corn—which drives corresponding variations in the rest of the market.*

Economists sometimes attribute to market forces a supremely stabilizing influence. They speak in terms of market equilibrium: a stable state in which all economic indicators are at an ideal and balanced level. Supply adjusts itself to match demand precisely, so that the markets are always "cleared." There is no waste and the goods are distributed to the greatest benefit of society.† This state, called the Pareto optimum after the Italian sociologist and economist Vilfredo Pareto (1848–1923), would allow the capitalist system to

---

*Jevons may have been maligned unfairly for this idea. It is now known that a conspiracy of astrophysics and meteorology does enable the solar cycle to influence the weather, and thus potentially to affect crop yields. And there is actually evidence of a fingerprint of the solar cycle in the wheat prices of seventeenth-century England. This was surely not the dominant factor behind the business cycle at that time, but neither does Jevons's theory now appear absurd.

†Adam Smith's laws of supply and demand, which provide the "hidden hand" that is supposed to engineer this equilibrium, were challenged in 1958 by arguably the first "econophysicist," M.F.M. Osborne. He questioned the fundamental assumption that prices depend on consumer demand in such a straightforward manner. This issue is still debated today, although empirically there is evidence that prices do fall as demand grows, just as Smith believed.

run smoothly and efficiently if it were not for the disruptive influence of "exogenous" factors and the meddling of governments. Even in the absence of such perturbations, however, achieving the Pareto optimum in conventional economic models demands assumptions that bear little relation to the real world.

In a mathematical tour-de-force, the French economist Léon Walras (1834–1910) put the concept of market equilibrium on a formal footing, and his work gave rise to the "general equilibrium theory," which dominated economic thinking for most of the twentieth century. The trouble is that this theory relies on suppositions about how trading is conducted and how traders act that are patently absurd, and the theory collapses if these constraints are relaxed. But faith in the Holy Grail of equilibrium is still endemic in economic practice, and it has engendered a strong belief, particularly among recent U.S. economic-policy advisors, that deregulation—freeing the market of all legal hindrances and burdens—is the best way to keep an economy healthy and to enable it to grow and create wealth.

A prominent economist who echoed Marx in arguing that fluctuations are intrinsic to the economic system was John Maynard Keynes (1883–1946). In the 1930s he attempted to explain the recurrence of boom and bust by analyzing the flow of income. Keynes pointed out that the vigor of an economy was characterized not by how much wealth it represented but by how much money was changing hands. Even in a depressed economy there may be a small proportion of fabulously wealthy people coexisting with a mass of workers on the breadline and a high rate of unemployment. But in that situation, the rich tend to hold on to their money rather than investing it. As long as money keeps flowing, an economy may remain buoyant: businesses expand when there is investment capital to draw on, and employment levels and wages stay high. That in turn encourages more saving, investment, and growth.

Keynes suggested that if people "freeze" part of their income by hoarding it rather than investing it, money is siphoned out of this cycle, until eventually the economy spirals into depression. In Keynes's picture, businesses must constantly borrow and expand to keep money flowing and maintain a healthy economy, which means that production and consumption must also increase. In this view, an economy's stability depends on its dynamism: like the Red Queen in *Alice's Adventures in Wonderland*, it must constantly keep moving in order to stay in the same place.*

RANDOM WALKS

Irving Fisher was not the first to suspect the presence of chaos in the so-called business cycle. In 1900 a Frenchman named Louis Bachelier proposed that the fluctuations in the prices of stocks and shares, and thus the underlying structure of the market economy, are effectively random.

Bachelier's name does not tend to feature in economics textbooks, because he was not an economist. He was a physicist, studying for his doctorate at the École Normale Supérieure under the eminent mathematical physicist Henri Poincaré, whose work supplied the foundations of modern chaos theory. Bachelier's thesis was unusual: called "Théorie de la spéculation," it presented a model of economics based on ideas from physics. For his contemporaries this was simply too strange, and Bachelier made no subsequent impact either on science or on economics.

Yet what Bachelier achieved in his thesis was remarkable. To construct a mathematical description of random fluctuations, he had to devise what amounted to a theory of the random-walk problem—five years before Einstein began to secure fame with his own treatment of the matter in his celebrated study of Brownian motion (page 43).

The direction of motion of a particle undergoing a random walk fluctuates unpredictably, and Bachelier assumed that stock prices do the same. The fluctuations are a kind of "noise." We saw in Chapter 2 how this background of random "static" due to erratic particle motions pervades everything, and that its amplitude is a measure of temperature. The hotter a gas is, the more pronounced are the fluctuations of its constituent particles. In other words, there is a characteristic "scale" to the fluctuations of particles undergoing random walks; there is a typical size of the deviations.

---

*The hydraulic metaphors of flow in Keynesian macroeconomic theory were not lost on some of his contemporaries, who even attempted to construct mechanical devices that, by appropriate manipulation of water flows, could generate predictions about the economy. They were the first "simulations" of economics: cumbersome, quixotic, supremely Newtonian, and later rendered obsolete by the computer.

A random walk has a very well-defined mathematical signature that emerges from the statistics of the process. It is impossible to say, at any instant, how big the next random change in direction will be. But if we keep a tally of the size of these fluctuations over a sufficient time span, a pattern becomes clear. Let's say we draw a graph of the size of the fluctuations against the number of times such fluctuations appear.

What we find is our familiar bell-shaped curve: de Moivre's error curve, championed by Adolphe Quetelet as the hidden "regularity" of social statistics, and now generally known as a Gaussian curve. As the statisticians of the nineteenth century discovered, any set of quantities whose values are randomly determined will fit onto a curve like this.

Bachelier thus assumed that stock prices fluctuate in accordance with Gaussian statistics. It seemed a fair guess, especially in view of the prevalence of Gaussian behavior elsewhere in the social sciences. He went on to construct an economic model that incorporated this random element in prices. Bachelier made no attempt to explain where the fluctuations came from, but simply accepted them as a given feature of the data.

In Bachelier's day the concepts of noise and fluctuations were new ones in physics. Most scientists were interested in what happened above the level of the background noise. They needed to know only the pressure that a gas exerted on a confining wall, not the tiny fluctuations in pressure from microsecond to microsecond caused by the very small differences in the number of molecules striking the wall at successive instants. Indeed, these fluctuations were generally too small even to detect using the techniques of that time, as Maxwell implicitly acknowledged in his kinetic theory.

Today it is recognized that noise and fluctuations are among the most subtle and significant aspects of statistical physics. One of the most important considerations to emerge from these studies is that not all noise is Gaussian. Just because a series of data points looks wildly erratic and unpredictable, it is not necessarily governed by Gaussian statistics. Had Bachelier carefully measured the stock market fluctuations, he would have discovered that his assumption was wrong. We can see this in the raw data, once we know what to look for. Figure 8.2 shows a typical record of stock price fluctuations over time (a), along with a graph (b) of Gaussian noise. Clearly market variations are not Gaussian—they are punctuated by occasional large peaks, whereas all the Gaussian fluctuations tend to

8.2   Fluctuations in "returns" for the Standard & Poor's 500 market index (*a*), a common measure of the state of the U.S. economy (see page 198). A return is the difference between two values of the index separated by a certain time interval; here the interval is ten minutes. So a return of zero at a particular time indicates that the S&P 500 Index has not changed over the past ten minutes. Gaussian fluctuations (*b*), which would be expected for purely random changes, look quite different.

stay within a limited range: they have a typical "scale" equal to the width of the band of spikes. For the real data, there is no such well-defined scale.

Another way of showing the distinction, which allows us to be rather more precise in our comparison, is to draw a graph of how the probability of a fluctuation of a particular size depends on its magnitude: the probability distribution function (Figure 8.3). The error curve is the probability distribution function of Gaussian fluctuations. For the non-Gaussian market variations of Figure 8.2*a*, we can see at a glance that small fluctuations are more likely than large ones. The probability distribution function tells us exactly how much more likely they are.

Even for small fluctuations, these statistics aren't very close to a Gaussian distribution. And for large fluctuations the differences are pronounced: a Gaussian distribution significantly underestimates their likelihood. In Bachelier's Gaussian model, big fluctuations—stock market crashes or sudden booms—were so improbable that they'd virtually never be seen. The

8.3   The probability distribution function for fluctuations in the S&P 500 market index. Here the time interval for the returns is one minute rather than ten minutes, as in Figure 8.2a. But the general shape of the curve is the same for time intervals between one minute and at least a day. For comparison, the probability distribution for Gaussian fluctuations (a random walk) is shown as a dashed line. The solid line shows the probability distribution function for a Lévy flight—see page 196.

market would just waver around a more or less steady state. But of course crashes do happen.

These "extreme" events are said to represent the "tails" of the probability function—the parts of the curve that tail off to zero probability. You might wonder whether we should really care whether a theory matches the data way out here, where we are talking about fluctuations that rarely happen. Is it not good enough that the bulk of the fluctuations, the smaller ones, trace out something kind of close to a Gaussian curve?

No, it is not. For the extreme events are the very ones that bother economic forecasters most: the market crashes. Trying to understand economics with a model that cannot describe crashes is like planning river management for small, normal variations in water height and not bothering about the occasional big changes that cause catastrophic flooding. And the deviations from Gaussian behavior don't just apply to big crashes like those in 1987 and 1997. The statistics of even moderately sized "spikes" in economic fluctuations deviate markedly from a Gaussian distribution. The simple fact is that market behavior is not random.

FAT TAILS

If the fluctuations in the market are not Gaussian, what *are* they like? In the 1960s the mathematician Benoit Mandelbrot, popularly known as "the father of fractals," looked at the fluctuations in cotton prices and recognized that Bachelier's random walks would not suffice to describe them. He proposed that the probability distribution of these fluctuations was "fat-tailed": close to Gaussian for small fluctuations, but with pumped-up tails encompassing the big fluctuations.

This demanded a big shift in the way that the dynamics of the markets were analyzed and modeled. In 1964, Paul Cootner of the MIT School of Management said:

> Mandelbrot . . . has forced us to face up in a substantive way to those uncomfortable empirical observations that there is little doubt most of us have had to sweep under the carpet up till now. He has marshalled evidence of a more complicated and much more disturbing view of the economic world than economists have hitherto endorsed.[12]

Mandelbrot proposed that instead of following a random walk through the range of possible values, price variations executed a Lévy flight, named after the French mathematician Paul Lévy, who introduced this notion in 1926.

A Lévy flight is rather like a random walk punctuated by occasional big leaps.* Some animals forage in this way. They explore a small area by wandering at random; but if they find no food, they will move rapidly to a new area and do the same there. This can be more effective than merely following a random walk over the entire terrain, since it gets one out of unpromising areas more efficiently.

Systems showing this kind of dynamic behavior are said to be Lévy-stable processes. Mandelbrot suggested that the fluctuations of economic markets are governed by such a process, and that the occasional big leaps account for the fat tails of the probability distribution functions of the fluctuations. The

---

*Technically speaking, a Lévy flight is a random walk in which the step size has a power-law probability distribution function (see page 229): the steps can be of any size, but their probability decreases as the steps get bigger. In a normal random walk all steps are the same size.

important thing to appreciate about this claim is that it is *descriptive*, just like Bachelier's formulation of random walks. It is a way of describing the fluctuations without proposing to explain how they come to be that way.

Purely descriptive models were alien to mainstream economics at that time. The approach of most academic economists was to work not from data but from "first principles." They would construct models based on assumptions about how the market operates, and see what they predicted. The model predictions were rarely compared against hard data. The economist Paul Ormerod argues that the scientific tradition of gathering data against which theories can be tested did not permeate to economics until the past few decades. So handling real data was a notion foreign to many economists; *starting* with it would have seemed bizarre.*

THE SHAPE OF CHANGE

Since the mid-1960s, however, Mandelbrot's Lévy-stable description of market fluctuations has gradually been accepted by many academic economists. Practitioners, on the other hand—traders and their advisors, who seek to use economic models to make real forecasts—have tended to persist with the idea that fluctuations are Gaussian. This is partly a matter of practicality: purely random Gaussian noise is relatively easy to understand intuitively and to handle mathematically in economic models, whereas Lévy flights are not. These hands-on economists trust that the precise nature of market fluctuations will not matter too much to their calculations.

But the fact is that even Lévy flights provide an imperfect description of real market fluctuations. In 1995 the physicists Rosario Mantegna and Gene Stanley at Boston University performed an analysis of more than one million records of a standard market index—five years' worth of economic data—to uncover the market's statistical behavior. They looked at the statistics of the "returns": the differences in the value of the index at times separated by some specified interval: from one day to the next, say, or from one hour or

---

*By no means all economists have shared this attitude. Indeed, two years after Mandelbrot the economist Eugene Fama proposed independently that price fluctuations follow a Lévy-stable probability distribution.

one week to the next. The returns are a measure of the market fluctuations: when the index remains steady from one step to the next, the return is zero. Mantegna and Stanley used the Standard & Poor's (S&P) 500 Index: the sum of the market capitalization of 500 top U.S. companies, selected for the size of their market share, their state of liquidity, and the diversity of industries they represent. The index is supposed to offer a single-figure summary of how the U.S. economy is faring.

Figure 8.3 showed what Mantegna and Stanley found. It was certainly no random walk. The smallest fluctuations fit the probability distribution of a Lévy-stable process rather well. But the larger fluctuations deviate from this curve, falling instead somewhere between a Gaussian and a Lévy process. In other words, whereas Bachelier's description of the fluctuations under-estimates the frequency of large events, Mandelbrot's description of the fat tails *over*estimates them. There seems to be a crossover from a Lévy-stable process for small fluctuations to some other kind of behavior for large ones.

This statistical distribution of market fluctuations seems to be the same over a wide range of different time intervals between successive data points. Mantegna and Stanley found that the statistics of the returns in the S&P 500 Index from minute to minute are just the same as those from hour to hour and from day to day. In other words, the behavior of the market looks the same at different levels of "magnification" in time (within limits, as I explain below). This is a way of saying that there is no characteristic size of the fluctuations: they are *scale-free*, unlike Gaussian fluctuations, which, as we saw earlier, have a certain typical size. If we were to magnify a part of a graph like Figure 8.2*a* (page 194), we'd see a wiggly line that looks just as jagged as the original one. We can magnify just one day of a week's fluctuations, and then just one hour of that day, and then just one minute of that hour, and in each case we see the same kind of jumpy graph.*

This similarity of form or shape at different levels of magnification is a

---

*More precisely, the *magnitude* of the variations does depend on the duration of the time intervals; a large return is more likely to accumulate over a day than to happen all at once. To allow for this, we need to *rescale* plots with different time steps. This is just like rescaling two probability distri-bution functions of children's heights—for a single school and the whole country, say—so that they both lie on the same Gaussian curve (see page 93). Such rescaling doesn't affect the mathematical form of the curves—that is, whether they are Gaussian, Lévy-stable, or whatever.

characteristic of the fractal structures mentioned in Chapter 5. It is a feature of many natural shapes, such as mountain ranges and coastlines. Mandelbrot gave fractals their name, and he was the first to deduce that the ups and downs of an economic market over time have fractal properties. But with his Lévy-stable distribution he oversimplified the precise mathematical form of those properties.

Just what is this form, then? The harder one looks, the subtler the question seems to be. No single kind of curve appears to fit the entire statistical distribution of market fluctuations. It depends, for example, on how large a variation (in percentage terms) you are looking at, and on what timescale. Although the distribution shown in Figure 8.3 seems to hold over a wide range of timescales, for very lengthy periods between each data point (several months, say) the distribution begins to look more like a Gaussian curve. (That's one reason why we should be skeptical about any single theory that claims to explain how the market works.) Yet this statistical behavior, however slippery, is consistent across the markets: the S&P 500 Index shows much the same behavior as the Japanese Nikkei Index and Hong Kong's Hang Seng Index. There does seem to be a universal statistical signature of the way capitalist markets work.

What remains contentious is whether major crashes are typical or atypical of market fluctuations. Clearly they are atypical in the sense that they don't happen very often. But that's exactly what the probability distribution of fluctuations implies: very big events are rare. The question is, do crashes fit onto the same curve as smaller fluctuations? Because crashes are uncommon, it is not easy to give a definitive answer. Some say that these extreme events fit the distribution perfectly, which means that crashes, while rare, are not *anomalous*—they are an inherent feature of the way markets behave, *arising from the same fundamental processes that cause smaller fluctuations.* Others are not so sure. For example, Mantegna and his colleague Fabrizio Lillo have shown that the *short-term* statistical distributions of returns on days when a market crashes, or when it rallies after a crash, differ from those on "normal" days. There seems to be something unusual about the way the market is functioning on those days.

But whatever the precise statistics of market fluctuations, it is indisputable that they are not fully random (Gaussian) but are relatively more prone to attaining large values. What does that mean for economic theory?

THE RIGHT WIGGLES

One of the dreams of entrepreneur economists is to develop a theory that can predict the ups and downs of the market before they happen. This would allow traders to make profits indefinitely, buying and selling exactly when their calculations tell them the time is right.

Real traders—at least, those with a serious understanding of market dynamics—have long since given up on this idea. They know that it cannot be achieved. Indeed, the impossibility of making completely accurate forecasts is enshrined in one of the central tenets of economics: the "efficient market hypothesis." One way of stating this hypothesis is that it is not possible to predict future stock prices from previous values. Note that it is couched as a hypothesis—a belief, if you will—for there is no rigorous proof that it is true.

Analyzing the statistics of market fluctuations has shown that *empirically* the efficient market hypothesis holds water. The question of predictability hinges on the idea of *correlations*. To make an accurate forecast of the price of a stock based on its previous price, there would need to be some mathematical relationship between the two prices: technically speaking, they would have to be correlated. Imagine, for example, that a declining price over a day invariably meant that the price would continue to fall at the same rate over the next day. Sometimes, of course, this relationship does hold; a stock can devalue steadily over the course of many days. But equally obviously, it is not a *general*, invariant relationship—if it were, then a stock that began to fall in price would inevitably decline until it was worthless. There is no telling whether a stock that is falling on one day will recover or continue to fall on the next.

Physicists have a mathematical tool for measuring correlations, called a correlation function. They use it, for example, to deduce to what extent the motion of a particle in a fluid is predictable on the basis of the motion of other particles. If the correlation function between two particles has a high value, then the trajectories of the two will be closely related. For example, a mother and child holding hands and moving through a crowd are highly correlated in their motion. We can predict where the child goes by tracking where its mother goes.

The *auto*correlation function of a particle characterizes how its motion at any point in time is related to its *own* motion in the past. The autocorrelation function of a particle in a liquid has a high value over very short times because the particle continues to move along its previous trajectory. But this value drops rapidly to zero over longer times because collisions with other particles quickly randomize the motion, destroying all trace of the earlier trajectory.

It is possible to calculate the autocorrelation function of an economic index or stock price, which tells us whether its future value depends at all on its past. Typically this analysis shows that the correlation falls to zero within about five to fifteen minutes, confirming that for times longer than this, the efficient market hypothesis is valid: the price has completely "forgotten" its past.

One might imagine that by buying and selling in the few minutes during which correlations exist, one could use the past to forecast the immediate future and operate at a sure profit. Practically, however, this is not feasible, partly because of the finite time needed to make a forecast and then complete a transaction, but also because even the small transaction costs that are imposed will quickly erode any potential profits from exploiting very-short-term correlations.

Thus there is no magic formula that will let one play the market without risk. This fits with experience and with common sense. Yet it is a telling comment on the state of conventional economic analysis that at least one celebrated theory suggests the contrary.

Robert Merton and Myron Scholes were awarded the 1997 Nobel Prize in economics for the theory of options pricing that they developed with Fischer Black, who would surely have shared the award had he not died several years earlier. The Black-Scholes model, extended by Merton, more or less defines the thinking of dealers in options.

Options are basically a kind of insurance contract that protects dealers against the vicissitudes of the market or of world events. The buyer of an option pays a small sum for the right to buy or sell something at a specified price some time in the future. If that price falls short of the actual value at the time of the transaction, the person or company that underwrites the option makes up the difference. Options are a way of diffusing risk, or of passing it on to someone else. At least, that is the theory. Options can be used to insure against almost anything: variations in the weather (which might ruin

a crop or an outdoor event), changes in exchange rates, or whatever might
bring the buyer future losses through ill luck or acts of God. They are one
form of the class of assets known as derivatives, which depend on the value
of something else. The type of derivatives known as futures are agreements
to buy or sell something at a given time for a price agreed to now. They are
basically a form of gambling.

Derivatives have developed notoriety in recent years because of their
high-risk behavior, which has led to massive losses incurred by organizations and institutions such as Barings Bank and Procter and Gamble. In
1997 alone, derivatives are estimated to have cost traders $2.65 billion
worldwide.

Options, however, are supposed to be relatively tame derivatives, thanks
to the Black-Scholes model, which has been described as "the most successful theory not only in finance but in all of economics."[13] Black and
Scholes considered the question of strategy: What is the best price for the
buyer, and how can both the buyer and the underwriter minimize the risks?
It was assumed that the buyer would be given a "risk discount" that reflects
the uncertainty in the stock price covered by the option he or she takes out.
Scholes and Black proposed that these premiums are already inherent in the
stock price, since riskier stock sells for relatively less than its expected future value than does safer stock. On the basis of on this idea, the two went
on to devise a formula for calculating the "fair price" of an option. The theory was a gift to traders, who had only to plug in the appropriate numbers to
get out the figure they should pay.

There was just one element of the model that could not be readily specified: the market volatility, or how the market fluctuates. To calculate this,
Black and Scholes assumed that the fluctuations were Gaussian. Not only do
we know that this is not true, but it means that the Black-Scholes formula can
produce nonsensical results: it suggests that options writing can be conducted in a risk-free manner. This is a potentially disastrous message, encouraging a false sense of confidence that can lead to huge losses. The
shortcoming arises from the erroneous assumption about market variability,
showing that it matters very much in practical terms exactly how fluctuations
should be described.

The drawbacks of the Black-Scholes theory are known to economists, but
they have failed to ameliorate them. Many extensions and modifications of

the model have been proposed, yet none guarantees to eliminate all the problems. It has been estimated that the deficiencies of such models account for up to 40 percent of the 1997 losses in derivatives trading, and it appears that in some cases traders' rules of thumb do better than mathematically sophisticated models. Econophysicists such as Jean-Philippe Bouchard in France have proposed new models of options pricing that go beyond the Black-Scholes theory by taking into account the non-Gaussian nature of fluctuations. These models banish the illusory notion of risk-free pricing. Bouchard's confidence in the role physics can play in economics is reflected by his having set up a consultancy company, Science & Finance, exposing his theories to the exacting demands of the marketplace.

Options pricing is just one area in which a better understanding of the statistics of the marketplace could transform economic theory. Another active area is the improvement of schemes for optimizing portfolios. Physicists use their methods to search for correlations between different stock prices and thus to find the least risky way of spreading investments over a market. If one invests in strongly correlated stocks, for instance, a downturn in one of them could sink the lot.

For some of these purposes it is enough to have a more accurate description of market fluctuations. But in other cases, particularly for devising models of economic behavior, one also needs to know where these fluctuations come from in the first place. It is to this question that we now turn.

NINE

AGENTS OF FORTUNE

WHY INTERACTION MATTERS TO THE ECONOMY

If economics could become a true branch of science, it would enormously increase our capacity to predict the course of events, as well as the outcome of attempts to change that course . . . Economic science . . . would increase our ability to foresee the consequences of changing the workings of the economic system, and thereby to choose the most favorable course of action.   —Robert Heilbroner (1999)

Napoleon was fond of telling the story of the Marseilles banker, who said to his visitor, surprised at the contrast between the splendor of the banker's chateau and hospitality, and the meanness of the counting-room in which he had seen him,—"Young man, you are too young to understand how masses are formed,—the true and only power,—whether composed of money, water, or men, it is all alike,—a mass is an immense center of motion, but it must be begun, it must be kept up"; and he might have added, that the way in which it must be begun and kept up is by obedience to the laws of particles.   —Ralph Waldo Emerson (1860)

Once, long ago, the economist Kenneth Boulding asked me, "What would you like to do in economics?" Being young and brash, I said very immodestly, "I want to bring economics into the twentieth century." He looked at me and said, "Don't you think you should bring it into the eighteenth century first?"   —W. Brian Arthur

If Maxwell's statistical physics borrowed from social science, economists of the late nineteenth century had no qualms about returning the favor. But could economics really aspire to the accuracy and certainty of physics? Surely the precision possible in the world of atoms seldom extends to the sphere of fickle human behavior? Was the phrase "economic science" destined to become oxymoronic? When the historian Thomas Carlyle called economics the dismal science,[1] he did not mean that it was bad science but that often it led to unpalatable conclusions. Economists, however, soon enough found plenty that was dismal in the discrepancies between their models and the real world of trade and industry.

Francis Edgeworth was undeterred, proclaiming cheerfully that economists should not set their sights too narrowly. The goal, he said, was "not so much to hit a particular bird, but so to shoot among the most clustered covey as to bring down the most game."[2] In his *Mathematical Psychics* (1881), he confessed that belief in a "social mechanics"—a mathematical economics—was partly an act of faith, which drew moral strength from the evident successes of physics:

> "Mécanique sociale" in comparison with her elder sister [*mécanique céleste*, celestial mechanics], is less attractive to the vulgar worshipper in that she is discernible by the eye of faith alone. The statuesque beauty of the one is manifest, but the fairy-like features of the other and her fluent form are veiled. But mathematics has long walked by the evidence of things not seen in the world of atoms (the methods whereof, it may incidentally be remarked, statistical and rough, may illustrate the possibility of social mathematics). The invisible energy of electricity is grasped by the marvellous methods of Lagrange; the invisible energy of pleasure may admit of a similar handling.[3]

"Pleasure," according to Edgeworth, was the force that propelled his hedonistic "charioteers," the individual agents of society who (in his characteristically colorful imagery) interact like so many atoms in the void. He began to glimpse an economics that treated people like "the multiety of atoms which constitute the foundations of the uniformity of Physics."[4]

In this respect, Edgeworth resembled his contemporary Alfred Marshall, John Maynard Keynes's tutor at Cambridge, in seeking an understanding of the economy from "first principles." Marshall did for economics what Maxwell and Boltzmann did for thermodynamics, holding up to it a magnifying glass and attempting to reveal what individual particles were doing on the microscopic scale. This was the start of microeconomics, which underpins most economic theory today.

"Economics, after all," says Robert Heilbroner,

> does concern the actions of aggregates of people, and human aggregates, like aggregates of atoms, do tend to display statistical regularities and laws of probability. Thus, as the professoriat turned its eyes to the exploration of the idea of *equilibrium*—the state toward which the market would tend as a result of the random collisions of in-

dividuals all seeking to maximize their utilities—it did in fact elucidate some tendencies of the social universe.[5]

In other respects, Edgeworth and Marshall were very different. Marshall
was a practical man, whereas, in Keynes's evaluation, Edgeworth was concerned not so much with using theory to formulate policies or to determine
how we should conduct our affairs as with discovering "theorems of intellectual and aesthetic interest."[6] Some would say that this tendency is alive
and well today.

Economists in the early twentieth century were perfectly prepared, even
eager, to accommodate analogies from the atomistic statistical mechanics
devised by Maxwell, Boltzmann, and Gibbs. The trouble was that they were
the wrong analogies. As we saw earlier, physicists focused on the idea of
equilibrium states. Likewise, economists wanted to believe in a steady market, gently ruffled by random noise. But the economic system is very clearly
not in equilibrium. Its theorists are still coming to terms with this distinction.

Nevertheless, by invoking images of particles in motion, economists
were implicitly suggesting that they might emulate physics and formulate a
description of their object from the bottom up. That is the subject of this
chapter.

THE RATIONAL TRADER

Throughout much of the twentieth century, the scientific community was
said to be afflicted by a disease called physics envy. Scientists in disciplines
other than physics wished that their own subjects could boast the intellectual profundity, the mathematical agility, and the foundational rigor that was
evident in physics. Economists, failing perhaps to appreciate that theirs was
the more difficult task, were not immune. Paul Krugman tells of an Indian
economist who explained to his students his unorthodox theory of reincarna-
tion: "If you are a good economist, a virtuous economist, you are reborn as a
physicist. But if you are an evil, wicked economist, you are reborn as a sociologist."[7]

Edgeworth's dazzling mathematics may be seen as a compensatory ploy, as if aiming to show that even if economics could not attain the precision of physics, it could dress in the same clothes. Marshall's approach to economics, although rather different, also sounds a lot like physical science: theories are developed from fundamental postulates. But he cautioned against taking this analogy too far. "Economics," he said, "cannot be compared with the exact physical sciences, for it deals with the ever-changing and subtle forces of human nature." Heilbroner expands on this:

> There is an unbridgeable gap between the "behavior" of [subatomic particles] and those of the human beings who constitute the objects of study of social science . . . Aside from pure physical reflexes, human behavior cannot be understood without the concept of volition—the unpredictable capacity to change our minds up to the very last moment. By way of contrast, the elements of nature "behave" as they do for reasons of which we know only one thing: the particles of physics do not "choose" to behave as they do.[8]

There are several problems with these assertions. Heilbroner is, of course, right to say that humans make choices whereas particles do not (although quantum physicists sometimes speak as though they do). Yet the resulting element of indeterminacy need not in itself prevent social science from devising models of mass behavior, as we have seen already. The early statisticians realized that regularities can arise in large populations even when their individual motives are unknown.

Heilbroner also forgets that, in many situations, people are faced with only a very limited number of choices and can therefore generate relatively predictable long-term patterns of behavior even when each individual makes those choices quite freely. If the pedestrians in Dirk Helbing's "human trail" simulations (page 136) were to arrive at random at any point on the perimeter of an open space and cross to any other random point, no clear trail pattern would emerge. It is because the entry and exit points are constrained that order appears. The same is true in economics: traders on the market can exercise free will, but their choices are restricted to buying or selling a limited range of commodities at any point in time.

Heilbroner's caution about whether economics can be an exact science is quite understandable, and to some extent perfectly justified, for we have seen how haphazardly the markets fluctuate. Yet ironically, conventional economics has gone to the opposite extreme. Rather than throwing up their hands in despair at ever capturing the capriciousness of human behavior in an economic model, theorists have assumed that humans act like perfectly predictable rational automata. A part of the agenda of physicists is to inject a *more* realistic and less idealized description of human behavior into the field.

How can it be that economists came to regard humans as automata? If I say this is the only way they could make their models work, I do not mean to sound cynical. It is simply the truth: there was no obvious means, until relatively recently, of dealing with the indeterminacy of human conduct.

But mathematics also played a part in squeezing the human element out of economics. The early economists, from Adam Smith to John Stuart Mill to Karl Marx, regarded themselves as moral and political philosophers. They mined their studies of the capitalist market system for implications for social and political behavior. Many of the most eminent twentieth-century figures, such as Keynes and Schumpeter, did the same. But mainstream academic economists, following Edgeworth's lead, began to construct ever more elegant and abstract mathematical models, in which there was no room for the messiness and clamor of the real world. Some of them asserted that to include noise in their models was to insult the intelligence of the market.

By what legerdemain, then, was free will spirited out of conventional economic theory? This theory has long assumed that individual market agents—the traders who buy and sell stocks and shares—are Hobbesian to a man or woman. They are out for all they can get, seeking only to maximize their own gains, whether this be the profit they make from their transactions or some more generalized well-being that economists call *utility*, which includes factors such as risk and security.

This seems reasonable enough. Why trade on the stock exchange if not for gain? The problems become apparent, however, once we consider how these agents go about maximizing their utility. The traditional view makes them preprogrammed, omniscient computers. It assumes that each agent has a fixed objective and pursues it in a wholly rational way. He or she assimilates, at every instant, all the information available about the state of the market and uses it to calculate the next best move, like IBM's Deep Blue computer playing chess with Garry Kasparov. The agents are, in other words, perfectly informed "rational maximizers."

The attraction of this assumption is that it makes it easy to see what agents will do next. Their behavior is predetermined by a set of rules. Traditional models then go on to assume that the factors influencing prices are unpredictable and external to the market itself, so that the fluctuations are random (that is, generally Gaussian) and beyond the control of the traders. The traders simply respond to these imposed ups and downs in whichever way allows them to maximize their utility at that particular moment.

Although prices fluctuate unpredictably, this does not mean (within the conventional theory) that the underlying factors influencing prices are wholly mysterious. Rather, changes in asset prices are assumed to reflect the (unpredictable) changes in the "fundamentals"—the causative forces that determine value. The fundamental value of a company, for example, is the value of all future dividends it will pay. The price of a company share is assumed to reflect this value.

Does that seem a little strange? How on earth can one know how the company's fortunes, and thus its dividends, will fare in the future? Of course, one cannot know. This is one of the leaps of faith that traditional microeconomic theory is forced to take to keep the math tractable. A second such leap can be identified in the assumption that information about fundamentals is simultaneously available to all agents in the market and is incorporated instantly into asset prices.

These beliefs underpin the efficient market hypothesis (page 200), a precondition of which is that asset prices change only when fundamentals change—that is, when new information becomes available. And because this information is available to all and because all agents know exactly how best to maximize their utility, no one can exploit it to the detriment of others. So there is no surefire way to secure a market advantage.

Two other aspects of this picture are worth mentioning, since we shall revisit them later. First, all agents are identical: all are rational maximizers playing by the same rules. Second, all have fixed preferences: they never change the beliefs that govern their trading decisions.

No one will deny that these assumptions are all simplistic. The question is whether their simplicity matters. Finding ways to simplify complex problems is the hallmark of good science. Often scientists make approximations that would seem absurd to outsiders, yet find that the resulting theories work surprisingly well. Einstein once said that scientific theories should be made as simple as possible, but not simpler. That, in many respects, is the test of a good theory. So we should not be too quick to disparage traditional micro-economics for presenting this cartoon view of the market. Cartoons often carry all the information one needs in order to follow a story. But the critical test is whether the theory measures up against reality. And the answer to that depends on what you are measuring.

The concept of rational individuals maximizing utility and profits and operating with perfect foresight is one that dominates conventional micro-economic theory, which is called neoclassical microeconomics. This picture is enshrined in the so-called real business cycle (RBC) theory, which seeks to explain the ups and downs of the market.

It is a rather bland, not to say dissembling, picture that RBC theory offers. It proposes that the causes of the business cycle are exogenous; they are imposed from outside the market system in the form of a series of random shocks delivered by technological progress. The market simply responds to these shocks.

The most unsatisfying aspect of the theory is that it seems designed to give us back only what we put in. The statistical properties of price fluctuations depend heavily on what one assumes about the randomness of the shocks. Inevitably, perhaps, the usual assumption is that they obey Gaussian statistics. This then generates economic fluctuations with particular characteristics, which can be compared with market data. Superficially, RBC theory gives plausibly erratic price curves; but on closer inspection the statistics come out wrong. For example, the autocorrelation function (see pages 200–201) does not match that of the real data.

A deeper criticism follows from the demonstration by the economist Robert Shiller in the 1980s that asset prices are not necessarily tied to their "fundamental" value (the supposed "true" value that they would attain in an equilibrium market), as RBC theory assumes. The prices change more erratically than the fundamentals: in market terminology, they are more *volatile*.

If asset prices vary while the fundamentals do not, what drives this change? Why are traders paying different prices for goods even though their underlying value has not altered? Clearly this is at odds with the idea that agents always act rationally.

In fact, we hardly need a detailed analysis of the statistics to tell us that agents sometimes act irrationally. If prices rise above the level that the fundamentals should dictate, they are "overvalued," and common sense should tell traders not to buy. But buy they do. This prevents the market from regaining its instant equilibrium through the conventional laws of supply and demand. Irrationality, it seems, is rife.

Again, this comes as no surprise to economists. We should not imagine that they have somehow been deluded into a deep misconception about human behavior. They all know that markets are not complete, that people are not rational, that they make myopic decisions and lack full information, that they are different from one another.[9] The question is, what can one do about it?

IGNORANCE AND BELIEF

There is an unfortunate tendency for outsiders to deride economists' neglect of irrationality, as though to imply that they are psychologically unsophisticated. One recent critic (not a physicist) claims, "Mainstream economic theory, as set out in countless textbooks, stipulates that a perfectly knowledgeable man or woman will always make a rational and fully informed decision on what to buy and what to sell and at what price . . . This starting proposition is dotty."[9]

But to this accusation, the economist John Kay rejoins, "Indeed it is. That is why mainstream economic theory today does not stipulate anything of the sort. The economics of imperfect information has been an important research theme—perhaps *the* most important research theme—of mainstream economics in the last 30 years."[10]

This is a valid defense. Many economic models are now commonly couched in terms of "bounded rationality": rational decisions made in the face of constraints such as incomplete information which simply try to make the best of circumstances.

All the same, we must be careful to clarify exactly what we mean by "imperfection" or "uncertainty." You can still make a rational best decision on the basis of incomplete information, by weighing up probabilities about the nature of what is unknown. The calculus of trade and risk in the face of an uncertain future or incomplete information is central to the work of many leading economists, such as Nobel laureates Harry Markowitz, Joseph Stiglitz, George Akerlof, and Michael Spence. Yet much of the effort in this area has nonetheless gone into finding optimal (and thus rational) strategies in the face of the market's unpredictable fluctuations.

Some economists accept that the uncertainties faced by traders and businesses are so great that they give up the futile quest for an *optimal* solution and instead accept one that is merely *good enough*, judged against some set of criteria. They are said to "satisfice" rather than to maximize. Again this assumes that decisions are basically rational, albeit laced with a strong dose of reality. Some believe that the difference between maximizing and satisficing is in any case rather small.

Keynes, however, doubted that humans are readily capable of optimal or even particularly rational choices: "A large proportion of our positive activities depend on spontaneous optimism rather than on a mathematical expectation . . . a spontaneous urge to action rather than inaction, and not as the outcome of a weighted average of quantitative benefits multiplied by quantitative probabilities."[11]

In other words, you can devise all the complicated formulas you like, but in the end most of us, including traders, are guided by instinct and impulse, by what Keynes called animal spirits. Top-level management decisions—for example, about a company's policies on recruitment, investment, diversification, or specialization—rely a great deal on such subjective judgments and are based less on the mathematics of economic rationality and more on their chief executive officers' years of experience.

One of the prophets of this perspective on economics was the American Thorstein Veblen, whose thinking was shaped by the harsh, pugilistic climate of the economy of the New World. He envisaged the business world as irrational to the point of barbarity—a world where behavior was governed as much by custom and foolishness as by planning and logic. Entrepreneurs, he believed, positively welcomed the unpredictabilities of the market, since these offered opportunities for profit that a staid economy did not.

It is indeed rare to find the wolfishness demanded of most market traders unaccompanied by a degree of impulsiveness that pays no heed to cool calculation. But perhaps the most productive way of accounting for the irrational element in trading behavior is simply to concede that different individuals, armed with exactly the same information, will tend to act in different ways. These ways are by no means unconsidered. Rather, irrationality tends to manifest itself in terms of different *beliefs* about how to respond to circumstances.

Thus there is *heterogeneity* in the market: not all agents are alike. There is no consensus as to what the best course is. Several economists have recently begun to build this aspect into their models. In "discrete choice theory," for example, each agent chooses one of several alternative actions, each with a certain probability.

By and large, however, views about the role of rationality in economics diverge along the fault line between academic economists and market traders. The academics have traditionally preferred the idealized abstractions championed by Edgeworth. Their models describe a world quite alien to that inhabited by the traders themselves, for whom Veblen's hawkish and opportunistic picture rings true. Neoclassical economic theorists cleave firmly to the notion of an equilibrium market where prices always reach their proper level and there are virtually no opportunities to exploit market trends. For their omniscient traders, a strategy that works better than others would be adopted instantaneously by all, squeezing out any advantage it offers. Market traders, on the other hand, live by their wits and cunning, each believing they can discover an unforeseen avenue for profit. They speak as if the market has a personality of its own—sometimes nervous and jittery, at other times confident and brash. Some traders dismiss academic economics as fantasy. George Soros, an undoubted expert on market-floor "practical" economics, says, "It may seem strange that a patently false theory [conventional neoclassical economics] should gain such widespread acceptance."[12]

One of the first economic theorists to introduce belief and choice into modeling was Alan Kirman of the University of Aix-Marseilles in France, who, in the 1990s, proposed that the inclinations of traders could be encapsulated within two general categories. The *fundamentalists* are the traditionalists: those who adhere to the "rationalist" dogma that prices reflect fundamentals, and who will buy and sell accordingly. The *chartists*, on the other hand, take a more empirical approach. They think that future prices can be predicted from their past behavior. Some might use rules of thumb, or intuition; others will rely on complex formulas from probability theory. But all are essentially optimists, believing that they need not be at the mercy of a market fluctuating purely at random. Chartists are also sometimes called noise traders, reflecting their conviction that there is valuable information in the market fluctuations (noise) which can be teased out if you know how.

In order to model the irrational tendencies of agents, Kirman did not consider it sufficient simply to introduce heterogeneity into microeconomics. He needed something else as well. Once several options are open to us, we cannot ignore the fact that our own choice is inevitably influenced by what others do. We have to take onboard the factor that represents probably the most egregious omission of conventional neoclassical theory, and that brings us most firmly into the realm of statistical physics. That factor is *interaction*.

FOLLOW YOUR NEIGHBOR

The economic system is a supremely interactive one. Traders influence one another directly: a rush to buy or sell a particular asset can prompt others to do the same. It seems intuitively clear, simply from looking at pictures of faces on the trading floor, that major crashes are stampede phenomena in which individuals respond to the mood of the market in a herdlike and sometimes panic-stricken manner. Yet microeconomic models that ignore interaction insist on a different interpretation, one in which crashes are driven by some exogenous fluctuation that is beyond control, or in which agents all decide independently and simultaneously on the same course of action.

Moreover, agents influence one another indirectly. The choices they make have a direct effect on prices, which in turn affect the choices of others. As an engineer would say, there is strong *feedback* at work. Whereas traditional models assume that agents adopt their (rational) strategies in response to an externally imposed evolution of prices, in reality the agents help to make those prices as well as responding to them.

Here again there is a danger of caricaturizing the way economists think. John Kay points out, "The idea that the behavior of participants in markets is influenced by what happens in markets" has been explored in "literally thousands of books and articles written by economists."[13] What physics has to offer microeconomic modeling is not new insights into the factors that control the markets but new tools with which to accommodate them. Physicists have been dealing with systems of many interacting particles for more than a century. It would be foolish to assume that these tools can necessarily be translated directly into economic terms. But equally, it would be surprising if some of the phenomena already well understood in physics did not turn up in some form in economics.

The man who first introduced interactions into microeconomics was a mathematician familiar with both physics and economic theory. In 1974, Hans Föllmer of the University of Bonn in Germany concocted an "interacting agents" model of the economy which was based on the principles of the Ising model of magnets (page 90). As we have seen, in this simplified description of a magnetic material, the magnetic atoms all sit on a regular grid, and they "make choices": they align their spins either in one direction or the opposite. These choices are interdependent—the alignment of each atom depends on those of its neighbors, since their magnetic fields exert forces on one another. In Föllmer's model, each atom represented an agent faced with choices about how to trade. The same idea is now widely applied by economists and econophysicists who, like Alan Kirman, are seeking to extend traditional microeconomics by using "interacting agents" models. The predictions of these models depend on the rules that govern the interactions.

Föllmer found that his model generated more than one stable state— more than one macroeconomic landscape—just as the Ising model offers two magnetically aligned states. This was already food for thought for economists weaned on the idea that the market has a particular, unique equilibrium.

Föllmer's approach was extended in the 1990s by the U.S. economists William Brock and Steven Durlauf, both literate in modern physics. They formulated their model of interacting agents making binary choices in a manner that paralleled the way Pierre Weiss had modeled the magnetic phenomena studied by Pierre Curie (page 87). The Curie-Weiss model makes a so-called mean-field approximation: it assumes that each atom is influenced by the average effect of all the others, not just by its near neighbors. In physics this is something of an oversimplification, the consequences of which become particularly apparent near a critical point. But in economics, "global" interactions are more feasible. Traders on the London Stock Exchange are influenced by others on the same floor; but they can also, thanks to telecommunications, be aware of the trading patterns on the New York and Tokyo exchanges. So for economic modeling, a mean-field theory might be quite a realistic approximation.

The idea of a "mean field" that enables each individual to know and respond to what all the others do is also apparent in attempts by some economists to handle interactions using game theory—a mathematical description of how people make choices in competitive games, which we shall investigate in Chapter 17. In such models, traders try to second-guess what other traders might do. Game theory has yielded some useful insights into the nonrational aspects of market trading, but it has not furnished an explanation of why the market fluctuates the way it does.

Yet even in economics the mean-field assumption cannot be entirely sound. No trader can possess an instantaneous global overview of the market. There is bound to be some degree of localization of interactions between agents. In particular, the choices each agent makes are more likely to be influenced by others trading in similar assets, no matter where they are geographically, than by what a distantly related market is doing.

With such issues in mind, Alan Kirman has focused attention on the question of exactly how information propagates on the trading floor. The key issue, he says, is what the structure of the information network is like: "It seems to me that the question of how economic networks evolve is one of the most important if we are to begin to understand how markets come to be organized."[14] Several microeconomic models of interacting agents have assumed that the links between agents are random. That is to say, each agent has an equal probability of being "connected" to any other random member of the trading community. Individuals are influenced only by those others to whom they are connected. Kirman proposed this kind of communication network in 1983. He found that network effects in interacting-agents models can lead to clustering: groups of traders develop who buy and sell primarily among themselves and have very few interactions outside the group. We shall look at network structures in Chapter 15, where we shall see that this assumption of randomness is not the only, and not necessarily the best, way to describe many of the social and business networks that people form.

In addition to clustering, interactions among agents can lead to an effect known as *herding*: the tendency for traders to mimic one another. Episodes of apparently irrational economic behavior prompted by a collective mania that sweeps through the market are notorious in economic history, not least because they defy the basic assumptions of theorists. In seventeenth-century Holland, for example, tulip prices ballooned absurdly, while conventional economic theory says that the laws of supply and demand should have stabilized the price.

John Maynard Keynes again got here first. In the 1930s, he likened eco-

nomic markets to the kind of beauty contests that were then appearing frequently in the popular press. Readers would be presented with a selection of
"beauties" and asked to guess which of them would gather the most votes
from other readers. This is, of course, subtler than simply nominating the
"most beautiful"—it requires some intuition about how others will cast their
votes. Keynes supposed that this sort of competition would create herding
behavior, but he could not find a way to describe it theoretically.

In the 1980s, Robert Shiller considered how herding behavior might influence market dynamics in a quantitative way. He was interested in what
controls the moment-to-moment variations in the amount of trading. Economic data commonly show a property called volatility clustering, in which
big fluctuations come in bursts separated by relatively quiescent periods.
During the bursts, the market is very active. It seems possible that these
bursts are the result of herding behavior, which prompts increasing numbers
of traders into frantic buying or selling.

But the underlying question remains: Where do the fluctuations come from?

NOTHING FUNDAMENTAL

We have seen how modern theories, as represented by the real business cycle theory, remain resolute in their determination to banish fluctuations from
the model and place them outside as a "given"—the result, say, of changes
in technology. Interacting-agents models, on the other hand, can supply an
endogenous explanation for the fluctuations, showing that the "spiky" behavior evident in share prices or economic indices (see Figure 8.2a on page
194) can arise even in a system disturbed by nothing more than Gaussian
noise. As Alan Kirman puts it, "Models which take account of the direct interaction between agents allow us to provide an account of macro phenomena which are caused by this interaction at the micro level but are no longer
a blown-up version of that activity."[15] In other words, the fluctuations in an
economic index due to the activities of thousands of traders need not be simply a scaled-up version of the random forces to which each individual trader
is subject.

This has been demonstrated by the economist Thomas Lux, formerly of
the University of Bonn and now at Kiel in Germany, and the physicist

Michele Marchesi of the University of Cagliari in Italy. In 1998 they used Alan Kirman's model of fundamentalists and chartists to probe the origins of variations in asset prices. The chartists were divided into two groups: optimists, who will buy additional units of an asset in hopeful anticipation of rising prices, and pessimists, who sell more units than their charts dictate in gloomy expectation of sinking prices. Both groups of chartists take the behavior of the other traders into account when they make their calculations. Chartists can switch between an optimistic and a pessimistic outlook, displaying herdlike tendencies which are dictated by the majority opinion: when most chartists are optimists, for example, the pessimists are more likely to convert. Moreover, chartists of either persuasion can become fundamentalists, and vice versa. The agents make these choices by considering which strategy happens to be the most profitable at the time: if another camp is doing better, there is a certain probability that they will join it.

The changes in asset prices in this model are determined by what the traders do, according to the normal laws of supply and demand. The driving force for change, meanwhile, is the variation in fundamentals. Lux and Marchesi assumed that these fluctuations are Gaussian. Whether that is so in reality is immaterial for these purposes; what the researchers wanted to test was the idea, central to conventional microeconomic theory, that fluctuations in prices reflect those in fundamentals—the efficient-market hypothesis. If this idea holds, the prices that emerge from the model should also show Gaussian variations.

It was entirely possible that the model might merely generate economic nonsense. But in macroeconomic terms it turned out to be perfectly well behaved. Over the long term the market was "efficient"—prices and fundamentals both varied considerably but stayed more or less in step (Figure 9.1a). In the short term, however, the story was quite different. The returns, which reflect the price fluctuations, varied in a decidedly non-Gaussian way (Figure 9.1b). In other words, interactions between the traders were converting a Gaussian "input" (the fundamentals) into an "output" (prices or returns) with quite different statistical properties, in which extreme fluctuations were much more common than the input alone would seem to require. Moreover, the non-Gaussian probability distribution for short time steps changed into a Gaussian one for long intervals, just as is found in real market data.

And Lux and Marchesi found just the kind of volatility clustering also seen in reality. Periods of high volatility (large fluctuations) coincided with periods in which most traders were chartists. In other words, a predominantly chartist strategy destabilizes the market, causing bursts of trading. But the model contains a compensating mechanism that restores stability. When fluctuations are large, prices can differ substantially from those dictated by fundamentals. A chartist will simply trust to whatever trend happens to be prevalent at the time, whereas a fundamentalist will exploit this divergence as an opportunity to profit. Thus the fundamentalist strategy will fare better and will come to be seen as preferable, prompting chartists to change camps. This sets a limit on how wild the market can get.

Thus Lux and Marchesi showed how the characteristic fluctuations of the market can arise *endogenously*: from inside the system. Paul Ormerod has constructed a similar "interacting agents" model in an attempt to account for variations in the industrial productivity of nations, typically defined by their gross national product (GNP), the total output of all the nation's businesses. In this model the autonomous agents are not individual traders but businesses. Just as market traders keep an eye on each other's activities, so do

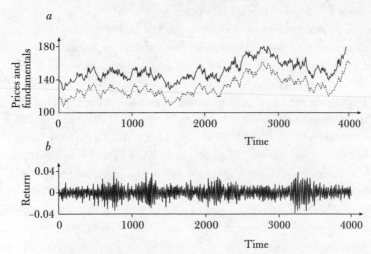

9.1   The economic model developed by Lux and Marchesi shows how random variability of the "fundamentals" that drive price changes gets converted by trader interactions into nonrandom price fluctuations. The *long-term* variation of prices (solid line) shadows that of fundamentals (dashed line), showing that the model is "properly behaved" (*a*). But the variation in *short-term* returns (for a definition of returns, see Figure 8.2) has a non-Gaussian statistical distribution (*b*)—that is, the fluctuations are nonrandom, with more large spikes.

businesses. They adjust their output for each economic period (generally a quarter year) according to short-term forecasts of demand which take into account the perceived mood of the markets. Fluctuations in productivity are essentially a surrogate for the business cycle, the erratic series of booms and slumps reflected in swings in financial indices such as the Standard & Poor's 500. We saw earlier that the conventional real business cycle theory predicts the wrong statistical properties for these fluctuations. Ormerod's model, in which variability is driven by nothing more than Gaussian noise, produces a much more realistic result: the fluctuating total productivity displays statistics close to those seen in reality. If Ormerod modified his assumptions about how firms are distributed in size, the fluctuations retain the same statistical character but may change in amplitude: the "mountains" get flatter or more rugged. The amplitude gets larger if the distribution is skewed in favor of a few big firms, while the variations are smoothed by an industrial profile composed of many small firms. This is a sober message in the light of the tendency for small companies to get swallowed up by a handful of big ones: in such an economy, we must expect deeper, more severe recessions. In this sense at least, a "healthy" market is a diverse one.

There is no unique way to model the shifting and sometimes irrational beliefs of market traders or businesses. The psychology behind these beliefs, no doubt a complex mixture of blind faith, wishful thinking, careful data analysis, past experience, and much else, is probably impenetrable, or at least impossible to express in mathematical terms. But that need not matter. As the economist Brian Arthur and his co-organizers of a recent workshop on "complexity" in economics have said, "*How* individual agents decide what to do may not matter very much. What happens as a result of their actions may depend much more on the interaction structure through which they act—who interacts with whom, according to what rules."[16]

What most of these interacting-agents models reveal is that once rational maximizers are jettisoned, the myth of an equilibrium economy vanishes too and is replaced by something that looks much more like the real world: a market that fluctuates wildly and is prone to crashes. Arthur and his co-workers, for example, have developed a model in which heterogeneous agents trade on the basis of a whole range of strategies and expectations, which the agents revise continually. Successful strategies are retained; those that perform badly are discarded. If the agents rarely switch between alternatives, the economy resembles that predicted by rational neoclassical theories. If the agents revise their behavior with a frequency that the researchers considered to be more realistic, the market descends into the messy, ever-changing, and unpredictable state that it seems to inhabit in real life. In other words, the economy does not appear to be guided very much by rationality.

LET IT BE?

If interactions between traders or between firms can provide an explanation for the furious ups and downs of the market, it is sheer folly to imagine that the market can be engineered to iron out blips on timescales of days or weeks, or perhaps even longer. Yet many governments continue to believe that this kind of manipulation is both possible and desirable. Most probably it is wasteful of resources, if not positively harmful to the economy.

It goes without saying that market fluctuations, particularly recessions, have serious social as well as economic consequences. Economic growth is linked intimately with employment: in a slump unemployment is high, and in a boom it is low. Keynes advocated that during a major, extended slump like the one that followed the Wall Street crash, governments should inject money into the economy to stimulate it back into a growth phase. Keynes feared that, left to itself, the economic system might spiral into a decline that would ultimately leave it frozen, unable to bounce back through the normal business cycle.

There are good reasons to believe that Keynes was right: government intervention may be the antidote to a recession. Certainly, it seems that by pumping money into the economic system after the 1987 crash, the U.S. Federal Reserve may have helped to avoid a slump. But the questions about precisely how Keynesian intervention might (or might not) work are complex, hinging on such issues as the lags between increases in monetary supply and their effects on the market, the question of how to define monetary supply in the first place, and the expectations generated by interventionist policies. These complexities are beautifully explained in Paul Krugman's *Peddling Prosperity* (1994).

Recessions, however, are extreme events that call for extreme measures.

The microeconomic models we have discussed suggest that governments worrying about short-term, moderate fluctuations in unemployment might be like meteorologists fretting about a wet week in July. Such variations are simply part of the system, and the most effective response is not to try to microengineer the economy but to alleviate the temporary miseries of unemployment. If there were greater public understanding of the inevitability of such fluctuations, governments would no longer have to fear the opportunism of opposition parties seeking to make currency from transient economic downturns (while ignoring the occurrence of the same phenomena the last time they were in power). This is not defeatism, but a proposal for effective focusing of resources. By the same token, we would do well to be skeptical when governments claim spurious credit for short-term economic improvements which, in all likelihood, have nothing at all to do with their policies.

On the other hand, there is perhaps a greater danger today of falling prey to the belief that the market should not have to endure any interference whatsoever. As I indicated in the previous chapter, free-market fundamentalists argue that total noninterventionism is the best way to let the economy reach equilibrium. In that blissful state, all prices are supposed to reach their "true" value as determined by Adam Smith's laws of supply and demand (the "market clearing" price), and all goods are assumed to be allocated efficiently to their optimal social uses.

The appeal of laissez-faire economic policies is long-standing. Decades before *The Wealth of Nations*, Charles Davenant concluded, "Trade is in its Nature free, finds its own Channel, and best directeth its own course"[17]—a hydrodynamic metaphor that carries with it the notion of prices finding their own level. While Smith himself felt that trade should not lose sight of elementary justice, he nevertheless asserted that "we may often fulfil all the rules of justice by sitting still and doing nothing."[18] If that is not laissez-faire, what is! To the conservative Edmund Burke, all trade regulation was "senseless, barbarous and, in fact, wicked."[19] That is a view by no means more extreme than those of contemporary right-wing economists.

We can now see that the idea of an equilibrium market is absurd. No process or system that fluctuates in such an undisciplined manner can be considered to approach anything like equilibrium. What is more, it no longer seems possible to blame these disturbances on "external" influences which perturb an otherwise well-behaved system. The interacting-agents models of microeconomics suggest that the economy is intrinsically unstable.

All the same, some might say, the market responds so flexibly and so rapidly that it is instantaneously optimal even as it changes. In other words, it still distributes goods with maximum efficiency from moment to moment.

But even this pervasive idea, the last refuge of Smith's alluring and enduring notion of an "efficient market," has now been challenged. A highly complex and sophisticated interacting-agents model called Sugarscape, devised by Joshua Epstein and Robert Axtell of the Brookings Institution in Washington, D.C., and described in Chapter 14, reveals that once trade is confined to realistic agents lacking infinite knowledge and access to all other traders (that is, if they are not omnipotent, omnipresent, and omniscient), the distribution of goods is inevitably less efficient than it could be. In other words, some agents are unable to procure goods that they both desire and can afford.

Moreover, the Sugarscape model suggests that while trade increases a land's "carrying capacity"—the free exchange of goods makes a region able to sustain a larger population—it also skews the distribution of wealth toward greater inequality. In other words, trade seems inevitably to place most wealth in the hands of a few—a situation discussed in the next chapter.

This may be the price we must pay for capitalism, because experience tells us that a rigidly controlled economy (like that of the former Soviet Union) works against the interests of both general welfare and efficiency. This is borne out by econophysics. Sorin Solomon and colleagues at the Hebrew University of Jerusalem have shown that attempts to enforce a worldwide equality of wealth are apt to make that wealth dwindle uniformly to zero. And an interacting-agents model of the economy devised by Zdzisław Burda at the Jagiellonian University in Kraków, Poland, and his co-workers suggests that "socialist" economies that severely restrict trade are more apt to collapse into a state in which a large proportion of the wealth ends up in the hands of one person. In other words, they are more prone to corruption. Corruption is generally regarded simply as a human failing; but it seems that some economic systems are more likely than others to create the preconditions for it.

Yet extreme capitalism is perilous too. After all, cautionary tales of megacompanies such as Enron and WorldCom hardly encourage the belief that corruption is unique to socialist economies. Solomon and his team have shown that once markets become global, there is an increased risk of all the wealth getting concentrated in one place. This is not only morally intolerable but dangerous for the market, which is then much more liable to collapse catastrophically than if such "wealth condensation" can be spread over many different places. "Extreme capitalism and extreme socialism," they say, "are equally counterproductive and possibly disastrous."[20]

Findings like these are based on models that can justifiably be called schematic, even crude. They should not be seen as anything more than suggestive. Should we believe them at all? Let me pose the question a different way. Can those on the political right provide any comparable justification, based on theory rather than ideology, for their advocacy of wholly unfettered markets? Doubtless such markets work out very nicely for some players, but they do not appear to be the utilitarian ideal the (libertarian) right might have us believe. "Certain economists," say Epstein and Axtell, "ascribe nearly miraculous powers to markets," insisting that all government intervention simply prevents the market from realizing its self-regulating, beneficial qualities. This "is also, unfortunately, a position frequently promulgated in policy circles, especially when there is no econometric or other evidence upon which to base decision making."[21] It is time to recognize such claims for what they are: expressions of faith, unhindered by facts and based largely on predetermined views about the role of governments, taxation, and legislation.

So deeply entrenched is the free-market philosophy in American economic theory today (I am talking here about the pundits who exert a real influence—the TV analysts, the *Wall Street Journal* op-ed columnists, the think-tankists, and all too often the White House advisors—but not the academic economists*) that the supporters of this creed are hoping even to ride out the catastrophic stock market collapse that is proceeding at full throttle at the time of writing. They place the blame on a few corrupt CEOs, on gov-

---

*All these groups are typically labeled "economists," but they have little in common. Paul Krugman prefers to call the non-academics "policy entrepreneurs," and the academics "professors." The professors, he says, include plenty of right-wingers and market fundamentalists, but their ideas remain rooted in theory rather than political dogma.

ernment policies, on fickle small investors, on labor unions, on left-wing critics who spread doubt and negative thinking—anywhere but on the market itself. If only all these people would *behave*, say the free-marketeers, stocks would keep rising forever.

This transparent nonsense will probably always find a voice when times are good—when the economy is buoyant, extreme libertarians will consider themselves vindicated. Slumps will be touted as a thing of the past, until the next bubble bursts. The truth is that economic ups and downs are an intrinsic feature of the capitalist game and cannot be conjured away by political ideology. They are a part of the natural rules of the economy; for such rules surely exist, even if we have not yet fully uncovered or explained them. If we wish to do so, it seems likely that physics-based models can assist, and so help prevent us from Canute-like posturing before the waves instead of rescuing those in danger of drowning.

TEN

UNCOMMON PROPORTIONS

CRITICAL STATES AND
THE POWER OF THE STRAIGHT LINE

. . . the reader will appreciate the orderliness of the lines . . . and he will see how this orderliness points to the existence of a fundamental governing principle.

—George Kingsley Zipf (1949)

The aesthetics of natural science and mathematics is at one with the aesthetics of music and painting—both inhere in the discovery of a partially concealed pattern.

—Herbert Simon (1996)

Much of the real world is controlled as much by the "tails" of distributions as by means or averages; by the exceptional, not the mean; by the catastrophe, not the steady drip; by the very rich, not the "middle class." We need to free ourselves from "average" thinking.　　　　　　　　　　　　　　—Philip Anderson (1997)

One might imagine that the best way to make a fortune on the stock market is to devote one's life to studying it. For the economists John Maynard Keynes and David Ricardo, knowledge paid: both men made fortunes as speculators. (It is not clear, however, whether their theoretical ideas played a stronger part in their success than did their shrewd instincts.) By contrast, Ricardo's friend Thomas Malthus, although a professor of political economy, never developed an aptitude for playing the market. By letting excessive caution guide his investments, he missed his chance to profit fabulously from Wellington's defeat of Napoleon.

It would certainly be a splendid way to gauge economists' faith in their own ideas if we were to demand that they test them using their own pockets. A cynic might expect to see a substantial thinning of the literature on market forecasting if this were a criterion of publication.

In 1995 the French scientist Jean-Pierre Aguilar had the rare courage to stake his money on the proposal that there is physics in economics. He was persuaded by a physics-based model of market crashes to buy options on the account of a fund management company that traded on the basis of such models. The model predicted a crash in Japanese government bonds in May of that year. It never happened, and Aguilar needed to engage in some delicate countertrading to avoid losing his stake.

Not surprisingly, Aguilar was among the skeptics when the claim was made in 1998 that a similar econophysics technique had retrospectively "predicted" the crash of October 1997. This predictive method suggested an intriguing idea: that crashes are closely akin to critical points in statistical physics.

The notion that market dynamics are somehow redolent of phenomena at critical points has a broad constituency. In order to see the connection, we need to look again at this strange location in the landscape of many-particle systems; many of the most powerful and startling insights in statistical physics stem from investigations of this unique place. The critical point motivated van der Waals's exploration of "continuity" between liquids and gases. Many physical systems, such as magnets and superconductors, must pass through this portal as they undergo global changes of state. Critical points are like the black holes of statistical physics, for everyone in the field gets drawn to them sooner or later. But usually they emerge on the other side with a far deeper and richer appreciation of how this branch of physics unifies the physical world.

Indeed, it is now fashionable to see critical points everywhere: in earthquakes, evolution, forest fires, even in hospital waiting lists and world wars. The persuasiveness of such claims varies enormously, but it is fair to say that many of the characteristics of critical-point phenomena—such as extreme sensitivity to fluctuations, "scale free" events, and, in particular, a special form of probability distribution—are found widely both in nature and in the world of human affairs. Critical points are at least a good metaphor, and sometimes rather more than that, for the strange combination of the unpredictable and the rule-bound that governs much of our lives.

PHYSICS ON A KNIFE EDGE

Van der Waals's theory explained the existence of a critical point at which the liquid and gaseous states of a substance become indistinguishable. But the theory could not account for some of the strange phenomena found there. For one thing, fluids close to the critical point become cloudy (this is known as critical opalescence), and no one knew why.

One experimental peculiarity that the theory did embrace was the extraordinary sensitivity of the critical point. A system near its critical state becomes extremely responsive to disturbances. If you squeeze a substance, it shrinks in volume. The resistance it offers to this compression is a measure of its so-called *compressibility*. A rubber ball is more compressible than a steel ball, and a gas is typically much more compressible than a liquid— one can squeeze it more easily. At the critical point of a liquid and gas, the fluid becomes absurdly compressible—in fact, more or less infinitely so. In principle, the gentlest squeeze is sufficient to collapse a critical fluid into invisibility. This sounds absurd, and experimentally one can never observe such extreme behavior, because maintaining a substance exactly at its critical point is too difficult—the critical state is too unstable. But one can see the compressibility start to increase very rapidly as the critical point is approached.

The same is true of a substance's response to heat. To raise the temperature of a system, you have to put energy in: you have to heat it. The amount of heat energy needed to raise the temperature by one degree is called the *heat capacity*. Water has an unusually high heat capacity, which is why it takes a long time—a lot of heat input—to boil a kettle.

At a critical point, a substance's heat capacity becomes infinite. This means that it becomes a kind of infinite heat sink: you can heat it as much as you like but the temperature doesn't change at all. A critical point separates ultra-cold liquid helium from the weird, inviscid state known as a superfluid (page 95), and the sudden increase in the liquid's heat capacity at around two degrees above absolute zero is the telltale signature that it is approaching its phase transition to a superfluid.

These bizarre behaviors are called *divergences*: some property of the substance diverges off to infinity at the critical point. Van der Waals's theory predicts divergences in the compressibility and heat capacity of fluids at the critical point. It explains why things get out of hand there.

There is a convenient quantity called the *critical exponent* that specifies the rate of this divergence. By measuring how rapidly a quantity such as heat capacity increases as the temperature approaches the critical temperature, scientists can easily calculate the critical exponent. The astonishing thing is that this exponent is the same for *all* fluids. The critical exponent that characterizes the divergence of compressibility is different from that for the heat capacity, but again it has the same value for all fluids. Liquid-gas critical exponents seem to be "universal."

To understand the meaning of critical exponents requires some mathematics, but only a little. The exponents define a mathematical relationship between two quantities that is called a power law. There is no hidden Hobbesian significance here in the word *power*; it is just a mathematical term. If the value of some quantity y depends on the value of another quantity x according to a power-law relationship, this means that each time you double x, y increases by some constant factor. The exponent of the power law is a number that tells us how big this factor is. The bigger the exponent, the more rapidly y increases for each doubling of x. If the exponent has the value 2, for example, then y increases fourfold each time x is doubled. If the exponent is 3, y increases eightfold.

This is perhaps a simpler concept than you might think. There is, for example, a power-law relationship between the length of a cube's sides and its volume—and the exponent in this case is 3.* If you double the length of its sides, the cube's volume increases by a factor of eight. You can fit eight cubes of width 2 inches into one cube of width 4 inches.

Each property of a fluid that diverges at a critical point does so at a pace set by a critical exponent that is the same for all fluids. Some quantities don't increase to infinity at a critical point but decrease to zero. The density difference between liquid and gas does this, for example, as does the magnetization of a magnet at its Curie point (page 87). Here again, there is a characteristic rate at which these things fall to zero. They too can be assigned critical exponents, but the exponents have negative values.

Van der Waals's theory of the liquid-gas critical point predicts these power-law divergences, but it miscalculates the values of the critical expo-

---

*In other words, the volume depends on the length of the side raised to the power 3: the length cubed, or $(\text{length})^3$. The exponent is the superscript in this relationship. A general power law has the mathematical form $y = x^n$, where n is the exponent.

nents. In effect, the theory tells us that the divergences loom like mountains at the critical point but fails to say how steep they are.

This became apparent in the 1890s when Jules Verschaffelt, in van der Waals's old laboratory at Leiden, analyzed very accurate measurements of the critical behavior of a liquid hydrocarbon called isopentane. He found that the critical exponent of the vanishing of the density difference seemed to be about −0.343, whereas van der Waals's theory predicted a value of exactly −0.5.

That is a small shortcoming, one might think, and others at the time thought so too. But if the critical exponents are *universal* for all fluids, they must be encoding something pretty fundamental about the nature of matter. So it is well worth asking what is missing from van der Waals's theory that prevents it from telling the full truth about the critical point.

THE SHAKY BALANCE

What van der Waals's theory could not tell him about critical points was that the key to their oddness lies in their fluctuations. A critical point represents a forked path: a place where choices are made. This is how, you may recall, a critical phase transition differs fundamentally from the first-order transitions of, say, freezing and melting. When a liquid teeters on the brink of its freezing point, every part of it stares the same fate in the face: to become a solid. But if we cool a fluid* through its critical temperature, suddenly it can exist in either of two states, both equally appealing: a liquid or a gas. The same is true of a magnet: you will recall that below the critical temperature (the Curie temperature), the atoms in Ising's model (page 91) can align all their magnetic needles (spins) in either of two opposed directions. *And there is nothing to choose between them.*

So a critical point is like a ball poised on the top of a hill between two identical valleys. It must roll down toward one or the other—but which? The choice is determined by fluctuations.

---

*To physicists, a fluid is simply a substance that flows: it could be gaslike or liquidlike. Above the critical point, we can't really speak of a "gas" or a "liquid" at all, so the term "fluid" then becomes indispensable.

A supercritical fluid has, in theory, the same density throughout. But chance fluctuations—the random motions of the atoms—will make it momentarily denser in some places and less dense in others. As the fluid is cooled through the critical phase transition, the denser regions will be inclined to become liquid, since they are already on the way there. The less dense regions tend to opt for the gaseous state. Either of these random choices is self-perpetuating. This is easier to see in the case of a magnet, where a domain in which all the spins are aligned will encourage spins at the periphery to align the same way too, via interactions between one spin and its near neighbors.

As a consequence, the system becomes exquisitely sensitive to random fluctuations. A tiny chance preference will tip the balance. This instability means that the critical state is highly precarious, constantly on the verge of rolling down into one valley or the other. Holding a substance at its critical point is like trying to balance a needle on its tip by blowing on it from all directions at once.

A fundamental peculiarity emerges from this extreme sensitivity to perturbations: events in one part of the system can have a virtually instantaneous effect in any other part. The orientation of one spin can affect that of another spin a long way away, even though the two are too far apart to feel each other's influence directly. In the parlance of statistical physics, there are long-ranged *correlations* in the system. The range of the correlations—the typical distance over which one particle can affect another particle—is another of the quantities that diverges to infinity at a critical point.

This hypersensitivity is a collective effect. The range of the force of interaction between two particles or spins in the system does not suddenly get longer at the critical point; in the Ising model, for instance, it extends only to the nearest neighbors. But these interactions can be transmitted from particle to particle over long distances without being overwhelmed by the randomizing effects of thermal motion. At the critical point, then, the particles can all act together.

The trouble is that each particle "wants" all the others to follow its own choice. As a result, the critical state fragments into regions that make one choice or the other, each triggered by localized imbalances that arise by chance. These regions can be of any size, from a single particle to a group big enough to span the entire system (Figure 10.1). There is no characteris-

10.1 At a critical point, fluctuations of all sizes can occur. In this computer simulation, black regions could represent a liquid and white regions a gas. Or the two could denote regions in a magnet where the spins point in opposite directions. The size of these regions spans all scales, from the size of a single particle to the size of the entire system. There is no typical scale to this unevenness: it is *scale free*.

tic scale to these domains: they appear at all possible length scales, and are constantly forming and dissolving, growing and shrinking. The critical state thus conjures up its own special kind of fluctuations from the thermal noise,* and they are *scale free*, of all conceivable sizes.

This behavior is the cause of critical opalescence: the milkiness of fluids near their critical point. Under these conditions a fluid separates into regions of liquid and gas of many different sizes. Some of these regions are about the same size as the wavelengths of visible light (a few hundred millionths of a millimeter), and this coincidence means that they will scatter light strongly, like the microscopic fat globules in milk. So the fluid takes on a pearly, opaque appearance.

Van der Waals's theory gets the critical exponents wrong because it doesn't acknowledge this microscopic picture of critical-state fluctuations. Rather, it regards the critical state as, on average, the same everywhere—as though we were to squint at Figure 10.1 from a distance and see only gray, not separate regions of black and white. According to that picture, each par-

---

*Thermal noise arises from the fluctuations in particle energies caused by collisions. It manifests itself as tiny variations in temperature at any location in the system. These fluctuations are random, or Gaussian: like those shown in Figure 8.2*b*, they have a characteristic size.

ticle would feel not the "whiteness" or "blackness" of its discrete neighbors but only some average "gray" influence of all the other particles. This is what is meant by a mean-field approximation (page 215).

The mean-field theory of van der Waals does a surprisingly good job of describing the critical state, falling down only in the details. As we've seen, Pierre Weiss's analysis of the magnetic Curie point (page 87) is a mean-field theory too, and it predicts the same slightly incorrect critical exponents for the magnetic transition as van der Waals's theory did for the liquid-gas critical transition.

Lars Onsager's analysis of the two-dimensional Ising model (page 90) went beyond the mean-field approximation, allowing him to calculate the exact values of the critical exponents. But to predict these values theoretically for a real fluid, we need to reproduce Onsager's success for a 3-D model. This, as we saw earlier, has so far proved impossible.

There is a way around the difficulty: a trick that allows theoretical physicists to sneak up on the "true" values of critical exponents in models like the 3-D Ising model. It is called renormalization, and it takes advantage of the scale-free nature of the critical point. The technique was developed by Kenneth Wilson of Cornell University in the 1960s, for which he was awarded the Nobel Prize in physics in 1982. In effect, renormalization is a mathematical way of squinting at the critical state and *selectively* eliminating fine details: the small blobs blur to gray while the big ones remain. By conducting this renormalization process over successively larger scales, one can calculate the true values of the critical exponents. Applied to the 3-D Ising model, this method supplies values very close to those measured experimentally for real fluids.

The Ising model of fluids, with its filled or empty boxes on a grid, is a pretty crude approximation of a real fluid. But the model's critical exponents in three dimensions are, as far as we can tell, exactly the same as those measured experimentally for liquids and gases. This is again a manifestation of the notion of universality: as far as critical behavior is concerned, the details don't matter. Only the broadest-brush properties of a system affect its critical behavior—not whether it is nitrogen or isopentane, or a magnetic metal, or a crude model of either, but whether it is two-dimensional or three-dimensional, and whether its particles interact through long-range or short-range forces. Differences of this kind are enough to put two systems in different *universality classes*, each of whose members share the same critical exponents. Unless there are such differences, seemingly different systems look the same at the critical point.

CRITICAL CRASHES

A paper coauthored by Jean-Pierre Aguilar in 1999 opens thus: "It is rather tempting to see financial crashes as the analogue of critical points in statistical mechanics, where the response to a small external perturbation becomes infinite, because all the subparts of the system respond cooperatively."[1]

This is the temptation to which Aguilar succumbed in 1995. It had been proposed that financial crashes correspond to a special kind of critical transition: one that shows so-called log-periodic behavior. Such critical points arise within certain models in statistical physics, and they have a distinctive signature. This kind of system is prone to oscillatory, periodic fluctuations; in an economic context they would be analogous to periodic business cycles. But log-periodic variations are not like the regular oscillations of a light wave or a tuning fork. Instead, the peaks and troughs of the waves get steadily closer together. At the critical point itself, they pile up on top of one another. The approach to such a critical transition is therefore signaled in advance by peaks and dips that follow one after the other at ever shorter intervals—a series of accelerating wobbles that herald catastrophe.

This precursor to the critical point in such systems has led some physicists to believe that if financial crashes really are log-periodic critical points, they can be predicted by identifying the telltale oscillations in an economic index and extrapolating them to the point where they coincide with each other. In other words, it should be possible to calculate when a big crash will come. In 1998, physicist Marcel Ausloos and his colleagues in Belgium used this approach to study the October 1997 market crash. They analyzed the fluctuations just before the crash and claimed the fluctuations would have allowed them to predict when the crash was going to happen.

Their work made headline news. If it was true, it was revolutionary. No longer need investors be ruined by unpredictable collapses of the market. Simply by keeping a careful eye on its ups and downs, they could anticipate the future. It was not a foolproof way to play the market, since the oscillations showed up only near a crash; at other times, prices "lose all memory" of the past within minutes and prediction is impossible. But it would be a fantastic tool for investment companies.

The French mathematical physicist Didier Sornette, based at the University of California at Los Angeles, is convinced that market crashes are log-periodic, and he claims to show how several of them in the recent past fit the model of accelerating wobbles (Figure 10.2). In this picture, an impending crash is foreshadowed in the way a market behaves not just weeks but months or even years in advance. But some other econophysicists remain unconvinced. Aguilar and his co-workers in France have argued that the way Ausloos's group matched the curve predicted by the log-periodic model to the real 1997 data involved a selective interpretation of what constituted a dip. They say that the technique doesn't seem to work when applied to other crashes. And they consider it highly improbable that a crash can really bear any relation to what the market was doing several years beforehand. Good fits between the theory of log-periodic critical points and real data (as in the October 1997 crash) are merely fortuitous, they conclude.

Of course, "hindcasts" are questionable in any event. Since what *did* happen is already known, is it possible to remain truly objective in "pre-

10.2   In the log-periodic model of market crashes, oscillations in economic indices pile up at a critical point, whereupon a crash occurs. Some researchers claim they can fit such a log-periodic curve to several crashes. Here are Didier Sornette's fits for the Standard & Poor's 500 Index leading up to the crash of October 19, 1987 (*a*), and for the Hang Seng Index before the Hong Kong stock market crash of March 1994 (*b*).

dicting" it? Sornette admits with some chagrin that trying to make a genuine prediction of a future crash is a thankless task.* There are, he says, at least three possible outcomes:

1. No one believes the prediction, but the market crashes anyway. Then critics will say it is just a lone, lucky correlation with no statistical significance. Besides, what good is a warning if it fails to avert a crash?

2. Many investors believe the prediction, are triggered into panic buying and selling, and thereby cause a crash—that is, the prediction becomes self-fulfilling.

3. Many investors believe the prediction and take careful compensatory action so that a crash is averted—that is, the prediction becomes self-defeating.

That is the problem with the dream of predictability in economics: future market behavior depends on what traders and investors believe that behavior will be, so the act of predicting the future (if it is taken at all seriously) is likely to change it.

THE SELF-ORGANIZED MARKET

Despite skepticism about the log-periodic model of crashes, the idea that market dynamics are governed by critical-state-like behavior has found wider advocacy. It is not hard to understand the attraction of this idea. We saw in the previous chapter that the statistics of economic fluctuations are not Gaussian. Instead, the fluctuations seem (at least for short to moderate timescales) to be scale free: variations of all sizes are seen. Statistically, the economic data in the "fat tails" of the probability distribution follow a power-law distribution: a characteristic feature of critical behavior.

The power law tells us about the probability of fluctuations of a given size. Suppose we peruse the time sequence of an economic index like that in Figure 8.2*a* and keep note of how often we get a return of a certain size. The

---

*Perhaps we shall see. At the time of writing, Sornette and his colleague Wei-Xing Zhou have bravely predicted a crash (or at least a pronounced change) in the U.K. housing market, arriving around the end of 2003. Sornette boldly advertises his economic predictions on his Web site at http://www.ess.ucla.edu/faculty/sornette/, which is a risk few economists would be likely to take.

erratic graph hovers on either side of zero—the most probable return is zero, and we can see that large deviations are comparatively rare. If we measure the relative number of fluctuations of successively increasing size, we find that the probability decreases according to a power law.

Just as a system at its critical point is apt to respond to a perturbation by undergoing a fluctuation of any size, so, we might suppose, this distribution of fluctuation probabilities in the economic market is a sign that it too is poised, in some sense, in a critical state, and is sent hither and thither in unpredictable lurches by the random factors that affect it.

But a critical state is highly precarious and liable to collapse one way or the other at the slightest provocation. How, if the market is critical, can it have stayed that way for so long?

In 1987 a group of physicists working at Brookhaven National Laboratory on Long Island in the United States stumbled by chance onto a system that had the strange, seemingly miraculous property of constantly reorganizing itself into a critical state. They called the phenomenon *self-organized criticality*.

These researchers—Per Bak, Chao Tang, and Kurt Wiesenfeld—hadn't set out to investigate critical points. What they were trying to do was make sense of a recondite problem in solid-state physics, about the way in which electrons sometimes move through crystalline solids in a series of waves called charge-density waves. This turns out to be a kind of *correlated* motion. The electrons don't move independently, as they do in a normal current-carrying metal wire. Instead, the movement of one electron strongly affects the others: they are coupled.

Bak, Tang, and Wiesenfeld began to think more broadly about coupled systems of many interacting particles. As a crude model of charge-density waves, they imagined many swinging pendulums connected by springs. The two systems do not sound similar at all; but mathematically there is an analogy between them, just as a single pendulum serves as a good model for any process that involves simple periodic motion.

When the researchers studied the behavior of this model by writing down its Newtonian equations of motion and solving them on a computer, they found a curious thing. If one pendulum in the coupled array was set swinging, it could set others moving as energy was transferred along the springs. But this influence could extend over any range. Sometimes a single pendulum would do no more than make its near neighbors oscillate. On other oc-

casions an "avalanche" would sweep through the system, setting thousands
of pendulums into motion. There was no telling how big the avalanche would
be. When Bak and his colleagues drew the probability distribution of these
avalanches—the dependence of the probability of an avalanche on its size—
they found a power law. The signature of power-law graphs is that when they
are plotted on a logarithmic scale (see caption to Figure 10.3), the probability distribution curve becomes a straight line.

The Brookhaven team then devised a new and more intuitively appealing
model of this coupled behavior. They replaced their pendulums and springs
with a pile of sand. Imagine dropping grains of sand one by one onto a table-
top. The pile builds up slowly into a little sand mountain. Once the slopes
reach a certain steepness, dropping new grains on top can trigger an ava-

10.3   The power-law probability distribution of avalanches in a mathematical model of a
sandpile. The best way to show the power law is to plot the logarithm of the avalanche size
against the logarithm of its probability: the graph then becomes a straight line. The slope of
the line is equal to the exponent of the power law. Here the slope is close to −1, which is characteristic of self-organized criticality. Large events, to the right of the graph, are less frequent,
so the statistics are less reliable here: the line becomes more jagged. The straight line that fits
the data most closely is shown as a dashed line.

lanche. Before this point, the grains are held in place on the slopes by friction, which prevents them from sliding. At some well-defined angle, friction is no longer able to restrain the motion, and the avalanches commence.

Once one grain starts to move, it collides with others, triggering a kind of chain reaction: the grain motions become coupled by collisions. This process might run out of steam after only a few dozen grains have been set rolling. Or it might continue to propagate until virtually the entire slope slides in a catastrophic avalanche.

Here's the crucial thing: there's no telling, when we add a new grain to a pile poised on the brink of an avalanche, whether it will set just a few other grains in motion, or trigger the collapse of the whole slope, or something in between. When Bak, Tang, and Wiesenfeld devised a simple mathematical model of this sandpile experimental and studied it on a computer, again they found that the size distribution of avalanches followed a power law (Figure 10.3). Big avalanches are less frequent than little ones, but avalanches of all sizes are possible. The fluctuations of the pile, in other words, are scale free. The pile can be considered to exist in a critical state.

Each avalanche releases "tension" in the pile, lowering the angle of the slope and restoring stability. But only just. The avalanche merely returns the pile to the brink of sliding, so that the very next grain that is added could trigger another landslide.

So the sandpile hovers around this state of precarious stability, never deviating far from it. Rather than the pile being liable to collapse irretrievably, the critical state is constantly resurrected after each avalanche. This is why the critical state is said to be self-organized. In contrast, the critical state of a liquid or gas can be regarded as self-destructive, ready to transform itself into a wholly different, stable state at the slightest disturbance.

The sandpile model describes a nonequilibrium situation. The critical state here is a "stationary state," in the sense that the system never strays far from it. But it is not an equilibrium state, because it is not unchanging: new grains are constantly being added. This dropping of grains onto the heap is the driving force that prevents equilibrium from being attained. Self-organized criticality is a property of nonequilibrium systems.

Self-organized criticality is one of the few genuinely new discoveries to have been made in statistical physics over the past two decades, and it has proved an astonishingly fertile idea. Per Bak has identified a wide range of

natural phenomena that seem to undergo power-law fluctuations suggestive
of self-organized criticality, or something like it. Earthquakes have long
been known to obey a power-law probability distribution: that is, the probability of an earthquake diminishes as its size increases according to a power
law. This was first discovered in the 1940s by seismologists Beno Gutenberg
and Charles Richter from the California Institute of Technology. They consulted a world catalogue of earthquakes and made a graph of the number of
events of each size. At the time, they had no way of interpreting their power
law. But Bak and others have suggested that it reflects self-organized criticality in geological fault systems. The motions of the Earth's crust build up
stresses in rock formations until slippage finally occurs. This releases the
stress—but only just enough to restore stability. Then the pressure begins to
build up again. Usually the stress is released in small doses, causing minor
tremors. But every so often there is a catastrophic release, and disaster
strikes a city such as Los Angeles or Tokyo.

Bak also saw self-organized criticality in forest fires. Many large forested
areas are prone to wildfires, but most fires are of only limited extent. Sometimes, however, a fire can go on spreading from tree to tree until it engulfs almost an entire forest. Volcanic activity, solar flares, the "starquakes" thought
to occur on exotic celestial objects called neutron stars, even the pattern of
species extinctions in the fossil record—all, said Bak, show the power-law
fingerprint of self-organized criticality.

Ironically, it appears that the sandpile itself might not be strictly in a
self-organized critical state at all. The experiment sounds simple in principle, but it is rather hard to measure sandpile landslides unambiguously in
practice, and to gather data over a big enough size range of avalanches so
that one can distinguish a genuine power law from something looking very
much like it. Various experiments have produced conflicting results. But it
seems that the original mathematical model constructed by Bak, Tang, and
Wiesenfeld neglects some important features of real sandpiles, such as an
accurate description of how the energy of a tumbling grain is dissipated in
collisions. True self-organized criticality (SOC) appears to be a rather elu-
sive thing in piles of grains—it crops up, perhaps, in some cases but not in
others. For instance, SOC has been claimed for piles of *rice* but not of sand:
the shape of the grain matters.

This rather undermines Bak's assertion that SOC is the key to "how nature works": it appears to be an insufficiently general and robust phenomenon to aspire to such a degree of universality. Nevertheless, there is little doubt that the basic characteristics underlying SOC—a power-law probability distribution of fluctuations, and catastrophic events that "release tension" while returning the system to the verge of instability—seem to provide a powerful framework for understanding a wide range of phenomena. We should hardly be surprised, then, if these phenomena include aspects of human social interactions.

HARD-LINE ECONOMICS

In 1988, Per Bak was working at the Santa Fe Institute in New Mexico—the center of the intellectual universe for any researcher interested in complex interacting systems, whether in physics, biology, geophysics, social science, or anything else. He was approached by two economists from the University of Chicago, Michael Woodford and Jose Scheinkman, who had heard about self-organized criticality and wanted to investigate whether the economy behaved in this way. It was an astute intuition, especially as the concept of SOC itself was barely a year old. Some errant economists had already become interested in applying ideas from chaos theory to economics—Scheinkman was one of them—but that notion was decidedly out of the mainstream. Woodford and Scheinkman saw more promise in Bak's theory, since it seemed to allow for precisely the kind of extreme events—big "avalanches"—that traditional economic models neglected.

As a physicist, Bak's introduction to economic modeling was revealing. Physicists generally deal with inanimate objects whose behavior has nothing of the irrational in it, and even when they make all manner of approximations and simplifications (as they invariably do), the math becomes too complicated to solve without a computer. Bak anticipated that economical modeling would be even more complex; so he was astonished to find that his economist colleagues preferred to look for "models that can be solved analytically with pen and paper mathematics." It was the beginning of what Bak called "a very productive, though rather painful, collaboration."[2]

The result was an agent-based model of the economy along the lines of those discussed in the previous chapter. The theory was published in 1993 by Bak and his fellow physicist Kan Chen, along with Woodford and Scheinkman. "Our conclusion," said Bak,

> is that the large fluctuations observed in economics indicate an economy operating at the self-organized critical state, in which minor shocks can lead to avalanches [crashes] of all sizes, just like earthquakes. The fluctuations are unavoidable. There is no way that one can stabilize the economy and get rid of the fluctuations through regulations of interest rates or other measures.[3]

The idea of a self-organized critical economy is an appealing one. But sadly, like the sandpile model, it seems to be right in spirit but wrong in detail. The very essence of SOC is scale-free behavior described by a power law. But although economic data, such as the rise and fall of the S&P 500 Index, can appear to behave in this way within certain limits, these features do not persist in the big picture. The fluctuations observed for time steps of a few minutes look more or less like those on hourly or daily time steps if everything is "rescaled" appropriately (see footnote on page 198)—but in fact the statistics are not precisely the same. The larger the time step, the more the probability distribution of fluctuations approaches the Gaussian form. Over periods of years, meanwhile, prices grow more or less linearly (that is, proportional to the time elapsed), punctuated by sharp drops. So any model that assumes or predicts a single mathematical form for the statistics of price changes on *all* timescales cannot be right. By the same token, the probability distribution functions of fluctuations in prices obey a power law only over a limited range in the "fat tails"; outside this range, a different kind of relationship is observed. So although it provides a nice illustration of how extreme events can represent a natural, if rare, aspect of how a system fluctuates, self-organized criticality alone is not the key to how the economy works.

THE SPIRIT OF THE LAW

Power-law probability distributions, which do not discriminate (as Gaussian fluctuations do) against extreme events, may turn out to be a pervasive feature of the way people organize their affairs. The physicist Sidney Redner at Boston University showed in 1998 that the statistics of citations in the scientific literature obey a power law. Scientific papers include a list of references to earlier papers whose findings are mentioned or used in the research reported. Some papers—the SOC paper by Bak, Tang, and Wiesenfeld in 1987 is one of them—contain ideas that are very influential and stimulate scores of follow-up studies. These are widely cited. Other publications address highly specialized topics and few subsequent researchers have any need to refer to them, so they garner only a handful of citations. Redner looked at nearly 800,000 papers published in 1981 and found that their citation statistics obeyed a power law: most papers were cited only a few times, but a few appeared in a large number of reference lists. (In fact, about half the papers were never cited at all, a telling reminder of the modest nature of most research projects. Tennyson was right to say that "science creep[s] on from point to point.")[4]

One might argue that these power-law relationships are statements of the obvious. *Of course* big earthquakes are rarer than smaller ones. *Of course* some research papers are highly influential, while others make an impact on no one. That's true, but a power law says more than that. It describes a particular *kind* of way in which the probability of an event declines as the event gets bigger or more extreme. There is no a priori reason to suspect that each time you double the size of the event, its probability will diminish by a constant factor: that the probability of four citations will be an eighth of that of two citations, say, and the probability of eight citations will *also* be an eighth of that of four citations. The general message of a power law is indeed intuitively obvious; the precise mathematical relationship, however, is not at all inevitable.

In his book *Ubiquity*, the physicist and science writer Mark Buchanan made the intriguing suggestion that history operates in a self-organized critical state. His idea is that conflicts and wars are the result of tensions that hold international relations poised at the brink of catastrophe. As a consequence, wars of all magnitudes, from minor skirmishes to global conflagrations, are inevitable. Buchanan offers what he admits is tentative evidence for this proposal in the form of a graph that displays the relation between the number of conflicts and their size, measured according to the number of fatalities that resulted. Wars that left only a few hundred dead and world wars that have killed millions: all follow a single power law.

This behavior was first noticed by the British physicist Lewis Fry

Richardson. One of the early pioneers of a modern physics of political policy, Richardson used concepts taken from meteorology to develop mathematical models of the arms race between antagonistic nations. A Quaker who served as an ambulance driver in the First World War, Richardson hoped to promote international peace by elucidating the causes of war.* Between the 1920s and the 1950s he amassed data on the statistics of "deadly quarrels," in which he provocatively included war alongside other kinds of killing, such as murder in civilian society. He found that there was a kind of Gutenberg-Richter law (page 240) of conflicts in which all events from individual murders ("magnitude 0 conflicts") to the two world wars ("magnitude 7"), obey a power-law probability distribution.

The sobering corollary, Buchanan argues, is that there can be no telling how big a conflict might be sparked by the smallest disturbance. The First World War, after all, followed from the assassination of the Austro-Hungarian archduke Franz Ferdinand in Sarajevo in 1914, an event that hinged on an unhappy and unpredictable set of circumstances on that fateful day. One can immediately quibble with this idea. It seems highly likely, for example, that the Second World War was made inevitable by (among other things) the First World War and the Treaty of Versailles, rather than being triggered by some freak occurrence in 1939. Indeed, the First World War was not really fought over the archduke's murder at all. But that is not the point; Buchanan is saying that if there is "tension" in a complex system like this, small events can have disproportionate consequences. In any event, Richardson's data imply that Immanuel Kant's notion of natural laws governing the course of history has some mileage still. The question historians will ask is whether this can tell us anything about history *as such*: whether it helps us understand *why* things turned out the way they did. That is, in a rather different context, a question to which we shall return later.

---

*Richardson's interest in whether wars between neighboring nations depended on the lengths of their common boundary led him to realize that the measured length of boundaries and coastlines depends on the step size used to track these tortuous zigzag lines. This discovery prompted the later development of fractal geometry by Benoit Mandelbrot.

THE LEAST ONE CAN DO

The American social scientist George Kingsley Zipf (1902–1950) helped revive the Enlightenment belief in natural laws of society and the possibility of a truly scientific sociology with his book *Human Behavior and the Principle of Least Effort*, published in 1949. It is a curious document, both phenomenally prescient and very much a product of its times.

Zipf's central notion was that people act in ways that require them to make minimal effort. This seems commonsensical enough, and Zipf regarded it as the sociological equivalent of the physical principle proposed in the nineteenth century by the Irish mathematician William Hamilton. The "law of least action," said Hamilton, determines how any entity moves under the influence of forces. He demonstrated that underlying the laws of mechanics adduced by Newton was the tendency of an object to follow a trajectory that generates the smallest possible "action"—a mechanical quantity that depends on the path taken. There are many trajectories that take a ball at rest on a table to a position at rest on the floor; but only one of them minimizes the action of the motion, and that's the path the ball follows when it rolls over the edge.

It is straightforward to calculate the action accrued by an object moving along a particular trajectory. But determining the "effort" involved in human activities is far from trivial. This is not simply the energy expended; as Zipf pointed out, different people might choose different ways to complete a given task, depending on how much they *think* it will cost them, in energy, time, discomfort, money, or whatever. One engineer will link two towns by boring a tunnel through a mountain; another will build a road over the peak. According to Zipf, both act according to their estimates of how much "effort" will be involved; it is simply that their estimates differ.

This is clearly a shaky principle, given the subjectivity of any assessment of "effort." The idea sounds good, but its central parameter is all but unquantifiable. Nonetheless, Zipf attempted to explain a bewildering variety of human traits and behavioral patterns this way, including the properties and development of language, the structure of music, human demography, the distribution of industries, travel statistics, marriage data, international and civil conflicts, and income distributions.

Zipf's main contribution to the physics of society was, however, empirical. He collected data in all these areas and showed that they all featured probability distributions characterized by power laws, with their distinctive straight-line graphs. At that time social scientists seldom looked much beyond the statistics of Gaussian (random) distributions, and the significance of Zipf's data has come to be appreciated only through the recent emergence of power-law behavior at the heart of statistical physics. Whereas Per Bak (who died in 2002) believed that self-organized criticality is in some sense "how nature works," Zipf considered that his power-law graphs showed "how society works." He believed that the social sciences differed from the natural sciences in that they were dominated by power-law rather than Gaussian statistics. We now know that power laws are common in the natural world too.

Be that as it may, Zipf uncovered something of fundamental significance. In 1983, Benoit Mandelbrot wrote that "the failure of applied statisticians and social scientists to heed Zipf helps account for the striking backwardness of their fields."[5]

Zipf asserted that one power law in particular—one in which the straight-line plots have a slope of −1 (see the caption to Figure 10.3), now seen as diagnostic of self-organized criticality—was characteristic of phenomena in which people act in groups rather than as individuals: that is, phenomena in which people interact. This was more or less the full extent of his appreciation of the role of interactions in social phenomena, which we can now see to be crucial.

As the physicist Philip Anderson has pointed out, power-law distributions in social phenomena destroy the idea that what matters is the "average" behavior, or Quetelet's "average man." In effect, a power law raises the odds of extreme events, which a Gaussian probability distribution relegates to the status of negligible aberrations.

The sociologist Vilfredo Pareto was perhaps the first to introduce power laws into social science, and he did so before anything of the sort had been discovered in physics. He claimed in 1897 that incomes toward the wealthier end of the social spectrum are distributed according to a power law (Figure 10.4). This implies that much of a nation's wealth is held by a few individuals. For the United States, some estimates attribute 40 percent of the wealth to 1 percent of the population, and over half of the wealth to 5 percent of the population.[6] This inequality has been growing since the 1970s— a trend that has been identified in other nations too.

10.4   National income distributions typically follow a power law at the rich end of the distribution, as first observed in the nineteenth century by Vilfredo Pareto. He claimed that the slope of the straight line is always −1, but later studies have shown that it is usually large. The larger the slope, the more the economy is impoverished. Shown here is the wealth distribution for the population of the U.K. in 1996, according to figures collected by the Inland Revenue. This is a cumulative distribution: each data point shows the percentage of the population with a net capital wealth greater than the corresponding amount on the horizontal axis. The Pareto law appears in the data for the wealthiest 10 percent or so of the population.

Pareto expressed this imbalance in terms of the so-called 80:20 rule: 80 percent of the wealth is possessed by 20 percent of the people. He observed that this income distribution held true for many countries, regardless of their political system or taxation regime. The 80:20 principle has come to be regarded as a rule of thumb for management decisions: 80 percent of your benefit will come from 20 percent of your outlay, 80 percent of your results will be due to 20 percent of your workforce, and so forth. But focusing on these particular figures rather obscures the main point, which is not simply that there is inequality between, say, effort and reward, but that the *distribution* follows a power law. The exact figures—that's to say, the slope of the straight-line power-law graph—may vary: it's not always an 80:20 ratio. But the straight line remains. This kind of wealth distribution has been inferred even in ancient Egypt of the fourteenth century B.C., from an analysis of the distribution of house sizes in the ruined city of Akhetaten.

Pareto's income distribution later came to be regarded as a fundamental aspect of society, possessing much the same mystique as the Gaussian distribution had in the early nineteenth century. In 1940 the economist Carl Snyder said that "Pareto's curve is destined to take its place as one of the great generalizations of human knowledge."[7] In truth, however, it is still not entirely clear how rigidly the "law" applies. For one thing, income distributions are surprisingly hard to determine accurately (one way is to look at tax returns; another is to infer levels of wealth from payments of death duties). In 1935 the economist George Findlay Shirras questioned the validity of the Pareto law, saying that there was no place for it in studies of income distribution. But there seems good reason to believe that a power-law distribution does hold at least for the highest-earning bracket of any society, as Figure 10.4 shows. The "steepness" of such a power-law slope reflects the degree of disparity between incomes: the steeper the slope, the more unequal the distribution of wealth and the more poverty there is. In extremely unequal economies, such as those of Haiti and Zaire (or indeed, ancient Egypt), a very few have gotten rich at the expense of a vast majority of very poor people. This is one good reason never to trust the average statistics cited by politicians unless you know the underlying distribution on which they are based.

Econophysicists Sorin Solomon in Israel, Jean-Philippe Bouchard in France, and their co-workers have proposed models to explain how the Pareto law might arise, drawing on ideas developed to explain the movement of the chainlike molecules in polymers. The researchers compare these motions with the movements of money in investment markets. Solomon, working with Zhi-Feng Huang of Cologne University, has shown how there is a tendency for trade to increase the steepness of the Pareto slope in unregulated markets, creating an ever-increasing disparity between rich and poor. One consequence of this is that market fluctuations also increase: the market becomes less stable. So, the researchers argue, a social policy that aims to raise the wealth of the poorest members of society "is not just a humane duty but also a vital interest of the capital markets"[8]—a form of enlightened self-interest.

UNIVERSAL ORDER?

George Zipf hoped his observations would be used as a guide to policy making. His dream sounds, in these post-Marxist times, a little unnerving: "a systematic social science will make possible an objective social engineering."[9] But he was right, if by *social engineering* we mean no more than informed planning.

His ambitions went even further, however. In the true spirit of the Enlightenment, Zipf felt that the discovery of order (by which he meant the ubiquitous power law) at the center of human affairs revealed a kind of natural design operating beyond human volition. In an age of declining religious conviction, Zipf imagined that this might encourage a rational social science to take its place: "For we are finding in the everyday phenomena of life a unity and orderliness and balance that can only give faith in the ultimate reasonableness of the whole whose totality lies beyond our powers of comprehension."[10]

From there it is perhaps but a short step to Pythagorean mysticism. But that need not distract us from responding with a certain wonder at the universality that organizes many aspects of society in the same way as it directs the properties of atoms. We need not turn this into a "religion of science," any more than we need regard the whorls of a flowing stream as evidence of divine planning. We can simply celebrate the fact that there are indeed "laws of large numbers," and that they let us divine order and regularity in an otherwise terrifying diversity.

ELEVEN

THE WORK OF MANY HANDS

THE GROWTH OF FIRMS

I met a vineyard owner in California. He was, he told me, trying to squeeze every cent of margin out of his product "in order to develop my winery as fast as possible." "Do you plan to get very much bigger?" I asked. "Oh, I don't want to grow bigger," he replied, "I want to grow better." To do that, he needed better equipment, better vines, better people—not more.   —Charles Handy (1976)

Large human organizations always lead to oppression unless there is a democratic machinery to prevent it.   —J.B.S. Haldane (1949)

What scale is appropriate? It depends on what we are trying to do. The question of scale is extremely crucial today, in political, social and economic affairs just as in almost everything else.   —E.F. Schumacher (1974)

Three years ago, Brecon was a Welsh market town on the verge of becoming—like many other communities today—a supermarket town. Safeway, the giant food retailer, was coming. When environmental campaigner George Monbiot went to the town, he found locals expecting the worst.

"Safeways say they're bringing choice to Brecon," said Brian Keylock, a butcher. "But there are nine butchers in town already. You've got nine choices. If we all disappear, you'll have no choice. The fishmonger has already closed down. Clothes shops have closed down . . . Go to Leominster. Since that thing arrived there, the town is dead. They'll do the same to us."[1]

Frankly, Brecon was lucky to have survived for so long. British towns have been succumbing to the death of the main street (the "high street") and the demise of the local shop since the early 1980s. Now chain stores, banks, and real estate brokers colonize town centers, and it's hard to buy a pint of milk there. To do that, you need to go to the outskirts of town, where super-markets offer milk in ten different varieties, along with all the other items that once took people from shop to shop.

The problem—if you are inclined to see the disappearance of small shops as a problem—is not just that supermarkets and chain stores exist but that we all use them. Increasingly we have little choice: the out-of-town Sainsbury's supermarket in the Somerset town of Axminster was the only place I could get a cup of tea on a Sunday afternoon as I cycled through town. The small tea shops could not compete. And that's the trouble: big companies can afford to do things that small ones cannot.

. How many local bakers, for instance, could sell a loaf of bread for eight pence (about fourteen cents), as Safeway did in 1999? This is way below cost price; but supermarkets gamble that by undercutting competitors on staple products, they can create an impression of fantastic cheapness that will draw in customers who will then buy more expensive items too. The Lidl super-markets in Britain were recently charging *negative* prices (yes, deducting money from your bill) for cans of baked beans.

Capitalism's evangelists tell us this is all for the best. A 1978 British gov-ernment Green Paper (a policy document) claimed, "Unrestrained interaction of competitive forces will usually result in the best allocation of our economic resources, the lowest prices, the highest quality and the greatest material progress, while at the same time providing an environment conducive to the preservation of our democratic, political and social institutions."[2]

Brecon's butchers might beg to differ.

Capitalism's opponents, meanwhile, often portray the free market as a system in which big fish inevitably swallow up small ones, leading to a ho-mogeneous world of commerce dominated by a few major players. It can cer-tainly look that way. Between 1990 and 1996, British shops that sold less than a hundred thousand pounds' worth of goods a year declined by 36 per-cent, while the number of supermarket branches almost trebled from 1986 to 1997. In the broader picture, big companies have the economic muscle of entire nations: 52 of the 100 largest economies in the world are corporations.

There is ample reason to worry about this. Monbiot argues that the growth of megacorporations threatens not just the cohesion of local commu-nities but the concept of democracy itself. That seems reason enough to won-der about the factors that drive the expansion and contraction of companies.

In this chapter we shall explore whether there are fundamental laws that govern the way firms grow. Many economic theorists have suspected that there are; but if this is so, no one yet agrees on the form these laws take or how they operate. Nevertheless, signs are beginning to emerge of a kind of "universality" principle in business which implies that whether you are looking at steelmaking, publishing, or bread baking, the growth of firms follows an "iron law" that tells us what kind of distribution of small, medium, and large we must expect.

And that's the strange thing. For all the signs that the business world is becoming top-heavy, the undoubted inequities of globalization are not the whole picture. While a few big corporations exert an inordinate influence on product choice and advertising, there is a far greater number of small firms than big ones. Most of these tiny acorns never get much bigger; few indeed are destined to become mighty oaks. Yet the acorns are not dying out—not, at least, if we are to judge from today's figures.

Small businesses know only too well how strongly success depends on chance. But from the fluctuating fortunes of many individual companies, robust laws seem to arise that govern the way firms grow. Whether we can tailor the world of commerce to the shape we prefer (and whose preference?) is an open question; but surely we cannot hope to do so until we can characterize and understand these laws. If we wish, for instance, to limit the size of big corporations, or to help small firms survive without their feeling compelled to expand, we need models of company growth that allow us to predict the consequences of the legislation or trade rules we might apply. If we were to discover (and this seems likely) that any market permitting a modicum of free trade will inevitably spawn a few very large companies, then at least we know what our choices are.

KEEPING COMPANY

World trade does not *need* firms and companies. In Roman Britain one could buy goods made from Chinese silk. Conversely, Rome exported to the Orient: Roman coins have been found in Vietnam. The Tojaidi temple at Nara in Japan, built in A.D. 752, contains Byzantine glass. The awesome trade network of the ancient world was sustained by individual craftspeople selling their wares to merchants who passed them on, at a profit, to those who could afford exotic imports from far-flung lands.

So what are firms for? Artisans in the Middle Ages discovered some of the benefits of collective organization. The trade guilds protected their members from exploitation and were able to enforce wage standards for the contracts the guildsmen undertook. But the Industrial Revolution revealed the greatest advantage of collective labor: economy of scale, made possible by mechanization. The advantage was, of course, primarily to the few, not the many—industrialization benefited the capitalist bosses much more than the workers. Even so, a worker on subsistence wages with a big firm was better off, by and large, than the traditional craftsperson, who could not hope to compete against industrialized rivals. The former was on the breadline; but unless the latter signed up for the same, he starved.

Economies of scale are achieved in several ways. Big firms are able to purchase expensive machinery which is beyond the means of individuals. They gain efficiency by bringing their workforce together under the same roof and by apportioning tasks among specialized workers. "The division of labour," said Adam Smith, "occasions, in every art, a proportionable increase of the productive powers of labour."[3] And big firms reduce transaction costs by striking a few big deals instead of countless smaller ones.

The coercive employment market of the Industrial Revolution eventually softened in the latter half of the twentieth century into what could be a reassuring environment for the worker. The paternalistic big company offered job security and employee benefits such as health insurance, transport, or accommodation, as well as camaraderie and, in time, the promise of a pension.

This, perhaps more than anything else, undermined Marx's "inevitable revolution," based as it was on a nineteenth-century model of exploitation and Smithian market principles that compelled employers (if they needed compelling) to pay workers only the bare minimum necessary to keep them alive. In Japan in particular, the transaction between worker and boss became no longer just that of work for money but of lifelong loyalty for lifelong protection. Whether or not one still believes in this rosy picture of capitalism in action, there seems little doubt that most people still prefer to be part of a collective workforce than to be on their own.

The traditional theory of the firm has had rather little to say about the complex relationship between managers and their workforces. Developed by economists, this theory has kept things simple by assuming that agents exhibit the calculated rationality to which the discipline has become accustomed.

Firms, it says, exist to maximize profits. They aim to establish a revenue that is as far as possible above costs. Primed with this clear goal, companies then use a well-developed calculus of supply and demand that enables them to chart their optimal choices. They will seek to expand to the point where further increases in revenue are exactly balanced by rising costs, so that no more profit accrues from increased production.*

This is all very neat in the short term. Unfortunately, the conventional theory of the firm also predicts that, as a result, in a completely free market no one can make any profit at all in the long run. Capitalism, the theory says, is bad for capitalists.

The reason is simple, and Adam Smith foresaw it. In a market governed by perfect competition—where all firms are equally capable and can charge whatever they like for a product—there will always be advantage to be gained over other firms by cutting back on profit margins and selling more cheaply. Thus no firm can increase prices above the break-even point at which revenue is exactly balanced by costs. If it does, someone else will sell more cheaply and no one will buy the more expensive product.

The problem of where profits come from was therefore a major concern for nineteenth-century economists, as we saw in Chapter 8. Marx argued that they result from the "surplus value" that a worker gives to his employer by working longer hours than his wage warrants: the employer gets some labor "for free," and this is where he makes his profit. This suited Marx's image of the exploited workers; and in our present climate of shrinking workforces working even longer hours, one has to suspect that Marx was onto something.

---

*The ideal level of production is that for which the *marginal revenue* (the change in revenue per unit increase in production) equals the *marginal cost* (the change in total cost per unit increase in production). Because any given production plant has an optimal level of usage, trying to expand its output beyond this capacity eventually produces *diminishing* returns of scale (without capital input to make a bigger plant). When that happens, the marginal cost rises steadily as output increases, and this is what eventually curbs the profitability of increased production.

Economists look for this compromise between marginal revenue and marginal cost by drawing theoretical curves that show how they relate to output. It turns out that these idealized curves rarely look much like those derived from real data, which adds to the (often overlooked) problems of the conventional theory of the firm.

But it is not hard to think of many other reasons companies make a profit despite the operation of free-market forces. Advertising clearly distorts people's buying preferences: they will not necessarily look for the cheapest product but will be drawn toward those they believe to be the "best" or the most prestigious. People will pay a high premium for fashionable brand names.

That is not the whole story. The assumption of perfect competition is a way of sidestepping the awkward fact that companies are interdependent: the behavior of any firm can affect the pricing of goods and the profitability of the others. Economists know this is so but are not sure how best to deal with it.

The conventional, so-called neoclassical theory of the firm is comfortable only with two extreme scenarios. Under conditions of perfect competition, each firm can be considered to operate within a market that has fixed, externally imposed features. The relationships between manufacturing cost, selling price, and quantity of a product are assumed to be immutable: they arise from the average behavior of all the firms in the market. All a company has to do is juggle with these imposed conditions to seek profitability.

The same is true of the other extreme: monopoly, in which the market contains only one firm. By definition, interaction can then be neglected because there is no one else to interact with. But in practice monopolies are clearly special cases, for which specific legislation usually exists to regulate the way the market operates. (Otherwise the firm with the monopoly can charge whatever it likes for its products.)

The truth is that a free market rarely if ever operates under conditions of perfect competition between a large number of more or less equivalent firms. Instead there are usually many companies of varying size, each with overlapping but not necessarily equivalent product profiles.

That is where the conventional theory breaks down. But it does have a decent shot at describing a more restricted situation: the oligopoly, in which just a few major players dominate the field. This is not uncommon; in the U.K., for example, five supermarket chains account for three quarters of all grocery sales, and there are only half a dozen or so major national newspapers.

Interaction among companies is critical to pricing and profitability in an oligopoly. Undercutting operates in much the same way as it does in a perfectly competitive market, but with fewer players the feedback between firms is more extreme. This can lead to the kind of instabilities and sudden changes that, in physics, are often the fingerprint of strong interactions. For example, prices can stay steady for long periods of time and then go abruptly into free fall as a price war escalates. Such hostilities have flared in the recent past between manufacturers of personal computers, airlines, and national newspapers. These battles can work to the apparent advantage of consumers, who can find themselves paying way below cost price. But in the long term the struggle can eliminate players, reducing choice and letting the market slide toward a monopoly, as Rupert Murdoch's News International group in the U.K. threatens to become.

Economists have tried to build interaction into models of oligopolies in various ways. A crude solution is to recognize that rivalry may change the shape of the demand curve that relates price to quantity of goods. For example, a firm's rivals might be ready to match any price cuts it makes, but not price increases. But this simply modifies one of the imposed characteristics of the model: the relationship between price and quantity. It is really nothing more than an attempt to reduce the interactive market to one in which each firm maneuvers within externally fixed circumstances.

Oligopolies do not have to be competitive. It is clearly in the interests of companies to collude in fixing high prices, which consumers are then forced to pay through lack of choice. This creates cartels that can operate as de facto monopolies. Some countries outlaw cartels, but it is nevertheless very hard to prevent collusion at the expense of the consumer. "There can surely be little doubt," says George Monbiot, "that some of the big chains operate what are, in effect, local or regional monopolies. Tesco and Sainsbury [two major supermarket chains], for example, control 57 per cent of the grocery market in London and the south of England between them."[4]

Cartels rely on their members' playing the game and not defecting by lowering their prices and thereby grabbing a bigger market share. This situation of cooperative advantage juxtaposed with the temptation to profit by defection lends itself perfectly to the analytical methods of game theory, described in Chapter 17. Such methods are commonly used in economics to model oligopolies that are open to collusive practices.

None of these traditional approaches, however, takes much account of how the market is really structured: that is, how firms are really distributed in size. Perfect competition, monopolies, and oligopolies are all exceptions.

In the real marketplace there are generally firms of many sizes. Small firms may modify the behavior of oligopolies while big firms constrain the competitiveness of small ones.

Moreover, the motivations of firms are not well described by the conventional theory. Most firms are almost certainly *not* pure maximizers of profit. Each pursues its own agenda, and this can be a brew of many conflicting ingredients. Even the most hardened cynic cannot claim that all firms put profits before all else; remember, after all, that we are talking here not only about multinationals but about local bakeries with three employees. A company for which I once worked provides free lunches for its workers. If it did not do so, its profits would increase, and I doubt that anyone would quit as a result (although they might grumble). Of course, this is partly enlightened self-interest: it is worth a firm's while to keep its workforce happy.

Other companies act to maximize not profits but total revenue, figuring that market share is more important in the long run than profitability. Some make the growth of the workforce their objective. Some seek whatever course will keep their shareholders happy. Others aim, in the tortured terminology of economics, to *satisfice*: to achieve satisfactory (not necessarily optimal) performance according to a wide range of criteria (page 212).

Given these diverse strategies, one might anticipate that the growth of firms shows little or no common ground. But we must remember the importance of distinguishing individual from statistical outcomes. Just because people have many different motives as they move through their environment, this does not mean that their collective movement is devoid of pattern. Do firms, too, display statistical regularities beneath their individual idiosyncrasies?

BUSINESS LAW

One might think that a book entitled *Inégalités économiques* (Economic Inequalities) would be a Marxist denouncement of capitalism. But this was not at all the intention of its author, Robert Gibrat. In his book, published in 1931, this French economist developed the first universal theory of *how* firms get to be of unequal sizes.

Gibrat understood that any such theory must be statistical—it must ad-

dress the mathematical form of the probability distribution of firm sizes. It
was clear to him that this distribution is highly skewed: that is to say, there
are many more small firms than large ones. This basic fact remains as true
today as it was in the 1930s; the economic data from any industrialized
country confirm it. The economist Robert Axtell, of the Brookings Institution
in Washington, D.C., says, "The stability of this distribution over time
makes it . . . perhaps the most robust statistical regularity in all the social
sciences."[5]

Gibrat suspected that the factors underlying the way that firms grow include a considerable element of randomness, a conclusion he drew from the
work of the Dutch astronomer Jacobus Kapteyn. Around 1916, Kapteyn
showed that skewed distributions in population ecology can arise from Gaussian (that is to say, random) processes. Gibrat proposed that a firm grows at a
random rate, and that the rate at any given moment is amplified by the firm's
size at that moment.

This defines Gibrat's now-celebrated Law of Proportionate Growth, which
he envisaged as a kind of Newtonian law of the business world. It works as
follows: To predict how much a firm changes in size between now and some
future time, choose a number at random from between −1 and 1 (it could be,
say, 0.5, or −0.3528, or 0) and multiply it by the current size of the firm. Thus
bigger firms tend to change in size more markedly than small ones, but not inevitably so. (This element of chance is included because the various factors
that influence growth are hard to predict.) To put this growth rule another way,
the bigger a firm is, the more able it is to capitalize on whatever opportunities
come its way.

Gibrat's "law" leads to a size distribution with a mathematical form
called log normal.* He collected data on the plant sizes of French manufacturing firms for the years 1920 and 1921 and showed that their size distribution fitted his prediction rather well. And he didn't stop there. He presented
data suggesting that his law held for different times (applying just as faith-
fully in 1896 as in 1921), for different economic sectors (industry and agriculture), and for different industries within a sector. By the 1940s Gibrat's
model was highly regarded, and even today it is often used as a benchmark
for theoretical and empirical studies of firm growth.

---

*This means that a firm's size fluctuates over time in such a way that the probability distribution of
the *logarithm* of firm sizes is Gaussian.

Nevertheless, the model is basically wrong, and everyone knows it. Whatever Gibrat's model says about the distribution of firm sizes, it is clear, simply from observing how real firms grow, that they do not expand and shrink in steps of random size. And this picture is certainly not consistent with the neoclassical idea of firms rationally maximizing their profits (which implies that firms should respond similarly, rather than independently at random, to changes in market conditions).

Thus, while Gibrat's law serves as a useful idealization, no one expects it to come very close to reality. Indeed, in the 1950s, the whole idea of a universal law of growth fell out of favor. Instead economists became convinced that market structure and dynamics varied from one industry to another, perhaps because different manufacturing methods offered different economies of scale, or perhaps because of the varying roles of advertising or research and development. Also, more careful studies of size distributions began to suggest that Gibrat's "universal" curves were probably just a fluke produced by inadequate data. The economist John Sutton of the London School of Economics summarizes the prevailing modern view:

> There is no obvious rationale for positing any general relationship between a firm's size and its expected growth rate, nor is there any reason to expect the size distribution of firms to take any particular form for the general run of industries . . . empirical investigations from the 1960s onwards have thrown doubt on whether any single form of size distribution can be regarded as "usual" or "typical."[6]

A rejection of Gibrat's model is vindicated not only by its apparent lack of conformity with how firms really grow but by its failure to provide more than an arbitrary prescription for that process: his Law of Proportionate Growth is not justified by microeconomic principles. Yet the standard economic theory of the firm has nothing better to offer in its place. Robert Axtell points to "the inability of the neoclassical theory of the firm—with its U-shaped cost functions and perfectly informed and rational managers—to render a plausible explanation of the empirical size distribution."[7]

This appears to force economists into the frustrating business of having to devise industry-specific models of firm growth that are geared to the particular quirks of that trade. So much for Gibrat's universality. In Sutton's view, "the evolution of market structure is a complex phenomenon and the quest for any single model that encompasses all the statistical regularities observed is probably not an appropriate goal."[8]

FIRM PRINCIPLES

Others are unsure about that. In 1996 the physicist Gene Stanley, the economist Michael Salinger, and their co-workers at Boston University looked at the growth rates of all publicly traded U.S. manufacturing companies between 1975 and 1991. Encompassing around eight thousand firms, this was a larger data set than had been tackled in previous studies. The researchers found that the growth rates did not fit Gibrat's log-normal distribution but instead followed a power-law relationship, like that seen in critical phenomena. This means that plotting the logarithm of the growth rate against the logarithm of the probability of that rate produces a straight line. Or, to be more accurate, *two* straight lines: one for positive growth rates and one for negative, both with the same slope. This generates a tent-shaped graph (Figure 11.1). And the power law holds true for two different measures of firm "size": sales (revenue) and number of employees.

So there might after all be a general law of the growth of firms, albeit one different from Gibrat's. But there is more to it than that. Gibrat's log-normal distribution followed from the assumption that firms grow independently of one another—that the process is random, but "weighted" by the size of the firm. Physicists' long experience with power laws, on the other hand, leads them to believe that such laws are the universal signature of interdependence. A power law generally emerges from collective behavior among entities, through which local interactions can develop into a long-range influence of one entity on another.

This idea is supported by a microeconomic model of firm growth developed in the late 1990s by Robert Axtell, in which firms arise by the aggregation of many interacting workers ("agents"), each following its own agenda. The aggregation is driven by rules of the model that can make collective labor advantageous. Axtell's model is in the spirit of microeconomics in that it attempts to deduce the overall behavior of a system from the motivations of the individual agents that constitute it. But unlike many such "theories of the firm," it starts with no preconceptions about what firms are for or how they behave; indeed, the agents are not forced to aggregate into firms at all.

That they do so is an indication that it is in their best interests. These agents do everything with their best interests in mind. In that sense, they are

11.1 The probability distribution of growth rates for all publicly traded U.S. firms between 1975 and 1991. The growth rates are measured by increases (or decreases) in both sales (black circles) and number of employees (white circles). The growth rates are generally bigger for bigger firms, but the figures all fall onto a single graph if one considers the *relative* growth rate—essentially, the percentage change in a company's size. The probability of a certain growth rate decreases as the rate increases according to a power law. This means that the figures fall onto a straight line when plotted as logarithms. Decreases in size follow the same power law as increases, making the graph tent-shaped.

the "rational maximizers" so dear to the economist's heart. But they don't have perfect knowledge of what is happening everywhere in the system—far from it. And they are not the avaricious "profit maximizers" that some economists would assume. Rather, they are individualistic maximizers of their own happiness.

What makes these agents happy? That depends on the agent. Each of them pursues two goals: money and leisure. Unfortunately, in Axtell's world one can't have both at the same time, because money is made only through hard work. Money and leisure are conflicting demands. Where the balance is drawn varies from one agent to another: some settle for low wages if it means they can do a lot of lounging about; others diligently apply themselves at all hours in order to amass wealth. The relative preferences for money and leisure vary across the agent population. So what each agent tries to do is to find a job that will provide the desired compromise between these two things. If it likes to work, it will join a firm of hard workers. If it likes leisure, it might seek a firm that enables it to free ride, putting in relatively little effort without its being noticed. The wages might be low, but hey, who cares?

In effect, each agent is seeking to maximize its *utility* (page 208)—a measure of happiness, here denoting the preferred balance of labor and leisure. Jeremy Bentham invoked a similar concept in his philosophy of utilitarianism, in which he argued that society should seek out the state that maximizes the collective utility of the population: the state of greatest total happiness. But Axtell's agents are nothing near as community-spirited. They don't give a damn what the collective utility is, so long as their personal utility is as good as they can make it, given the choices open to them.

It's a fair question why agents in this model should wish to join a firm at all. If they like slacking, can't they do that just as happily alone? Conversely, why compromise your hard work by teaming up with potential slackers? This issue was pondered by Nobel laureate in economics Ronald Coase in a pivotal paper on firm theory in 1937, in which he wondered why the market did not simply consist of individuals trading with one another.

Coase's answer hinged on transaction costs: the inevitable costs of trading, such as the time, effort, and often the financial burden involved in negotiating, drawing up, and enforcing contracts, and so forth. Alfred Marshall, on the other hand, suggested in 1920 that the key to organized labor is to be found in the old adage that many hands make light work. That is to say, a certain amount of effort generally produces more returns if exerted collectively rather than individually. Division of labor, sharing of capital costs, and specialization enable the collective to achieve more than the sum of its parts. Economists call this an *increasing return of scale*: the bigger you are, the more productive you are. Axtell provides an incentive for his agents to form aggregates—firms—by building this property into the mathematical equation that relates an agent's efforts to the productivity of the group to which it belongs.

Standard theories of the firm assume that companies grow because of increasing returns of scale. Crucially, Axtell's model does not. It merely asserts that *individual agents* get more done for the same effort if they pool their resources. This might lead one to *expect* increasing returns as firms themselves get bigger, but the model does not guarantee that. (It does not, for example, specify just how hard each agent in a company actually works.) All it says is that it is in the interests of each agent to team up with others, no matter whether they are workaholics or slackers. In either event, they get more for less, and they all like that.

The agents, each with its own proclivities, are at liberty in the model to join and leave firms (groups of agents) in search of the best deal. Thus firms can grow and decline as agents come and go. But each agent actually has relatively few choices. It cannot see what is happening in the whole market; the rules of the model permit it to know only how it is faring relative to a small number of "friends," typically two. If one of these friends is doing better, the agent joins it; otherwise it stays put.

The key point about the model is that each agent has a choice about how hard to work. The decision varies depending on the situation. If a slacker can find its way into a firm of hard workers, it can do virtually nothing and still profit. Each group of agents constituting a firm gets paid in proportion to the total output, and each gets an equal share. If a single slacker joins a firm of 50 good workers and then puts in no effort, so that the total output does not rise proportionately, it gets 1/51 of the rewards earned by the other 50. Their reward, meanwhile, barely changes: instead of being 1/50 of the total, it is 1/51 of the same total. So if the firm is relatively profitable, its workers might tolerate this slacker in their midst. (No way is provided, within the simplest formulation of the model, for hard workers to gang up on slackers and expel them. All they can do is leave the firm for another, if that will give them better rewards for their efforts.) But if the slacker is on its own, it cannot afford to put in no effort, since then it will have nothing to live on—and even leisure-lovers have to eat. In a small company, the presence of an agent that is not pulling its weight will be more noticeable than it would be in a large company. So the other agents in a small firm may be more likely to leave, provided they can find greener pastures.

Each agent in the model makes a new decision at randomly chosen time intervals, as if waking up and thinking, "Do I want to be doing this?" On becoming "active" in this way, the agent readjusts its effort level to the best advantage of its utility. It also weighs the relative advantages of staying put, joining a friend's firm, or starting a new firm on its own.

One of the big questions for a model like this is whether there are stable states in which every agent is happy with its position. Economists call such states Nash equilibria, after the mathematician John Nash, who proposed them in 1949. They correspond to situations in which no agent can improve its circumstances by altering its behavior. Traditionally, economists are inclined to look for Nash equilibria in their models—they think these determine behavior in the real world.

But Axtell's model has no stable Nash equilibria. That is to say, it can never settle down into an unchanging state. There is constant flux as firms boom and go bust (Figure 11.2). The model is a *nonequilibrium* one, and this makes it quite different from most microeconomic models of firm growth.

This does not imply, however, that nothing certain can be deduced about the system. Rather, it means that we are forced to draw statistical inferences. We cannot say, for example, how long it will take for a firm encompassing 25 percent of the total labor force to appear. But we can, at any point, determine the probability of such a firm's existing. Indeed, a statistical sampling of model "runs" shows us the full distribution of firm sizes. It is a power law: a (logarithmic) graph of firm size against the probability of there being a firm that size is a straight line (Figure 11.3a). And that is just what is observed in practice, as Axtell deduced from the statistics on about 20 million U.S. firms in 1997 (Figure 11.3b). This is a striking result—no other microeconomic theory of the firm has correctly forecast the power-law nature of this distribution.

Another telling test of the model concerns firm growth rates. Gibrat, remember, argued that growth rates are randomly distributed. The real eco-

11.2   There is a constant turnover of businesses in Axtell's model of firm growth. The total number of firms (*a*) never attains a steady value, and even the largest firms (*b*) can collapse catastrophically.

nomic data show that, on the contrary, they are again distributed according to a power law, producing a tent-shaped plot on logarithmic scales (Figure 11.1). Axtell's model generates this same shape for the distribution of growth rates (Figure 11.4).

The skeptic might suspect that these power laws will arise no matter what we assume about the agents' behavior because power laws are somehow built into the model. But if we assume that agents switch firms at random (rather than to improve their utility), or that they choose levels of effort

11.3   The statistical distribution of firm sizes in Axtell's model is a power law, displaying a straight line on a logarithmic plot (*a*). The distribution of U.S. firms in 1997, compiled from census and Compustat data combined with self-employment data, also shows such power-law behavior (*b*). The largest sector consists of 15.5 million firms that have "no" employees—that is, they are self-employed individuals.

11.4   The rates of firm growth in Axtell's model show the same double-power-law, or "tent-shaped," distribution as that seen in the real world (Figure 11.1).

at random (rather than tailoring effort to circumstances), the firm size distribution is no longer a power law. So it seems that the power law is a consequence of *purposive* behavior on the part of the individual agents in the market. On the other hand, the power law remains even if we alter many of the details of how the agents make their purposive choices. For example, we can increase the size of each agent's friendship circle (and thus increase its "knowledge" of the labor market); or we can boost the factors that promote increasing returns of scale; or we can build in factors such as inducements to stay with a firm (loyalty bonuses). These things alter the slope of the straight-line plot; but the line stays straight.

No one is claiming that this is a perfect model of firm growth. It neglects all manner of important things, such as management structure, product specialization, and so on. Yet it seems that the model can make realistic predictions about the statistical properties of firms. How can that be? Axtell suspects that there might be a kind of universality at play here which parallels that in physics, whereby the overall behavior is insensitive to the details. He sees the origin of this universality in the fact that regardless of the relative weights of the different factors that influence an individual's decision, the actual options are quite constrained. Basically, an agent can do one of three things: join a firm, leave a firm, or join a rival firm. Tweaking the rules does not alter these choices. As long as these options, and the underlying motivations behind them, remain, the market will acquire certain collective properties regardless of how we specify the details.

RISE AND FALL

As far as one can tell from a comparison with real-world data, this looks like a good description of the way firms develop. But the model doesn't just allow us to predict these broad, cold (if important) statistical measures of firm growth. Because it is built from the ground up through interacting agents, it provides us with a history—in fact, a series of histories, one for each time we run the model on the computer. We can't expect any of these histories to mimic exactly what took place, say, in U.S. manufacturing in the 1950s. But we can search them for typical characteristics of the life cycle of firms, the careers of individual agents, and so forth.

The first thing to notice is that most firms are ephemeral. It is not immediately obvious that the real business world is like this, because we notice the firms that last, such as General Motors and Ford. But most do indeed go under (or get taken over), on a relatively short timescale. Of the largest five thousand U.S. firms operating in 1982, for example, only 35 percent still existed as independent entities in 1996. There is a high turnover of companies, which many economic theories of the firm do not acknowledge.

So why do firms fail? In Axtell's model there is a typical trajectory. First, a new firm grows more or less exponentially over time as increasing returns cause workers to flock to it. But at some point the firm reaches its peak, after which collapse is usually sudden and catastrophic. Reduced to a tiny fraction of its former size, the firm struggles on for a while with a handful of determined workers before eventually vanishing (Figure 11.5).

This collapse is a consequence of the firm's own success. Once it grows big enough, it becomes a haven for free riders who capitalize on the efforts of others. So the firm gradually becomes riddled with slackers, until suddenly the other workers decide they have had enough and jump ship. (Note that firms fail in this model because the workers leave for better jobs, not because the market for their products disappears or because there is a terrible warehouse fire, or any other reason. The failure is self-induced.) Tellingly, just before a firm collapses, the average effort of its workers plummets to zero.

Agents can't always do as well as they would like—their utility has ups and downs. But on the whole, the utility of agents in large companies rarely

11.5   The simulated firms have a typical history: exponential growth followed by sudden decline and gradual dwindling. Here is one example.

drops below the average for the whole system; big companies prove to be pretty good buffers against misfortune. Only occasionally, and transiently, do agents in these firms experience below-average utility, and that is because they sometimes get stuck in an ailing firm with no better options in sight.

If we believe that such a simplified model can tell us anything at all about the real world, then we learn some revealing things about firms. First, they are not maximizers. Firms as a whole maximize neither profit nor overall utility (as conventional theories would have us believe). Individual agents *do* try to maximize their utility, but this does not induce such behavior in the group as a whole.

The firms that do best are not those that aim to make the most profit. Rather, longevity in a company stems from being able to attract and retain productive workers. A firm fails not when its profit margins are eroded but when it is infiltrated by slackers.

The notion that putting profits first does not make a firm successful should not come as a surprise to those in the business community, although they have sometimes been reluctant to acknowledge it. Some market fundamentalists regard profit maximization not just as a principle of sound management but as a social obligation: it is the notorious "greed is good" paradigm. But as the British economist John Kay points out, it simply doesn't work. If the employees suffer from the profit motive, so does the firm: "The piece-rate systems of car factories were abandoned because they destroyed social relationships in the workplace, provoked endless negotiation and confrontation, and established a working environment in which no one cared about the quality of the product."[9]

Axtell's model does not allow for these complexities of worker interactions; but to the extent that it *does* imbue its agents with a kind of free will (how hard to work, whether to stay or quit), this worker choice comes immediately into play to determine the company's success or failure.

Axtell feels that traditional microeconomic theories of the firm have found themselves in a blind alley because of their neglect of this dynamic, ever-changing face of the labor market. The tendency, he says, has been to develop models that assume uniform intentions on the part of workers, and then to look for stable, equilibrium configurations in which each of them is happily in one firm or another. These theories, says Axtell, begin innocuously enough, with purposive agents in strategic environments of one kind or another, notionally similar to some known organization form (e.g., hierarchy). They then go on to derive the performance of the resulting firms in response to strategic rivals, uncertainty, information processing constraints, and so on. But these derivations are almost everywhere characterized by equilibrium theorizing, that is, inter-firm stationarity is seen as the result of intra-firm equilibrium and thus the homogeneity assumption is manifest.[10]

This quest for equilibrium, reminiscent of that of the founders of statistical mechanics, is transparently misguided in today's era of job-hopping. But even in former times, the idea of a "job for life" gave a misleading impression of the labor market, afflicted as it was by just as many unpredictable fluctuations and crashes. A theory of the firm must necessarily be a nonequilibrium theory in which the future is never certain.

By looking at real economic data, we can see that there most probably *are* general laws describing how firms develop and grow. Models like Robert Axtell's give us some hope that we can understand where these laws come from. But ultimately we must ask another question: What do we want? As is ever the case with a statistical social physics, we cannot expect or even hope that it will show us how to predict or control the fine details, such as whether Brian Keylock's butcher shop can survive when a big supermarket chain moves into Brecon. This is one of the major limitations of the approach. But we can perhaps hope to identify conditions under which shops like Brian's might have a fighting chance.

TWELVE

JOIN THE CLUB

ALLIANCES IN BUSINESS AND POLITICS

The science of society would have attained a very high point of perfection if it enabled us, in any given condition of social affairs—in the condition, for instance, of Europe or of any European country at the present time—to understand by what causes it had, in any and every particular, been made what it was.

—John Stuart Mill (1843)

The call for anti-fascist unity was, in some ways, likely to win the most immediate response, since fascism publicly treated liberals of various kinds, socialists and communists, any kind of democratic regimes and soviet regimes as enemies to be equally destroyed. In the old English phrase, they had all to hang together if they did not want to hang separately.

—Eric Hobsbawm (1994)

Why do wars and revolutions happen? We do not know. We know only that to produce the one or the other men form themselves into a certain combination in which all take part; and we say that this is the nature of men, that this is a law.

—Leo Tolstoy (1869)

I've written this book on an Apple Macintosh. That might mark me out, I suppose, as a radical, an iconoclast, a free thinker. At least, this is what Mac users like to think. Those who choose Microsoft-based PCs might instead see us as mugs who've picked the losing side.

Sometimes the marketplace presents a choice almost too dizzying to contemplate. (Ordering a coffee in the United States is always a brain-taxing task.) Sometimes, in contrast, there seems to be no option at all. (For standard mail, for example, there is only one postal service.) But sometimes the system bifurcates, and then the decision becomes momentous: Will you take the right-hand path or the left, knowing that there is no going back?

Life confronts us with many binary choices. We are then like a magnetic atom wondering which way to point our compass-needle spin, and we are likewise inclined to be influenced by the choices of those around us. The result can be a rich and subtle interplay of interactions, leading to outcomes that are impossible to predict without considering the dynamic of the group as a whole.

In this chapter and the next two I shall consider various social situations in which we are presented with stark and mutually exclusive choices. I begin here with a special class of situations commonly met with in big business and politics: the formation of mergers and alliances. Companies do not always grow in the manner outlined in the previous chapter—by a steady accumulation of employees or a gradual rise in sales. Instead they may merge with a competitor in the hope of gaining a synergistic advantage. Or they might choose to collaborate while retaining separate identities.

In the ruthless world of business mergers there may be little room for small fry. Players who do not secure a place as market leaders may have no option but to add their inferior muscle to the strength of the giants: to become engulfed by, or at least conform to the ways of, the big guys.

In multiparty political systems, alliances can decide the balance of power. As the number of parties increases, it may well be that no single party can do without the support of others. Minor players suddenly find themselves with make-or-break power. This has been the case for decades in the German political system, where, for example, the Social Democratic Party formed (West German) governments in coalition with the Free Democratic Party from 1969 to 1982.

History is alive with the consequences of coalitions between politicians and nations. Would Napoleon have lost at Waterloo without Prussian intervention? Was the United States liberated not by George Washington but by a makeshift union of European powers eager to see British influence diminished in the New World? How would Athens have fared against the Persians without the aid of its longtime foe Sparta? These may seem idle questions to many historians, who as a rule are interested only in what did happen and not what *might have*. But we shall see how, with some inspiration from physics, one can inquire further about that "might have" and start to explore the contingencies of history. While there is always but a single way the cookie crumbles, we can begin to systematize, rather than just guess at, the

alternatives, and thus make "counterfactual history" a more concrete and
objective pursuit.

SETTING STANDARDS

In the pantheon of businesses there is space at the top of the pile for just a
select few, their supremacy defined by size, market share, and revenue. But
even big firms are sometimes forced to cooperate rather than compete with
their rivals, for example when new technologies raise the issue of standards
and compatibility. Unless products sold by different companies are technically compatible with one another, each company risks being excluded from
some segment of the market.

This is often perceived as a modern concern, exemplified well by the
struggle between the VHS and Betamax video systems in the 1980s. But the
issue is an old one. It has been claimed that the inefficiency of the railroad
system in the southern United States during the 1860s, owing to the existence of three different track gauges, helped to ensure the Union's victory
over the Confederacy. (In the North, gauges were already largely standardized.) During the development of the typewriter, companies were faced with
the choice of whether to conform to the QWERTY arrangement of keys or to
innovate with some new configuration. QWERTY is now a petrified accident of
history and is far from being the optimal design, because the accessibility
of the keys does not correlate with their frequency of use. Some even suggest
that QWERTY was chosen in the nineteenth century *because of* its inefficiency,
which prevented the mechanical keys becoming jammed thanks to a typist's
excessive dexterity.* When the phonograph was invented, record manufacturers had to agree on the standard rotation speeds of 16, 33, 45, and 78 revolutions per minute (and later had to decide whether it was worth persisting

---

*The standardization of railway gauges and of typewriter keyboards are classic examples of so-
called path dependence and lock-in in the evolution of economic markets. In such cases, the eventual outcome is history-dependent (see page 108). Brian Arthur has developed an agent-based
model to investigate how railway gauges became standardized. He argues that path dependence
may result in the eventual emergence of a relatively inefficient or less-than-optimal standard, challenging the conventional (neoclassical) idea in economics that market competition always selects
the best outcome.

with the first and last of these). The conflict between metric and imperial units of measurement is notorious throughout the engineering industries, and the existence of the two standards still causes difficulties on either side of the Atlantic. (NASA was, to its cost, reminded of this in 1999, when a failure to convert between the two systems of units caused an error that doomed a $200 million space mission.) National grid voltages, the size of early compact discs, color coding for computer graphics—the list goes on.

The desire for agreed standards is obvious from the consumer's point of view, but for firms the best option is less clear. If the market is fragmented into many different user groups with functionally equivalent but mutually incompatible hardware, then any firm can capture only a small part of it. (Few consumers will be prepared to buy into more than one system.) If, on the other hand, a firm decides to standardize its product's operational specifications with those of a rival, it can sell to a broader market, but at the cost of sharing it. Moreover, the capital costs of altering a manufacturing process to fit a new standard present a significant disincentive. Yet if one standard begins to get an edge over the others, it might pay to adopt this, because while that may mean compromising with one rival, the company can enjoy an advantage over several others.

In some cases a technical standard is agreed on throughout an industry or a nation, perhaps even enforced by legislation. But this is rare. More often, market forces push an industry into adopting a few competing standards, and companies or national industries align themselves with one camp or another. Commonly, the competing camps are winnowed down to just two. This can be seen as a natural outcome of two opposing tendencies. If a company *has* to join one alliance or another, it might as well be a big one—so small alliances die out. At the same time, the company does not want to enter into collaboration with a firm with which it is in fierce competition. When there are only two alliances, each company can be part of as big a group as possible while still actively opposing its main rival.

The evolution of technical standards for computer operating systems in the 1980s was a classic example. A computer's operating system is the software package that controls the hardware and the way in which information is processed and passed around; it is, if you like, the computer's native tongue. Anyone with the right technical skills can write an operating system—can invent a language—and in the early days of computer technology, that is just what happened. The computer industry could have become a Tower of Babel.

In the late 1960s, however, two computer scientists at AT&T's Bell Laboratories in New Jersey devised a sleek and elegant operating system that became known as Unix. It was hugely popular, but AT&T could not profit from it because at the time the company was prohibited by U.S. law from entering the computer business. So it decided to give away Unix at a nominal cost to all who wanted it—which proved to be almost everyone. Yet those who acquired the nearly cost-free license for Unix were also entitled to alter and improve the language. And they did just that: by the 1980s there were around 250 different versions of Unix in use, each of them incompatible with the others. It was rather like speciation in nature: populations accumulate small mutations until they become genetically incompatible and can no longer exchange (genetic) information.

The market in computer workstations, most of which ran on Unix, had grown by then to enormous proportions—by 1990 it was worth $10 billion. There was a strong motivation for different companies to standardize their Unix operating systems. And so began a battle of wills.

The first move was made in 1987, when Sun Microsystems and AT&T agreed to use the so-called Unix System V, developed at AT&T in the 1970s. AT&T would license this system to others, but now at a price. The prospect of being in thrall to AT&T forced seven of its rivals, including the Digital Equipment Corporation and IBM, to form an alliance in May 1988 called the Open Software Foundation (OSF), which intended to develop a different standardized Unix operating system. In response, AT&T and Sun formalized their own alliance in late 1988 by establishing Unix International Incorporated (UII). The industry became polarized into two camps.

As it turned out, neither was really the winner. UII disbanded in 1993, but OSF had only a short time to savor its apparent victory before it too began to fragment, because of the inability of its members to resolve their competing interests. Happily, however, a consensus began to emerge in 1994, when a consortium of manufacturers called X/Open Company (which merged with OSF the following year to form the Open Group) agreed on the so-called Single Unix Specification. Supported by governments and commercial suppliers, all eager for a standard operating system, the major computer companies began to ensure that their products conformed to this specification.

The initial growth of OSF and UII forced computer companies to choose one way or the other. How should they have decided? Can we foresee the outcome of such alliance-forming in industry?

HEAD FOR THE VALLEYS

Political scientist Robert Axelrod at the University of Michigan says that yes, we can. And he and his colleagues have a theory to support that claim. They call it landscape theory, and it is a tailor-made form of statistical physics.

The players in this game are like gas particles on the point of condensing into droplets. They are drawn to one another by a kind of attraction, yet are also kept apart by repulsions. Configurations emerge out of this push and pull in which the particle-like agents aggregate into alliances.

The condensation of particles into discrete clusters is a common phenomenon in physics: think of water vapor in the air gathering into raindrops or the ornate filigree of snowflakes. Clusters also form if a mixture of two different liquids is suddenly "quenched"—that is, plunged to a temperature at which the two are no longer miscible, so that they begin to separate out. This is common in metallurgy, where two alloyed molten metals segregate as they cool, forming little blobs of the two pure metals. The ideal final configuration of the metal atoms—the one with the lowest energy—might be a complete separation of the two types, like a layer of vinegar sitting beneath a layer of oil in salad dressing that has been left standing for too long. But small clusters of each metal may get frozen in place before they can merge into a single layer, leaving the two metals intimately interspersed.

Often the growth of these clusters is self-amplifying: the bigger they get, the faster they grow, because of the increase in the area of the surface on which a cluster accumulates more particles. Small clusters are doomed to disappear or be swallowed up. In physics, this is known as Ostwald ripening. It is, if you like, a case of the rich getting richer and the poor getting poorer; in the business world it could serve as a metaphor for takeovers and globalization.

The picture makes intuitive sense when applied to business alliances: companies are naturally drawn toward those coalitions that look set to emerge as the strongest. But there are other forces at play too. A fundamental impulse is not to align with rival firms but to oppose them and try to squeeze them out of the market. Microsoft's controversial attempts to monopolize the personal computer market in the late 1990s demonstrate this stark Darwinian law of business. Arising out of necessity, alliances between firms can oppose such a one-sided takeover.

On the one hand, then, the force of attraction between two firms might be considered to increase in proportion to their size: the bigger firm A is, the greater the inducement for firm B to aggregate with it (it makes sense, for example, for a tiny computer manufacturer to align itself with a giant such as Sun Microsystems). But the counteracting repulsion between the two firms will depend on how much antipathy exists between them, which is likely to depend on the extent to which their products and markets overlap. Two computer companies that both specialize in PCs for the U.S. domestic market will regard each other as closer rivals than they will a third company that mostly sells software or mainframes to big businesses. And rivals are rarely charitable: firm A might choose to join one alliance not only because it is big but because by making it bigger, A increases the chance that the other alliance, to which rival firm B belongs, will be overwhelmed. Even if B's alliance is bigger, A might join the other side so as not to help B's cause.

In the landscape model of alliance formation developed by Axelrod and his colleagues, each firm is therefore like a particle that has an individually tailored force of interaction with every other particle. The attractive component of the force that A exerts on B depends on how big A is. The repulsive force depends on whether B is a close or distant rival of A. The model is like van der Waals's fluid, except that each particle is unique and there is typically only a handful of them. The principle that governs their final configuration is the same as it is in traditional statistical mechanics: What is the most stable way to arrange them? In other words, what is the equilibrium state?

To find this state, Axelrod and his colleagues defined a kind of "total energy" for a group of firms, which they calculated by summing up all the forces of attraction and repulsion between each pair when the firms are aligned in various coalitions. If close rivals are clustered together in the same alliance, the total energy is relatively high because of the firms' mutual repulsion. A better, more stable arrangement puts such competitors in different camps. In the lowest-energy, equilibrium configuration, no firm can bring about any further stabilization by switching from one camp to another. This is a Nash equilibrium (page 263), and it is the partitioning that we should expect to find in reality.

This is not to say that everyone is happy in the lowest-energy configuration. Indeed, almost inevitably some agents will not be. There are usually only two big alliances; yet all firms consider others to be rivals to a greater or lesser degree. So they will all have to share a bed with some rivals— perhaps even with close ones. There is no other option, other than the decidedly unwise course of going it alone. Thucydides understood in the fifth century B.C. how the threat of annihilation makes for unlikely bedfellows: "Mutual fear is the only solid basis of alliance."[1]

The most stable state, then, inevitably contains some "frustration" of each firm's desires. I use this term advisedly, for it is drawn from an analogous situation in physics. In the Ising model of magnetism, every magnetic atom sits on a regular lattice and points its needle (spin) in one direction or the other. In a so-called ferromagnet such as iron, the most stable state is the one in which all spins point in the same direction. But in some magnetic materials, the interactions are such that neighboring atoms prefer not to align their spins but to point them in opposite directions. These are called antiferromagnets.

If the spins are arrayed on a square lattice, this requirement of opposed alignments can be satisfied (Figure 12.1a). But if the spins are instead placed on a triangular lattice, complete satisfaction is no longer possible. Arranged thus, the spins can then be regarded as being grouped in threes, with each member of the trio equidistant from the other two (Figure 12.1b). Any two of them can have their spins in opposing directions, but then the third must be aligned with one or the other of them. This is called frustration: there is no way to satisfy the conflicting demands all at once.

It means that, for an Ising antiferromagnet on a triangular lattice, there is no unique and clearly defined most stable state: there is always some degree of imperfection or disorder in the orientation of spins, no matter how we align them. Such a system is called a *spin glass*.*

Instead of possessing a single equilibrium state, a spin glass has a mul-

---

*The word *glass* is used here by analogy with glassy materials, in which, unlike crystals, the positions of the atoms are not perfectly ordered and regular.

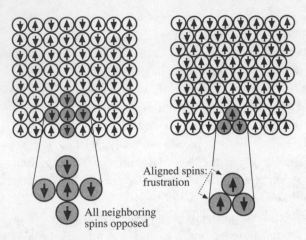

12.1 In an Ising anti-ferromagnet, the spins of neighboring atoms point in opposite directions (a). If the atoms are arranged on a triangular rather than a square lattice, it is impossible for each atom to orient its spin to oppose those of all its neighbors: there is *frustration* (b).

Aligned spins: frustration

All neighboring spins opposed

a   b

titude of different spin arrangements that all have very similar energies. One way of depicting this is to invoke the concept of an *energy landscape*: a map of all possible arrangements and their respective energies. We can see how spin configurations might be represented graphically by analogy with a chess game. The beauty, the art, and the enduring appeal of chess lie in the fact that, on an eight-by-eight grid, with sixteen pieces on each side, there is an astronomical number of different possible arrangements and ways of moving between them—so much so that even the world's most powerful computers cannot evaluate the best move by enumerating all possible subsequent outcomes. Yet, in principle, we could systematically work our way through every configuration of the pieces that can arise in the course of a game, and each of these configurations can be plotted as a point on a graph. For example, one graphical axis could represent the position of the white queen, running from 1 to 64, with each square assigned a number. (To be exhaustive, we might go from 0 to 64, with zero denoting absence from the board.) A second axis could represent the position of the left-hand red bishop. Then a configuration in which the white queen is on square 5 and the red bishop on square 42 is denoted as a point on this 64 × 64 graph (Figure 12.2a).

Of course, this isn't enough: What about all the other pieces? To fully specify the state of the board, we need a graph with 32 axes, one for each piece. Since we run out of spatial dimensions after three, we can't draw such a graph. But there is nothing in principle to prevent us from thinking in terms of such a multidimensional cube, as long as we don't worry too much about being able to visualize it. Every grid point on this hypergraph corresponds to a particular arrangement of all the chess pieces.

The same is true for the arrangements of spins in a spin glass, or indeed in any Ising model of a magnet. Each configuration of all those spins—whether each of them points "up" or "down"—can be represented as a grid point on a "hypergraph" with as many dimensions as there are spins. And each of those configurations has a particular total energy, which is calculated by adding up all the different interactions between pairs of spins—some favorable, some not.

Now we can add one further axis to the graph to denote these energy to-

12.2 Configurations of a chess game can be represented by points on a graph. A two-dimensional graph like that in *a* can show the locations of two pieces. To include all 32 pieces we would need a 32-dimensional graph. If we reduce the multidimensional graph of spin states of a spin glass to a two-dimensional surface, we can imagine an energy landscape in which the height corresponds to the energy of each respective state (*b*).

tals. It is asking a lot, admittedly, to visualize this; but the task becomes easier if we pretend that the gamut of different configurations is represented not
by a multidimensional space but just by a two-dimensional surface (Figure
12.2b). Each grid point on this surface corresponds to a particular arrangement of all the spins. And now we assign each of these points a *height* that corresponds to the energy of that configuration. Join these points up and we arrive
at the energy landscape (Figure 12.2b), a kind of topographic map that shows
how the energy varies as the configuration of spins is changed. The more stable a configuration is, the lower its energy. If a configuration cannot be altered
in any minor way (flipping just a few spins) without increasing its overall energy, then it is "locally" stable and corresponds to a depression in the landscape. Moving in any direction away from such a state means moving uphill.

For a spin glass, the energy landscape is typically very rugged. It is covered in dips, each of them corresponding to a relatively low energy state

12.3 The energy landscape of a spin glass (*a*) contains many dips and valleys, none of
which is significantly deeper than the others. Here, these dips and bumps are represented on
a simple two-dimensional graph, which could be regarded as a slice through the landscape—
like a topographic profile through a mountain range. For a ferromagnet (*b*), in contrast, there
are just two valleys of equal depth, corresponding to the equilibrium states in which the spins
are pointing all up or all down.

(Figure 12.3a). Some dips are lower than others, but none looks obviously much lower than all the others. For a normal Ising ferromagnet, meanwhile, the situation is quite different. There are just two deep valleys: one corresponding to the configuration in which all spins are "up" and the other to the configuration in which they are all "down" (Figure 12.3b). In the absence of anything to bias the choice, these two states are equivalent, with equal energies (equal "heights"). Small changes to either of these equilibrium configurations—flipping just a few spins to oppose the others—increase the total energy slightly by creating unfavorable alignments. Big changes—flipping a lot of spins—raise the energy a lot. So the landscape rises gradually and smoothly all around the lowest points.

The Ising model, remember, can also represent a fluid that can form a liquid or a gas. So we can imagine an energy landscape for a system of particles in which the grid coordinates correspond to different spatial arrangements of the particles and the contour heights correspond to the energy states determined by the forces between particles. This is the equivalent of Axelrod's landscape model of alliances, in which the "particles" are gathered into clusters that are forced to endure a certain amount of frustration because of their mutual "dislike."

BALANCE OF POWER

To find the most likely configuration of allied agents in their model, Axelrod and his colleagues surveyed the energy landscape to look for dips. Physicists do something similar when they seek the most stable arrangements of many interacting particles in computer models. Because they are typically dealing with thousands of particles, the number of possible arrangements is immense, forcing them to rely on special computational techniques to feel their way blindly around the landscape and look for the downward slopes. But if the number of agents is small, the search can be done exhaustively simply by calculating the "energies" of all possible aggregates and picking out the one with the lowest energy. This is what Axelrod's team did.

Because of the frustration in the landscape model, one might expect the energy landscape to be rugged like that of a spin glass. But the picture is simplified by the fact that the agents are not all identical, as they are in a spin glass. This imbalance means that some alliance configurations are much more stable than others. With two particularly big and powerful rival firms on the scene, for example, configurations in which all the smaller firms aggregate around one or the other of these two will tend to be much more stable than any other arrangement. So contrary to what intuition might suggest, the existence of a few big rivals can help create stable patterns of alliance. The partitioning of many competing medium-size firms, on the other hand, can be very sensitive to small changes in the forces between them, giving rise to a constant shifting of allegiance as the system jumps between one dip and another in a rugged landscape.

The big question is, does it actually work? Can the landscape model predict what happens in reality? To put it to the test, Axelrod and his colleagues used the example of Unix standardization. Would the model provide an accurate hindcast of how the industry split, given the state of play just before the alliances were launched? There were nine principal U.S. computer firms involved in the coalitions of the late 1980s, all of varying size and with different degrees of rivalry. Some, for example, were specialists in making technical workstations controlled by the Unix operating systems; others were computer-products generalists. The researchers assumed that the rivalry would be more intense, and thus the "repulsion" greater, between two specialists than between a specialist and a generalist.

There are several measures one could use for the "size" of a firm, such as the market share or the net corporate assets. Axelrod and his colleagues chose to represent size by the firms' share of the workstation market in 1987.

As for the relative strength of the forces of repulsion between close and distant rivals—well, there was no obvious and unique way of deciding that. But the researchers simply tried out a range of plausible values. They found that, except at extreme ends of this range, the outcome was not greatly affected by the precise strength of the forces.

They assumed that just two alliances would form. Who joins with whom? There are 256 possible ways of dividing nine firms between two camps. Yet the calculations of the energy landscape showed that in general there were just two stable configurations, and the one with the lowest "energy" was a very close match to the actual split between OSF and UII. In this configuration, only one company (IBM) was placed in the wrong camp ("UII" rather than "OSF"). As expected, the two alliances had similar sizes in both the stable configurations. Since the probability of getting this close to the historical reality by pure chance is about one in fifteen, it looks as though the landscape model does a good job.* Had it been available to the companies concerned in 1988, they could have used it as a fair guide for predicting how things would turn out, and for helping them decide which way to jump.

Industrial decisions like this are often made on the basis of all manner of long-term forecasts and cost-benefit analyses. But the landscape model invokes nothing of the sort. Instead, firms act on a decidedly myopic vision: in effect, they simply look at each competitor in turn and ask, "How do I feel about this firm?" And this feeling is assumed to be uninfluenced by the company the firm keeps: the interaction between A and B is the same irrespective of whether or not B has joined forces with C.† Moreover, the agents are assumed to converge on their final configuration by a series of small, independent steps, each of which takes the whole group further down the slope toward equilibrium. In the words of Axelrod and his colleague Scott Bennett, "The idea of descent need not be justified by an appeal to far-sighted rational decision making, but can easily be the result of a process in which each actor responds to the current situation in a short-sighted attempt to achieve local improvement."[2]

In effect, unless they change their attitudes toward their rivals, the firms are at the mercy of an inevitability.

It is with this in mind that we turn to a more dramatic and more far-reaching reason to conclude that it is worthwhile to search the landscape to see what the future holds.

---

*There is an important proviso, however. French physicists Razvan Florian and Serge Galam have shown that if the model is not constrained to produce just two alliances but is left to form however many clusters it wants from the nine companies, neither the configurations identified by Axelrod's group nor the one seen in reality emerges as a clear contender for the best arrangement. Instead, the average number of coalitions formed is between six and seven, and those arrangements aren't strongly preferred over a situation with no alliances at all. The unresolved question is then whether there are good arguments for insisting that scenarios like this will indeed tend to produce just two coalitions, as I discuss on page 275. Certainly, the model itself does not seem to insist on that.

†This needs to be stated quite carefully. The incentive for firm A to join an alliance containing firm B depends on the summed interactions of A with each firm in B's alliance. In this sense, whether or not A and B join forces depends on the company B keeps. But the interaction between A and B themselves is not affected by whatever alliance B is in.

EUROPE DIVIDED

If anyone has ever proposed that chess might be played as a game with more than two sets of pieces, it has never caught on. Neither has a version of football evolved in which more than two teams occupy the field. We do not play three-way tennis. This must be more than a matter of the limits to physical and mental coordination, for we all know what tends to happen in competitions between three players: sooner or later two of them will gang up to eliminate the third. There is something inherently unstable about competitions with more than two main players, which is one of the complaints often directed at multiparty politics. Even in the U.K., where the Labour and Conservative parties have for decades far outweighed the Liberals, talk of coalition has frequently been in the air.

This takes on an all too deadly significance when the game is not chess or football but war. Cromwell's army represented a complex mixture of different interests, but it disintegrated into fighting factions only after the common enemy, the Royalists, no longer threatened it. Twice in the twentieth century, Europe's many disparate nation-states formed up into two big alliances that set about destroying each other. If nerves or luck had not held out, the nations of NATO and the Warsaw Pact would have done the job much more comprehensively in the decades that followed the Second World War. The political scientist Kenneth Waltz sums up the bilateral nature of warfare: "The game of power politics, if really played hard, presses the players into two rival camps, though so complicated is the business of making and maintaining alliances that the game may be played hard enough to produce that result only under the pressure of war."[3]

As I intimated earlier, the Second World War is often portrayed as an inevitable consequence of the First. An embittered Germany, resentful of the indignity of enormous reparation debts, was primed for a leader who would restore Teutonic power and pride. There is surely truth in that; but Germany did not fight alone. Italy, Hungary, and Romania joined Hitler in the Axis alliance, which for a short but dangerous time seemed poised to overwhelm the Allies. Did Europe have to split this way? Did each alliance build up through a series of independent and entirely contingent negotiations? Or were broader forces at work?

There can be few more exacting real-world tests of a model coalition formation than the onset of the Second World War. Seventeen nations were involved; there are 65,536 ways of dividing them between two camps. Of course, many of these would be historical absurdities (it was never likely that the rest of Europe would unite against Estonia), but it is by no means trivial to anticipate the alliances that emerged in the late 1930s.

Axelrod and Bennett were bold enough to confront the landscape model with this challenge. The main difficulty was finding a way to quantify the interactions between the different nations—the forces that propelled them into one or the other camp. Traditionally, political scientists have taken the grimly named "realist" view that all states view all others as potential enemies, that all "repel" one another. There is ample historical justification for this rather bleak outlook; but of course states have also tended to respond to or anticipate very particular threats—for example, economic competition or ethnic or ideological divisions. Similarly, they might be prompted to form common cause with other states on the same grounds. The political scientist Glenn Snyder observes that alliances emerge from this mix of "conflicts and commonalities."[4]

Translated into the language of physics, conflicts and commonalities become repulsions and attractions. But how does one ascribe numbers to them? Most "neorealist" political scientists have no idea. Once the problem is expressed in general terms, however, rather than according to the specifics of historical contingency, one can start to see a way forward. Axelrod and Bennett classified the interactions between each pair of states on the basis of six factors, each derived from the political, economic, and demographic situation in 1936: ethnicity, religion, territorial disputes, ideology, economy, and past history. For example, some states were predominantly Catholic in 1936, others Eastern Orthodox, others atheist. In terms of ideology, some had democratic governments, others fascist, and so forth. The researchers considered that the existence of border disputes (like that between France and Germany) contributed to repulsion, as did a recent history of armed conflict between two states.

One can tick such boxes easily enough, but it is also necessary to assign a weight to each factor. Axelrod and Bennett made the simplest possible assumption and weighted each factor equally: +1 if there was commonality, and −1 if there was difference or antagonism. This is of course entirely arbitrary, but they had to start somewhere.

As for the "size" of each nation—again, many measures are conceivable, such as population or GNP. The researchers decided to use the "national capabilities index" assigned to each nation by the Correlates of War project conducted by U.S. political scientists since the 1960s. This index attributes a degree of "power" to each nation based on six measures of demography and military and industrial strength.

Then it was simply a matter of mapping out the landscape of 65,536 points, each corresponding to a configuration of all the nations divided into two groups. These groups were taken to represent nations that, at the outbreak of hostilities, had declared war on or been invaded by one or more of those in the other group.*

On the basis of this criterion the Allied powers in reality comprised Britain, France, the Soviet Union, Czechoslovakia, Denmark, Greece, Poland, and Yugoslavia. The Axis nations were Germany, Italy, Hungary, Estonia, Finland, Latvia, Lithuania, and Romania. Portugal was an anomaly: formally neutral, but with a defense agreement with Britain, which is why it should really be included with the Allies rather than excluded altogether, like Switzerland and Sweden.

The results generated by the landscape model are remarkable. There are just two broad basins in the terrain (Figure 12.4). One of them (the deepest) corresponds to a configuration in which the alliances match the list above almost exactly: the only wrong assignments are for Portugal and Poland, both placed in the "Axis" camp. The probability of getting a match as close as this by chance is less than one in two hundred.†

The basin that surrounds this configuration in the "energy landscape" is more than twice as large as the one that leads to the other low-energy configuration, so it is clearly predicted as the most likely outcome, reached from

---

*In other words, membership in an alliance here does not imply that there was some formal agreement to fight alongside the other nations in that camp. The Baltic states "joined" the Axis camp by default when the Red Army crossed their borders in 1940, making the Soviet Union their enemy. In fact, the massacres orchestrated by the Stalinist regime reportedly left some of those in Estonia, Latvia, and Lithuania praying for a Nazi "liberation."

†Again, relaxing the restriction to just two coalitions changes these conclusions, and degrades the model's ability to match the historical picture. But for the reasons given earlier, the two-alliance condition is defensible. Moreover, without this constraint the model produces three coalitions, with the Soviet Union split from the other allies: a situation which, as I shall argue, was quite likely, had it been politically feasible.

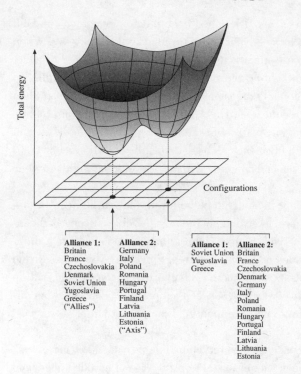

Total energy

Configurations

Alliance 1:	Alliance 2:		Alliance 1:	Alliance 2:
Britain	Germany		Soviet Union	Britain
France	Italy		Yugoslavia	France
Czechoslovakia	Poland		Greece	Czechoslovakia
Denmark	Romania			Denmark
Soviet Union	Hungary			Germany
Yugoslavia	Portugal			Italy
Greece	Finland			Poland
("Allies")	Latvia			Romania
	Lithuania			Hungary
	Estonia			Portugal
	("Axis")			Finland
				Latvia
				Lithuania
				Estonia

12.4   The two "energy" basins in the landscape of alliance formation immediately before the Second World War. The deepest basin corresponds closely to the observed split into Allied and Axis powers, with only Portugal and Poland placed in the "wrong" camp. The other basin predicts a very different history, with Europe united against the Soviet Union.

the majority of "starting points" in the landscape. But what is the alternative? Strikingly, the second basin speaks of a very different war, one in which almost all of Europe, including Britain, France, *and* Germany, is united against the Soviet Union, which is allied only with Yugoslavia and Greece. (These two countries had a recent history of antagonism with Germany, increasing their tendency to join the opposite camp.)

This may sound like a preposterous prediction. But there is nothing inherently absurd about the notion of a war, in the 1940s, between Western Europe and Stalin's expansionist state. After all, the Nazi-Soviet pact, which lasted until 1941—a ruse that in no way lessened the two states' aggressive intentions toward one another—all but made Britain and France enemies of

Stalin in the war's early years. There had been interventionist calls in both
Western countries when the Soviet Union invaded Finland in 1939. When
Britain declared war on Germany, it was only its lack of military capacity
that held it back from fighting the Soviets too. And when Churchill and
Stalin did eventually unite, the tension was immense (all the more so when
the United States entered the war). The Alliance created, in the words of historian Eric Hobsbawm, an "astonishing unity of opposites, Roosevelt and
Stalin, Churchill and the British socialists, de Gaulle and the French communists."[5] One could see cracks in the Alliance developing well before the
war ended.

Thus, if the landscape model is to be believed, the war between Britain
and Germany was the most likely but not the only possible outcome of the international situation in 1936. In this picture, the question of which "valley"
Europe entered depended on where it started. Since the Allies/Axis divide
has a much bigger basin, it was more probable that history would end up
here. But if the balance of relations between nations had been somewhat different, perhaps Britain would have identified Stalin, not Hitler, as its most
dangerous enemy.

What about the mistakes in the prediction, minor though they may seem?
Portugal is, as we've seen, an unusual case. And it is not hard to understand
how Poland could be placed in the "wrong" camp. It was antagonistic to both
of its powerful neighbors—Germany and the Soviet Union—in virtually
equal degree, and with good reason: the Soviet Union, after all, invaded
Poland just sixteen days after Germany, and Hitler and Stalin carved up the
country between them. Arguably Poland, as a twice-conquered nation, belongs in both camps, or neither.

Moreover, even this "error" is redeemable if one uses the data not from
1936 but from closer to the outbreak of war. The "size" of the nations changed
between 1936 and 1939, most dramatically in the case of Germany, as it rapidly expanded its military forces and occupied territory to its south and
southeast. On the basis of data for 1937 there are still two possible configurations, much the same as those above. But by 1938, there is only one: the
most realistic one. And by 1939, shortly before the outbreak of war, the increase in Germany's "size" has increased Poland's repulsion from it to such
an extent that Poland falls instead into the Allied camp in this single equilibrium configuration. Thus, just before war was declared (and so before the

alliances were cemented), the model predicts precisely the way things will go. This is a result that has a less than one in three thousand chance of arising fortuitously.

The skeptic might ask whether the Allies/Axis division was not a foregone conclusion, insensitive to the details of what we assume about international relations. Does the landscape model, for all its apparent success, actually do any better than a "realist" approach which assumes that everyone regards everyone else with equal suspicion? It is easy enough to run the landscape model under this set of conditions instead, whereupon the uniformity of interaction forces makes the landscape resemble that of a spin glass, with no fewer than 209 stable configurations instead of just two. None of them is an accurate picture of the historical truth.

REWRITING HISTORY

The landscape model does something more profound than provide a retrospective prediction of the course of history, impressive though that is. It gives us a picture of the historical landscape: a map of possibilities.

What are we to make of this? Many historians balk at the idea of discussing might-have-beens, what they call counterfactual histories. The task of the historian, they say, is to interpret what happened, not to construct stories about what might have but didn't. The philosopher Michael Oakeshott has pointed out, for example, the absurdity of attributing the spread of Christianity to St. Paul's escape from Damascus, as we seem forced to do the moment we start asking, "What would have happened if Paul had been caught?" "When events are treated in this manner," says Oakeshott, "they cease at once to be historical events. The result is not merely bad or doubtful history, but the complete rejection of history."[6]

In his book *Virtual History*, the British historian Niall Ferguson defends counterfactual history against such criticism. Some historians have argued that the past is an ever-branching tree, an "infinitude of Pasts, all equally valid,"[7] in the words of André Maurois—like Borges's Garden of Forking Paths. That may be true, opponents will say, but once each choice is made, the other branches stop short and we can say no more about them.

Some historians favor a determinist view which minimizes the role of chance in directing the course of world events. Modern determinists have often been inclined toward Marxism, with its tendency to invoke inevitability in human affairs. They and other "materialist" historians, taking their lead from Kant and Comte, regard history as being guided by laws analogous to those that govern the natural sciences, and so are generally opposed to counterfactual scenarios on the grounds that these laws, not fickle fortune, banish them. It might seem strange, then, to use analogies from physics to argue the case for counterfactual history. But the physics we are using here is statistical: it is concerned not with the narrative of how one thing led to another, like a succession of colliding billiard balls, but with the range of possibilities and the likelihood of each.

The landscape model promises something more concrete than flights of historical fancy. Not only might it offer an objective rationalization of the path that events took, but it can potentially draw the map of the terrain. This can help counterfactual historians to ground their discussions in the realm of the possible, and not to succumb to vague speculation. It gives concrete form to the range of influences that any historian knows must operate in shaping the course of real events. As the historian Hugh Trevor-Roper has said, "History is not merely what happened: it is what happened in the context of what might have happened."[8]

If we concede, then, that the successes of the theory are not flukes, we must grant that the landscape which provides them has some broader validity. We can talk in (somewhat) quantitative terms about worlds that might have been, and identify the factors that helped things turn out this way and not that.

An ability to do this, says Ferguson, can validate counterfactual history as a way of understanding the past. We must be concerned, he says, not with all things that *could* have happened, but with those that were most *likely*—"with possibilities which seem probable in the past": "By narrowing down the historical alternatives we consider to those which are *plausible*—and hence by replacing the enigma of chance with the calculation of *probabilities*—we solve the dilemma of choosing between a single deterministic past and an unmanageably infinite number of possible pasts."[9]

Here Ferguson seems to be implying that these "calculations" rely largely on the good judgment of the historian. Counterfactual scenarios, he says, should also take account of "those alternatives which we can show on the basis of contemporary evidence that contemporaries actually considered"[10] (such as a British assault on the Soviet Union in the 1930s). Fair enough—but how are we to assign quantitative values to any of this? Landscape the-ory shows how, in the particular case of alliance formation, it is possible to make a real calculation that produces numbers whose values, although highly approximate (and how could they be otherwise?), hold some degree of utility.

THE EDGE OF HISTORY

As war approached in the late 1930s, according to the landscape model, the European political terrain was thus divided into two realms of possibility: one in which the strongest democratic nations unite against Germany, the other in which they oppose the Soviet Union. That one scenario occurred and not the other is due to the conditions from which history began its "walk through the hills." This starting point, of course, itself follows from the his-tory of the earlier twentieth century, which in turn is shaped by the conflicts of the Victorian era. In one sense this is nothing more than a banal statement that the present depends on the past, except that we can, using this theory, imagine watching that development take place in the form of a fluctuating landscape.

As we have seen, the Ising model of a ferromagnet or a many-particle fluid also has a twin-valleyed landscape, corresponding to the stable config-urations of spin-up and spin-down states, or of liquid and gas. Which state prevails depends on which is lower in energy under the prevailing condi-tions. At the boiling or condensation temperature of a fluid, the liquid and gas "valleys" are equally deep and a transition can occur from one to the other. This is exactly what van der Waals described; couching the phase transition in terms of an energy landscape is just another way of looking at it.

It follows that something like a phase transition is possible between "al-liance energy" valleys: that it is possible, if two valleys become equal in depth, to jump from one to the other, from one alliance configuration to a to-tally different one. And just as, close to the liquid-gas transition, a small change in pressure or temperature can bring about a profound change in the state of the entire system, so too might small changes in political attitudes or in circumstances markedly alter the configuration of players, if alternative "energy" valleys are nearly matched in depth. Thus it becomes important to know not only which valley is deepest (since that is the one corresponding to the predicted outcome of pact making), but whether there are nearby valleys that are almost as deep, in which case the alliances may be prone to changing abruptly into a quite different configuration.

For a fluid of interacting particles, altering the pressure while keeping the temperature constant changes the energy landscape so as to alter the relative depths of the "liquid" and "gas" valleys. At the transition point these depths are equal. At a slightly higher pressure, the liquid becomes more stable than the gas: the gas is a *metastable* configuration (see page 163). In the same sense, the anti-Soviet alliance was metastable in 1936. We saw earlier that metastable states *can* exist if the system is prepared in the right way, even though they are constantly at risk of collapsing into the more stable state.

But if we continue to increase the pressure, the metastable gas state becomes ever less stable than the liquid state: the "gas" valley becomes shallower. At some point the metastable dip vanishes altogether: a gaslike configuration is no longer possible even in theory. This is called a spinodal point, and it represents the limit of metastability (Figure 12.5*a*).

In Axelrod's landscape model, altering the "pressure" might be equated with changes in the attitudes or the "sizes" of the various agents—that is,

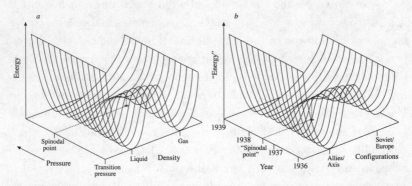

12.5   At a spinodal point, the less stable of the two possible configurations of a fluid (the metastable state) vanishes (*a*). Something analogous happens in the landscape model of national alliances between 1936 and 1939: history seems to pass through a kind of spinodal point at which the anti-Soviet alliance ceases to be viable (*b*).

changes in the strengths of the attractive and repulsive forces between them. We find that such changes alter the historical landscape between 1936 and 1939 from a two-valleyed terrain to one with a single stable state. This implies the occurrence of a kind of historical spinodal point: a set of circumstances under which the less stable alliance ceases to be viable. Between 1937 and 1938 the political landscape apparently passed through such a point (Figure 12.5*b*); after that, a war would inevitably involve the Axis and Allied coalitions.

It is not hard to find contemporary political scenarios on which to exercise the landscape theory. Indeed, Axelrod and his colleagues have used it to retrospectively "predict" that given the balance of power and international relations in 1989, just after German reunification, the desertion of the Soviet Union by its Eastern European allies was already in the cards. They predicted that only Bulgaria would stay with the former Warsaw Pact, while the others would align themselves with NATO. By the time the Soviet Union began to disintegrate, only Romania in fact remained in contradiction of this hindcast. Others have applied the landscape model to the breakup of the former Yugoslavia in the early 1990s.

Looking back is all very well, but the model might be of greatest value for looking forward. What can it teach us to expect of relations in the volatile Middle East, where, for example, Israel, Syria, Iran, and Jordan are locked into a frustrated mutual antipathy? Might religious similarities and fear of Western interference outweigh political differences in creating an alliance of Islamic states? Where would that leave Turkey? How is this picture affected by the new (and currently anarchic) political situation in Iraq?

It is neither likely nor desirable that anyone will decide international policy on the basis of a single and, let's face it, simple model. But there seems ample reason to believe that such an approach can help us foresee the general lay of the land. It might even assist in identifying the best points of leverage for engineering favorable alliances and avoiding conflict, although that is a challenge of another order: dishearteningly, most international alliances are forged in the run-up to or in the midst of war, not in order to avert it.

There are endless opportunities to explore the factors that affect the emergence of groups, coalitions, and organizations in many areas of society. It would be valuable to know whether characteristics such as race, class, religion, and ideology lend predictability to the way in which social groups (both small, such as office friendship circles, and large, such as national groups and communities) develop. Which of these characteristics tends to dominate the clustering process, for example? In organizations, it would certainly be helpful to know how best to subdivide employees. It is not always clear, for example, to which university faculty a particular department should be assigned (is geology an engineering discipline or a natural science?). If one had a means of determining how things should fall "naturally," such dilemmas could be more reliably resolved.

That we can even speak about such questions in the same breath as we speak about the fate of nations is a hint that there is something like a fundamental principle operating here. For a single model successfully to predict the allegiances of computer companies and the alliances during the descent into global war suggests that we have moved beyond the compartmentalized, case-dependent perspective of much traditional social and political science and hit a deeper seam in the order of things.

THIRTEEN

MULTITUDES IN THE VALLEY OF DECISION

COLLECTIVE INFLUENCE AND SOCIAL CHANGE

Only in a quite limited sense does the single individual create out of himself the mode of speech and of thought we attribute to him. He speaks the language of his group; he thinks in the manner in which his group thinks.

—Karl Mannheim (1936)

When I'd had my coffee this morning and went upstairs to get dressed for work, I never considered being a nudist for the day. When I got in my car to drive to work, it never crossed my mind to drive on the left. And when I joined my colleagues at lunch, I did not consider eating my salad bare-handed; without a thought, I used a fork.

—Joshua Epstein (2001)

To look closely at complex behaviors like smoking or suicide or crime is to appreciate how suggestible we are in the face of what we see and hear, and how acutely sensitive we are to even the smallest details of everyday life. That's why social change is so volatile and so often inexplicable.   —Malcolm Gladwell (2000)

Perhaps the best reporting on the U.S. presidential election in late 2000 was to be found in the satirical magazine *The Onion*, "America's finest news source":

In one of the narrowest presidential votes in U.S. history, either George W. Bush or Al Gore was elected the 43rd president of the United States Tuesday, proclaiming the win "a victory for the American people and the dawn of a bold new era in this great nation."

"My fellow Americans," a triumphant Bush or Gore told throngs of jubilant, flag-waving supporters at his campaign headquarters, "tonight, we as a nation stand on the brink of many exciting new challenges. And I stand here before you to say that I am ready to meet those challenges."

"The people have spoken," Bush or Gore continued, "and with their vote they have sent the message, loud and clear, that we are the true party of the people." With these words, the crowd of Republicans or Democrats erupted.[1]

This is no more farcical than the events that followed, as the man who would later lead the United States into two wars (and counting) came to owe his seat in the White House not to the nation's Constitution or to a majority of popular votes but to an escalation of the legal process. While the media grappled with the hitherto unknown concept of the "hanging chad" (an imperfectly punched hole on voting cards in Florida's notoriously error-prone "butterfly ballot"), the Florida Supreme Court ordered a recount of its pivotal votes. In overruling this decision by a single vote, the U.S. Supreme Court then decided the country's leader.

In December 2000, the treasured democracy of the United States fell prey to statistical variability.

While some methods of casting and counting votes are surely better than others, all incur inevitable statistical uncertainties. The result of the 2000 election was simply too close to enable that uncertainty to be discounted. When the presidency hangs on a hundred votes or so, there is no way of judging who has won that does not involve a strong element of the arbitrary. Trapped within the margin of error, we find that all outcomes are possible.

Democracy is supposed to be about making choices. The Bush vs. Gore election has highlighted how much contingency can be involved in the expression of those choices (for example, how clearly you make your mark at the polling station, and how voting ambiguities are resolved); but the principles at least are clear. All eligible voters participate in the selection of their leaders, and in theory no person's vote counts for more than any other's. In practice, each voter's power depends considerably on the details of how voting districts are defined and how the votes are counted; but the popular image of the democratic process is of millions of ballot papers accumulating on two (or more) great piles until the candidate with the highest one emerges victorious.

But leaving aside the technicalities of a balloting process, it is still not that simple. For one thing, not even the most ardent supporter of individualism could reasonably claim that our choices are truly independent. How, in a society flooded with mass advertising, can we hope to make decisions free from the influences of our environment? Today this applies as much to politics as it does to soap powders; in the 1999–2000 U.S. election campaign the Republicans and Democrats between them spent $300 million on advertising and marketing—more than the marketing budget of many multinational companies.

Few stones have been left unturned in studies of how we make our decisions, for after all, our political institutions hinge on it. And yet there is no consensus on this question, let alone any reliable means of prediction. Opinion polls led British voters to believe that the 1992 U.K. general election would be a close call: that Labour's candidate, Neil Kinnock, might narrowly end the thirteen-year reign of the Conservatives. In the event, there was no photo finish at all; the Conservative Party won a big majority, leaving the pollsters on that occasion thoroughly discredited.

There is a name for the business of election prediction: psephology. Behind that respectable-sounding word lurks a black art compared with which economic theory looks like a science of infallible accuracy. This is not necessarily the fault of psephologists, for they face human nature at its most slippery. People announce one intention in opinion polls and then go and do the opposite come polling day. Substantial changes in voters' affiliations may be triggered by last-minute events, such as a speech by a party leader. Single and sometimes quite trivial issues can mobilize large proportions of public opinion.

In the face of such psychological volatility, can we expect models of social behavior to reveal anything useful about the political decision making that characterizes any society with a degree of democracy? Economic traders may take risks, but they do not make their decisions lightly. Irresponsible choices by drivers of vehicles are strongly suppressed by heavy penalties—financial, legal, and medical. But some voters choose their leader because they like his hairstyle. Can science cope with *that*?

What physics can bring to the social science of decision making is not a fully fledged and precise theory but a deeper understanding of an ingredient that has often been ignored or clumsily handled in the past. By now we should know this ingredient well. It is the effect that one person has on another: the influence of interaction.

One of the features of collective behavior arising from local interactions is that it becomes impossible to deduce the global state of a system purely by inspecting the characteristics of its individual components. This is phys-

ical science's most important message to social science: do not be tempted
too readily into extrapolating from the psychology of the individual to the behavior of the group.

Throughout the twentieth century, sociologists have stressed that their
studies are concerned not with the individual but with the group. Yet frequently they have been able to do little more than embed their subjects within
a preexisting set of cultural norms: group behavior is postulated a priori.
Now, instead, we can begin to examine how, through the interplay of personal
choice and interpersonal exchange, such norms arise and change. We can
start to understand how a society creates its leaders, its customs, its fashions, and its problems through a mass of mutually interdependent decisions.

CAST YOUR VOTE

One of the issues that the U.S. election debacle has highlighted is the absurdity of wrangling over a few hundred votes in an election in which over
half the enfranchised population was silent. And that is hardly the worst of
it; U.S. county and city officials are typically elected by less than 20 percent
of the eligible electorate. Brazil allows no such apathy. It has compulsory
voting. In October 1998 over 100 million voters determined the Brazilian
president, Senate, and Congress. (The country's political structure is built
essentially on the U.S. model.) In addition, the voters selected a governor
and state deputies for each state in the country.

The physicists Raimundo Costa Filho and his colleagues at the Federal
University of Ceará in Brazil analyzed the election results. Since state deputies
made up the biggest group of candidates, Costa Filho and his colleagues
focused on the voting statistics for these 10,535 prospective minor officials,
looking at the proportion of the total vote that each candidate received.

What might one expect to find from such an exercise? If voting were a
purely random process—say, if the choice of each voter was determined by
the throw of a die—then we should see a Gaussian distribution. That is to
say, most candidates would receive some average proportion of votes, a few
would receive rather less, and a few rather more. The probability distribution—the number of candidates receiving some proportion p of the votes,
plotted against p—would then be a bell-shaped curve. Of course, voters
don't, on the whole, make their choices by rolling dice. But we might imag-

13.1   Voting statistics in the Brazilian elections of 1998. The number of candidates who gathered various fractions of the total votes follows a power law. The black circles show the results for São Paulo only, and the white circles are for Brazil as a whole. Both of the straight lines have essentially the same slope.

ine that the factors determining how individuals vote are so many and so varied that the result for a large number of voters is indistinguishable from their all having chosen at random.

But the researchers did not see a Gaussian distribution. Instead they found that the voting statistics fitted a power law (Figure 13.1). What is more, this particular power law seemed to have an exponent close to -1, as is seen in self-organized criticality (page 237): this means that the number of candidates receiving a fraction p of the votes is inversely proportional to p. The same power-law relationship also turned up in a state-by-state analysis of the voting patterns; it seems to be reliably reproduced in subgroups of the total electorate.

This tells us at once that the voting process does not consist of millions of independent decisions being made essentially at random.* As we saw ear-

---

*The conclusions I draw here about nonrandomness due to the interdependence of choices are in fact not at all airtight. Costa Filho and his colleagues pointed out that under certain conditions, a concatenation of independent random decisions by each voter can give rise to behavior that looks, within a certain window of parameters, like a power law. This emphasizes the point that any one data set is generally insufficient to allow us to uniquely identify the mechanism that produces it.

lier, a power law is typically a signature of some process governed by strong interactions between the "decision making" agents in the system. At the critical point of a magnet, a power-law distribution in the sizes of islands of oppositely oriented atomic spins arises from the influence of each spin on its neighbors. In a hypothetical sandpile, each grain "chooses" whether to participate in an avalanche on the basis of its interactions (via collisions and friction) with other grains. We might suspect, then, that a power law in voting statistics is telling us that voters are influenced by one another's decisions.

This is what Americo Tristão Bernardes of the Federal University of Ouro Prêto in Brazil and his co-workers believed when they set out to explain the observations by Costa Filho and his team. Bernardes and other researchers had previously devised a model of electoral voting in which each voter sways the opinions of his neighbors. The electorate is envisaged as a kind of grid of "spins," rather like the Ising model of a magnet. Each grid point represents a voter, and its "spin" may point in as many different directions as there are electoral choices. Just as there is a force that tends to align the spins of magnetic atoms, so social influence can align the opinions of neighboring voters. But this happens, said Bernardes and his colleagues, only if there is enough of a consensus: a "critical mass" of local opinion. Lone voters can't persuade their neighbors to vote the same way, but a cluster of several like-minded individuals can do so.

When they simulated this voting model on the computer, the researchers found that it produced a distribution of votes among the many candidates with precisely the power-law form seen in the real election results: even the slope of the power-law graph was the same.* Voting is, it seems, very much a group decision.

Needless to say, the idea that we are influenced by our friends, colleagues, and neighbors is hardly a revelation. Like most people, I have taken

---

*To obtain the right slope, however, Bernardes and his colleagues had to make a particular assumption about the "persuasiveness" of each candidate—that is, the way in which the ability of voters to persuade their neighbors depended on which candidate they were voting for. But they were able to do away with any such assumption by modeling the network of social interactions not as a grid but as a branching, "scale free" network of the kind discussed in Chapter 16, which seems to be a better description of real social interconnections. In this case, the model agreed with the real voting statistics without the researchers having to make any assumptions about how persuasive each candidate is.

vacations in places my friends have recommended, and I have gone to see movies and plays they have enjoyed. And we don't even need to know other people to feel a mutual influence. The reason organic produce has become popular, for example, is surely not because a significant fraction of the population independently decided that they prefer it. (There is an additional positive feedback at play here, since the law of supply and demand pushes prices down as more people buy organic goods, making them more attractive to others with marginal preferences.) There is no reason to expect our choice of political affiliation to be immune to such influences.

All the same, the notion sits uncomfortably with our sense of democracy and freedom of choice. Whatever else a democratic election is, it is apparently not a summation of so many million independent votes. The interdependence of voters' decisions means that it is far from easy to predict how public opinion will respond to particular events or inducements. There is, for instance, ample reason to suspect that small imbalances in the visibility of candidates (a result of differences in campaign funding, say) may not lead to correspondingly small differences in voter choices.

I have a strong suspicion that the outcome of the 1992 U.K. general election was at least in part a consequence of the same collective herding behavior that occurs in economic markets (page 216). Whether people confessed to one another that they could not after all bring themselves to trust Labour's Neil Kinnock or whether they simply sensed this in the prevailing mood of the nation, the large discrepancy between poll predictions and results does not seem a likely outcome of so many random independent decisions.*

Some political scientists have explored a model of political electioneering which has strong resemblances to models used in physics. It is a kind of landscape model akin to that developed by Robert Axelrod (see previous chapter), in which voters give their allegiance to one of several parties. The topography of the landscape is defined by the voters' preferences on a range of issues, and political parties are considered to rove across this landscape looking for a suitably attractive pinnacle on which to plant their flag—to erect their "platform" position on the various issues of the day. Thus, the

---

*The concerted scaremongering in the right-wing British press no doubt played a strong part too, though perhaps only in tilting the political landscape to favor one-way herding.

parties aim rather nakedly to tailor their policies so as to attract as many votes as possible, although they are at the same time conditioned by the different position in the spectrum of opinions from which they start.

This so-called spatial model of voting was first developed in the 1950s by the political scientists Anthony Downs and Duncan Black, who adapted it from a similar model of how businesses make economic decisions. The central idea, which aims to bring decision making within the embrace of quantitative, "scientific" analysis, is that each voter takes a particular position within a spectrum of views about each issue. That's to say, if we represent the range of possible opinions as a line between two extreme stances, each voter can be placed somewhere along that line.

We conventionally think about political opinions in such spatial terms. We speak of views that are on the "left," on the "right," or centrist. This terminology derives from a genuinely spatial division of the National Assembly (later the National Convention) of the French Republic just after the Revolution. The Assembly was split into two main camps: the Jacobins, supporters of Robespierre, who favored radical political and social change, and the somewhat more powerful Girondins, who were concerned to preserve the status quo. These two groups sat in the French parliament as far apart from one another as they could—the Jacobins on the left as one entered the chamber, the Girondins on the right. Those with more moderate views took their seats in the middle. Ever since then, the political left has been associated with change, the right with conservatism.

Political factions can, of course, rarely be divided along just a single axis. Elections are fought over a range of issues, many of which can be characterized by two mutually incompatible extreme views. The spatial model of voting broadens the one-dimensional picture of a political left and right into a multidimensional one in which there is an axis of opinion for each important issue. Every voter, in this model, can be placed somewhere along each of these axes. In other words, the political "space" has as many dimensions as there are significant issues, and each voter can be represented by a point somewhere in this space that indicates his or her position on each issue.

This is a rudimentary way of looking at how people define their political stance, of course. For one thing, not everyone has a precisely articulated preference on every particular electioneering topic. But the approach can be expected to capture some of the basic features of the election process. Certainly, it seems fair to suggest that political parties try to gauge the shape of the landscape of public opinion and to adopt policies they think will attract the greatest number of voters. This is not to suggest that all politicians are so cynical as to take up *any* viewpoint just because it will win votes: there are typically large regions of the landscape that are characteristic of one persuasion or another (liberal or conservative, say), and boundaries that politicians might prefer not to cross. But political parties manifestly do try to mold themselves to public opinion, a point emphasized by the recent fad for "focus groups" in liberal politics in the West. In many European countries, parties across the political spectrum have had to adjust their policies in response to strong public opinion on issues, such as immigration and crime, that have traditionally been the preserve of the political right.

The spatial model of voting has become one of the key tools for a political science that aims to understand how democracy works—and how it breaks down. It has been used, for example, to explore the factors that promote polarization or convergence of political stances, to contrast the differences between two-party and many-party systems, and to investigate the consequences of single-issue politics.

One of the most striking findings is that there is not necessarily a "best" winner in democratic elections. One might imagine that, in an ideal voting system, the winner will be the party whose policies lie closest to the preferences of the majority. But often it is impossible to determine just where that point lies in the space of opinions. Aristotle realized as much. Even in the simplest case, in which public opinion can be represented on a single axis, problems arise when the distribution of opinion has more than a single peak. Aristotle imagined a society in which there are two dominant constituencies with very different preferences: the rich and the poor. Is the ideal political leadership the one that adopts the policies of whichever of those two groups happens to have the loudest voice at any moment? Or is it better to elect a government whose policies lie in the middle between these two polarities and therefore match no one's preferences precisely? Aristotle concluded that a society like this can never be stable. For true stability, he said, it would be best to have a large "middle class" with centrist views.

Condorcet, in his pioneering treatise on the statistics of voting and public choice in 1785 (page 54), identified another problem with democracy. What, he asked, if the preference profiles are not single-peaked even for *in-*

dividuals? This might sound odd at first, for surely we always have a single preference? If, for example, we are considering how much we think the government should spend on public services, we are likely to have a certain ballpark figure in mind and look progressively less favorably on policies that propose spending increasingly more or less than that figure. But in some situations, if we can't have exactly what we want, then our second choice might be something quite different. When the United States was debating whether to intervene in the ethnic conflicts in the former Yugoslavia in the 1990s, some Americans felt that their country should either make a massive intervention that would end the fighting definitively or stay out altogether. Something in between—a peacekeeping force in which U.S. soldiers were exposed to risk but which could not end the conflict—was deemed less desirable than either extreme. Many had felt the same way about the Vietnam War two decades earlier.

Under such conditions, Condorcet concluded, majority rule (which is surely what we regard democracy as being about) may not be stable. He considered elections in which each candidate faces each of the others in turn in a one-to-one contest. In modern theories of voting, the contestant who beats (or at least ties with) all others in a series of such head-to-head contests is called the *Condorcet winner*. It is possible to show that in some elections the Condorcet winner is not necessarily the best compromise for reflecting the views of the electorate. Indeed, there are cases where the Condorcet *loser* (the candidate who loses each of those one-to-one bouts) is arguably the best overall winner.

Condorcet's voting procedure is by no means the only alternative to the principle of simple majority rule. The discipline called *choice theory* explores the consequences of different voting schemes. For example, the French political theorist Jean Charles de Borda proposed in the late eighteenth century that elections to the French Academy of Sciences should be made by the voters' giving each of the candidates a score denoting the number of candidates considered to be *less* preferable. The higher the score, the more desirable the candidate is. So the winner would be the candidate who accrues the highest total score from all the voters. This is a form of proportional representation, but it is far from ideal, since the "Borda winner" can depend on the rankings of candidates that *nobody* wants.

An ideal voting system would not be plagued by such apparent logical absurdities. Choice theorists look for a certain logical consistency in methods of voting. For example, the outcome should be *transitive*: if candidate A is preferred by the electorate over candidate B, and candidate B over candidate C, then candidate A should also be preferred over candidate C. Another desirable feature is that the outcome should not change under a contraction of the choices: if A is picked from A, B, and C, then A should also be picked from A and B alone. And so forth.

The trouble is, no voting mechanism satisfies all the logical criteria that one might reasonably demand from it. This is not because we haven't yet found the right system. It simply can't be done. That rather shocking result is proved in the "impossibility theorem" of the economist Kenneth Arrow. He argued that any truly democratic decision-making process should possess a particular set of logical features, and went on to show that the only collective-choice mechanism that satisfies all these characteristics *and* allows one set of preferences (that is, one candidate) to be unambiguously selected over the others is dictatorship: rule by a single person. Clearly, this is not a collective process at all! Arrow seemed to find himself forced, by a very different route, to the same conclusion reached by Thomas Hobbes three centuries earlier.

The implication of Arrow's paradox is that *there is no perfect alternative to dictatorship*. Either we need to accept that majority rule has some undesirable consequences (including instability) or we need to find an alternative to simple majority rule. Such alternatives have been thoroughly explored in theory, but it is very hard to find ones that are transparently fair and immune to manipulation. In other words, democracy is a very slippery concept. Western politicians and public alike frequently fail to acknowledge this crucial point: the argument for adopting a democratic system of government is not that it is the perfect or even in some vague sense the "fairest" system, but that it is (probably) the least susceptible to corruption.

And all of this, let's remember, follows from models in which each voter makes an *independent* choice. The statistical analyses of voting that physicists have conducted undermine this simplifying assumption, making the path to a workable and equitable democracy even murkier. It all drives home the message delivered by Otto von Bismarck to the Prussian Herrenhaus in 1863: "Politics is not an exact science."[2]

WHICH WAY TO TURN?

Joshua Epstein of the Brookings Institution points out that whereas many psychologists and sociologists have exercised themselves over how we make decisions, society is geared largely toward removing that need. Many social norms exist simply so that we no longer have to think about other options. If each day we had to choose afresh every aspect of our appearance, conduct, and activity, we'd never get anything done.

Thus, for example, the inhabitants of every nation have agreed to drive solely on the left-hand or the right-hand side of the road.* This agreement is enforced by law, of course; but, in general, laws simply consecrate preexisting social norms that evolved more or less spontaneously. In Britain, for example, driving on the left may be a legacy from the preference of passing an approaching horseman or carriage on the left-hand side, so that one could wield one's sword right-handed against any sudden attack. In continental Europe, in contrast, postilions (mounted riders guiding a wagon team) were mounted on the rearmost left horse and thus preferred to pass left side to left side (that is, to drive on the right). Today these conventions avoid accidents, of course, but they also free us from the need to guess, each time we get into the car, how everyone else might be driving today.†

If this seems like trivial common sense, consider how the same kind of mass conventions apply to many seemingly unimportant aspects of our lives. In Baroque England you would not think of appearing at the royal court without your wig; in the 1920s only ill-mannered and disreputable men walked the streets of New York City without a hat. There is no law (as far as I am aware) against shopping in your underwear, or shaking hands with your left hand, but few of us consider doing these things.

Epstein suggests that we structure society so as to keep the need for thought to a minimum. We are compelled to make a decision only when no

---

*Beware of exceptions! There is a 10–20 km stretch of Autoroute 20 in southwestern Montreal, Canada, on which one must drive on the left. And it was decreed around 1929 that vehicles in Savoy Court in London, a short stretch of road which gives access to the Savoy Theatre, should drive on the right so that vehicles queuing to drop people at the theater would not block access to the Savoy Hotel.

†One April Fool's Day a British newspaper reported that the U.K. was to convert to driving on the right to bring it in line with the rest of Europe. But to ease the tremendous administrative burden of this change, it would be phased in: for the first six months it would apply only to buses and taxis.

social convention prescribes it. The stronger the norm, the less we as individuals need to think about it. This urge to conform exists not just in society as a whole but within particular subgroups whose conventions, however alien to outsiders, are observed without a thought by group members.

Few if any social rules are engraved by nature; they have to be learned. That is why children and tourists are more likely than others to bump into people in the street (see footnote on page 134): they haven't yet learned on which side it is usual to pass people. As people familiarize themselves with the rules, says Epstein, they also learn *how much* to think about how to behave. Unless we are born into the aristocracy, we generally know that we have to think carefully about what to wear to a royal occasion.

Epstein proposes that this learning process involves checking what others are doing. If we find that they are all doing the same thing, it is clear what we should do. If their behavior varies, we may need to watch a larger sample before deciding what seems to be "normal." But in any event, our sample is usually limited: we don't ask the entire population of Manhattan whether it is safe to wander around the Lower East Side or how much to tip cab drivers.

To investigate how norms emerge from such a system, Epstein imagines a line of agents (individuals) arranged in a circle. Each agent makes a dual-option (binary) choice—this or that—based on a census of its neighbors. The circle isn't meant to mimic any particular social situation. It is just the simplest arrangement possible: a line, or what physicists would call a one-dimensional array. Joining the line at each end avoids anomalies at the ends, where agents would otherwise have fewer neighbors.

Individual decisions in this model are reached according to simple rules. The underlying philosophy is that agents want to do as little thinking as possible: they want to conform, but with minimum effort. This means that they look for the smallest feasible sample that will tell them how to make up their minds. Each agent updates its decision once every "round." It adds up the choices of all its neighbors within a given distance to each side, and conforms with the majority. But then it checks this decision by extending the sample by one agent in each direction, to see whether this produces the same recommendation. If it doesn't, the agent concludes that the initial sample wasn't representative. It continues to expand its field of vision until the consensus from one sample matches that of the next-largest sample.

Because these agents are lazy, they also look for ways to make the sample size smaller. Once they find a consensus view, they look to see whether they get the same consensus from a slight contraction of their field of view. If they do, they contract it further, and keep doing so until the sample is as small as possible while still reflecting the view of the "wider world." As Epstein puts it, the idea is, "When in Rome, do as the majority of Romans do— but with the smallest feasible definition of Rome."[3]

When a particular norm becomes entrenched across a wide span of the ring, the agents become small-minded: they look no further than those right next to them. You could say that they stop thinking about their choices.

Epstein finds that when preferences are initially assigned at random to a

13.2   Conformity to a norm eliminates the need to think too hard about decisions. In Joshua Epstein's model, agents placed around a ring consult their neighbors, out to a certain distance, in order to choose between two courses of action: whether, for example, to become "gray" or "white." They continue to extend their opinion-sampling range until they are confident of a local consensus. In these figures, each horizontal slice shows the affiliations and activity of each agent around the ring at each moment (so that in reality the left and right edges of the plots lie adjacent). Time advances from the top to the bottom of the plots. An initial random distribution of "gray" and "white" agents (barely visible here along the top edge of *a*) develops quickly into a stable global pattern of wide gray and white domains. Within each domain there is a clear consensus, and the sampling range for each agent dwindles to zero. This is shown in *b*, where the sampling range is indicated by a gray scale. Black regions are where this range has fallen to the minimum value (only nearest neighbors are consulted)—the agents here need no longer "think" about their choices. Only those agents near the boundaries of the gray and white domains need check much beyond their immediate neighbors in order to decide which camp to join.

large number of agents arranged in a ring, the ring becomes rapidly seg-
mented into regions where one or the other preference prevails. This is
shown in Figure 13.2a, where the two preferences are marked as white and
gray. To make the results easy to display, the ring is straightened out here
into a horizontal line, and successive time steps run down the page. Vertical
stripes in the diagram indicate regions of the ring where the agents' preferences stay fixed over time. In Figure 13.2b, a gray scale denotes the corresponding "search distances" of the agents. Away from the boundaries of the
domains, "thinking" quickly contracts.

If this system is "shocked" by briefly introducing a random element
(noise) into the decision making so that all agents make their choices purely
at random for one round, the ring settles into a new configuration (Figure
13.3). We might imagine that some new discovery or event has momentarily
shaken up people's preconceptions and forced them to rethink their position.
In a sense the outcome is not surprising, but it is a valuable reminder of the
fickleness of human nature: some people who were unthinkingly in the

13.3   The pattern of domains in Epstein's model is arbitrary yet stable. As in Figure 13.2,
this figure shows domains of "gray" and "white" agents (a) and corresponding variations in
their sampling range (b). When given a "kick" by randomizing the affiliations of agents for
several time steps (middle of plots), the model settles down into a new configuration. Agents
that were "confidently" gray before the kick (in the sense that they didn't need to "think"
about it) may be equally confident in their whiteness afterward.

"white" camp before the shake-up end up equally unthinkingly in the "gray" camp afterward.

Epstein suspects that many social norms and beliefs are determined this way. Who would doubt that the virulent opponents of the male fashions for long hair in the 1960s would have been any less ardently opposed to short hair in the seventeenth century, and for much the same reason? Epstein suggests that most American citizens believe the Earth is round not for any reasons they can articulate but because it is as much the popular received wisdom as a flat Earth was in older times.

Injecting a little random noise into this model (a little "confusion" or arbitrariness in the choices agents are making) doesn't prevent it from becoming resolved into distinct blocks of one persuasion or the other, but it makes the boundaries between the blocks more mobile. Majority views can alter over time in any region of the ring; but at any moment there are still broad regions encompassing many "minimal thinking" conformists, with the hardest thinkers prevaricating at their edges, where they sample other points of view. As a metaphor for social decision making, the model is absurdly simplistic—and irresistibly attractive.

WORLDS APART

This kind of polarization among ostensibly open-minded individuals was explored more than two decades ago by Thomas Schelling, a political scientist at the University of Maryland. In 1978, Schelling published a groundbreaking work in which he demonstrated how the collective consequences of many individual decisions can often defy intuition. Schelling's book, called *Micromotives and Macrobehavior*, is one of the cornerstones of the kind of social physics with which the current book is concerned. Schelling lacked access to the tools of modern statistical physics, but nevertheless he saw clearly how physical laws find analogies in the social sciences.

It is common in physics, for example, to speak of nonsentient systems as though they exhibit purposive behavior: to say that a soap bubble *tries* to minimize its surface area or that light *tries* to find the quickest path. What physicists mean is that the laws that govern such processes are determined by the tendency of the world to maximize or minimize some quantity. There is nothing teleological about this; it is simply akin to (indeed, formally allied to) the way in which, as we saw earlier, entropy increases in processes of change.

Of course, vehicle drivers, in contrast, really do *try* to find the quickest route through town (although they lack the apparent omniscience of light). Social behavior may also thus be guided by an impulse to minimize (effort, time, distance . . .) or maximize (profit, happiness . . .)—in short, to optimize. This was the basis of George Kingsley Zipf's "principle of least effort" (page 245).

In economics, Adam Smith's "hidden hand" is widely thought to optimize the market: to make it efficient at providing and shifting goods. "Somehow," said Schelling,

all of the activities seem to get coordinated. There's a taxi to take you to the airport. There's butter and cheese for lunch on the airplane. There are refineries to make the airplane fuel and trucks to transport it, cement for the runways, electricity for the escalators, and, most important of all, passengers who want to fly where the airplanes are going.[4]

And if economics works this way, why shouldn't similar principles apply to other human activities? Might society have some spontaneous organizational ability, unguided by laws or coercion, that allows it to run its affairs?

If economists have studied the matter for two hundred years and many of them have concluded that a comparatively unrestricted free market is often an advantageous way of letting individuals interact with each other, should we suppose that the same is true in all the rest of those social activities, the ones that do not fall under the heading of economics, in which people impinge upon one another as they go about pursuing their own interests?[5]

Here Schelling had in mind decisions about anything from when to turn on one's car taillights to whether to vaccinate our children, whether to wear safety helmets in sport, whether to carry towropes in cars, whether to break the law, where to sit in an auditorium, and whether to play golf on Fridays.

The crucial aspect of all these activities is that in conducting them we are influenced and affected by the choices other people make. Furthermore, we have to make our own decisions on the basis of imperfect knowledge about what others are doing or intend to do. This interdependence is what makes group behavior different from a trivial extrapolation of individual behavior.

Schelling's approach cried out for the tools and insights of statistical physics. In fact, he used physics-based concepts in a qualitative manner without knowing that he was doing so: he presented models that show phase transitions between different types of behavior, although he never explicitly used the term. But perhaps the most memorable and influential scenario that Schelling handed down to later generations of "micromotive" modelers is that of racial segregation in demographics.

The "multiculturalism" of Western nations was once celebrated, but increasingly it is becoming clear that behind this optimistic label often lies not integration but segregation. Neighborhoods in many big cities are demarcated by race and culture. The nonwhite inner cities become the ghettos of Baltimore, Chicago, and Los Angeles, while "white flight" creates the well-to-do suburbs. The result is all too well rehearsed: racial tension, mutual suspicion, and a recipe for explosive unrest.

It is not simply a black-and-white issue, nor white-and-brown or Hispanic. There are Chinatowns in North American cities from Toronto to Boston to San Francisco; London has its Greek and Irish neighborhoods; Grenoble has a Jewish quarter, Berlin a Turkish quarter. These ethnic neighborhoods enrich the cultures in which they are embedded and generally coexist amicably. But as Britain discovered in 2001, when riots exploded in the Asian Muslim communities of some northern cities, racism and resentment often smolder below the surface.

It is worth remembering that there is nothing "modern" about this kind of segregation. Before the dividing lines were drawn according to race, they were defined in Britain by class or, more rarely, religion. (Arguably, racial segregation is still in part a separation of rich from poor.) "A preference to mix with people similar to oneself is as old as humanity,"[6] says Paul Ormerod.

Schelling wanted to know how segregation happens. It is sometimes assumed that it reflects a high degree of racial intolerance. But it is certainly unfair to impute to all American citizens, say, an unwillingness to live next door to someone of a different race or color. On the other hand, we can imagine that, however regrettable it might seem, many people would not wish to remain in an area in which they are in a racial or cultural minority (assuming that they could easily move elsewhere).

Yet people move for all sorts of reasons, and one might suppose that the cultural stirring this generates, coupled with a reasonable degree of tolerance, should keep the population relatively homogeneous. Schelling devised a model—we'd now regard it as an interacting-agents model, like the ones we encountered in earlier chapters—which showed that, on the contrary, there is an unexpectedly strong *collective* pull toward racial segregation.

The model contained two different types of agent—two "colors," which could represent race or ethnicity or some other difference. Schelling specified the rule that in a mixed neighborhood, a family will move if more than one-third of its neighbors are of a different "color." Note that this is consistent with a desire to escape bigotry as well as to express it. But it nevertheless allows a degree of tolerance too: people don't mind if, for example, a quarter of their neighbors are of a different "color."

Schelling found that such a society quickly devolved into segregated enclaves, despite an initially uniform mixing of two types of agent (Figure 13.4). This situation has an analogy in physics: two well-mixed substances whose molecules have only a slight preference for others of their kind will gradually "phase separate" into distinct regions, like oil and vinegar separating in a salad dressing.* The segregation process is a collective effect because if one agent leaves a neighborhood, it becomes proportionately less likely that others of the same "color" will stay.

There are few better illustrations of how difficult and contentious it can be to separate the often straightforward predictions of a physical description of society from the interpretations and implications for policy makers. One way of interpreting these results is to say that we should not worry too much about racial (or cultural or class) segregation, since it is virtually inevitable. Or to look at it another way, we cannot infer that a segregated society is a highly prejudiced one (and therefore prone to confrontation). Such conclusions would certainly suit those who would rather worry about other things (U.S. senator Daniel Patrick Moynihan notoriously advised Richard Nixon in the early 1970s that race relations should be treated with "benign neglect"—a phrase understood, although apparently not intended, to endorse laissez-faire).

---

*Recent models of segregation taking an approach similar to Schelling's show more to the trained eye of a physicist. The Sugarscape model of Robert Axtell and Joshua Epstein (see the following chapter) generates segregation distributions that suggest a clear relation to critical points. Under some conditions, segregation produces domains with a "typical" size, as it does in phase separation. Under other conditions the segregated regions have a range of sizes, with no dominant scale, implying that they are nearer to a critical point. In this case the mixing is more intimate—the majority of the domains are very small, although a small proportion are rather large. So integration might be encouraged if one could find a way to make the demographic system "closer to criticality."

a   b

13.4   In Thomas Schelling's model of neighborhood interactions, the agents prefer to have a
majority of neighbors of the same color (here gray or black), but only marginally so: they will
move to free squares (white) only if more than some appreciable fraction (say, one-third) of
their neighbors are of a different color. Thus the agents have only a mild degree of prejudice.
("Color" here might represent a difference in class, race, religion, or such.) A random initial
distribution of the two kinds of agent (a) develops quickly into a highly segregated arrangement (b)—one that, at face value, might lead us to infer a much higher degree of prejudice
than individual agents display. The configuration in b here is the result of an average of only
two moves of each agent from the configuration in a. Notice that empty squares tend to appear
at the boundaries, since this alleviates the "tension" between adjacent domains. This reflects
the fact that the interfaces are relatively unstable—they have a kind of "surface tension."
(These results, based on Schelling's original model, were produced by Paul Ormerod.)

But would it not be more useful to ask how, if there is a strong likelihood
of segregation arising, we might want to respond? For example, it seems entirely possible that a separation of cultures will promote an increasing ignorance of, and thus fear of and hostility to, other ways of living—that
mild preferences could become transformed into entrenched prejudices.
(Schelling's model does not include such feedbacks on the agents' behavior,
but one can imagine ways in which this could be done.) So it might be profitable to focus available resources not on trying to suppress segregation but
on fostering close interactions between the distinct communities that result
from it.

    By the same token, we might benefit from knowing that the introduction
of choice and freedom of movement into social environments (such as
schools) that have previously had cultural mixing imposed on them is likely

to lead to rapid and extreme segregation. In policies on public services, "consumer choice" has become a mantra for some Western governments. Would they be so keen to promote it if they knew that unfettered choice begets segregation of, say, rich from poor, or of more able from less able? In other words, while this kind of modeling could be used to defend a laissez-faire attitude, a bowing to the inevitable, it need not carry that implication. Rather, it may force us to think more carefully about what kind of society we consider desirable, and help us identify realistic (as opposed to idealistic or simply naïve) means of achieving it.

Perhaps the most important general lesson of Schelling's work is that pronounced segregation does not necessarily imply a high degree of intolerance. In other words, individual tendencies *do not necessarily extrapolate to group behavior*. This cannot be emphasized enough, particularly to scientists seeking to make sociology more scientific. The Harvard biologist Edward O. Wilson argues that such efforts should take more account of the evolutionarily determined predispositions of human nature: rather than simply positing modes of individual behavior, one should look for the fundamental tendencies for which the brain has been hardwired by natural selection. This is an important objective—too many models, including many of those described in this book, begin with apparently commonsensical but ultimately rather arbitrary assumptions (or preconceptions) about human behavior. But Wilson fails to acknowledge adequately that group behavior may not be simply a scaled-up version of individual behavior. Rather, characteristics of the group appear that cannot be predicted from the nature of the brain's instincts alone. Schelling's mildly tolerant agents do not evolve into moderately mixed neighborhoods but instead display a high degree of segregation. Looked at from the top down, we might misread people's natures if we judged them on the basis of how they distribute themselves.

Michael Lind, a political scientist at the New America Foundation, puts it very aptly:

> A friend of mine who raises dogs tells me that you cannot understand them unless you have half a dozen or more. The behavior of dogs, when assembled in sufficient numbers, undergoes an astonishing change. They instinctively form a disciplined pack. Traditional political philosophers have been in the position of students of canine behavior who have observed only individual pet dogs.[7]

Today it is time to study the pack.

CRIME AND PUNISHMENT

Another reason people move is to escape crime. This can instigate an all-too-familiar downward spiral, for crime is indisputably linked to social deprivation. No one who is not a criminal wishes to live in a high-crime neighborhood (in fact, even criminals probably don't), but only those who can afford to move to more genteel areas will do so. This makes high-crime areas poorer, which brings more crime.

The equation is not at all simple, however. Some poor rural areas have lower crime rates than richer urban areas. Periods of high unemployment and lower standards of living in the past were not necessarily accompanied by higher crime rates than is the case today. We cannot say with confidence that there is a predictable relationship between social deprivation and the incidence of crime.

This makes it hard to identify the best solutions. The prevailing political fashion on both sides of the Atlantic is to be "tough on crime," with harsher penalties and longer prison sentences. This pleases certain sectors of the voting population, but there is no unambiguous evidence that it reduces crime rates. Rather, we can see it as a backlash against the more liberal attitudes of the 1970s, when crime was regarded as a disease whose cure was understanding and compassion rather than punishment. There is no reason, however, to believe that this earlier approach did any better either.

The hard-line attitude to crime control is supported by an "economic" model of criminality advanced in the 1960s by the American economist Gary Becker. He suggested that one can speak of a market for crime just as there exists a market for aluminum or bananas. The traders are the criminals, who enter into transactions (commit crimes) by making a cost-benefit analysis. If you get away with it, crime pays. If you don't, crime costs—either financially, for a minor offense, or through the curtailment of personal liberty at Her Majesty's pleasure, as we say in Britain. In Becker's model, criminals make a rational assessment of these pros and cons and decide accordingly whether or not to commit an offense.

If this is how crime happens, it stands to reason that it can be reduced by increasing the cost—that is, by imposing harsher punishments. That is a very Hobbesian outlook: man's will to power can be subverted only by a greater power, which he dare not provoke.

But many conventional criminologists regard this "rational decision" as pure fantasy. Crime, they say, is rarely undertaken after a cool weighing-up of advantages and disadvantages. Most criminals don't consider seriously the consequences of being caught. If that is so, tougher penalties have next to no deterrent effect, and one must look elsewhere for cures. Dostoevsky's Raskolnikov knew the inclemency of the penal system he faced; he even conceded that "almost all crimes are so easily discovered."[8] But this did not deter him from committing a frenzied murder for reasons that even he could barely articulate.

Dostoevsky has given us some of the most compelling insights into the criminal mind. If his portraits come anywhere close to reality (and as a convicted criminal who endured a mock execution in Siberia, he arguably knew whereof he spoke), there is no easy identification of the factors that deter and spur on the offender. But one issue that has perhaps not received the attention it deserves is social pressure: the effect of one's fellow people. Japan is by no means a country free of crime, but its streets and parks must surely derive some of their remarkable safety from the shame and social exclusion heaped upon criminals. In a culture steeped in social etiquette and peer pressure, this is perhaps a more powerful deterrent than any draconian penalties imposed by law. The same applies in some small communities in the West. If criminals know their neighbors, not only do they feel less able to depersonalize their victims, which seems to be a necessary part of most random crime, but they risk the disapprobation of everyone they subsequently encounter.

The economists Michael Campbell and Paul Ormerod have proposed a model of criminal activity that owes much to Schelling's approach, including as it does this element of interaction between its protagonists. Campbell and Ormerod assume that, simply put, there is a "cost" to crime which depends on how much it deviates from social norms. If nearly everyone on your street is a burglar, you aren't likely to risk social exclusion if you join them. Indeed, you'd be better off doing so, since otherwise you'd be in a minority of losers. If, on the other hand, everyone is a pillar of society, there is more pressure on you to abide by the law.

The choices here might be better regarded as a matter of calculated risk. One could argue that criminally inclined individuals have everything to gain by living among lawful citizens, since then they benefit from their crimes without suffering from those of their neighbors. On the other hand, that same temptation is open to everyone else. This is the kind of scenario for which game theory, described in Chapter 17, was devised. But Campbell and Ormerod take the view that social pressures make a potential criminal more likely to conform in a lawful neighborhood.

There is no reason to suppose a priori that these pressures are the major influence on the incidence of crime. Campbell and Ormerod simply ask what effect they have in societies subject to particular penal regimes and social conditions. They divide the population into three groups. The first is immune to the temptations of crime, come what may. Most women and pensioners are likely to belong to this group. The second consists of active criminals. The third represent the "floating voters," who are susceptible to becoming criminals under the right circumstances but who might instead decide to live lawfully. Individuals can switch from one group to another, and do so on the basis of a kind of peer pressure: the higher the relative proportions of each group in the population, the more likely it is that others will join it.*

Campbell and Ormerod assume a straightforward relationship between social deprivation and the propensity to become an active criminal. Hardship can be considered as a kind of overall "driving force" for criminality that subsumes a variety of social factors, such as wage rates and unemployment levels. One could imagine it being linked to a standard economic index.

On the face of it, increasing social deprivation would be expected simply to increase the proportion of active criminals in the population. But because of the interactions between individuals—the group pressure to conform— the relationship between social hardship and actual incidence of criminality is not simple. For a wide range of levels of deprivation, the model produces *two* possible levels of criminality: one high, one low (Figure 13.5*a*). In other words, there is no unique stable state for a wide range of conditions; the state of the system depends on its *history*.

If we start with a low level of crime and little deprivation (the lower curve, to the left of the graph), worsening social conditions lead to only a

---

*Changes from the nonsusceptible to the susceptible group are caused not by individuals changing their mind, which would contradict the very notion of nonsusceptibility, but by demographic shifts in the population.

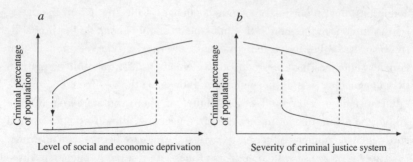

13.5   The incidence of criminality in a society can depend on social factors in ways that are not obvious or easy to predict. Ormerod and Campbell's model shows sudden jumps between "high crime" and "low crime" states that can be triggered by only small changes in social conditions. Here, changes in criminality are shown in relation to (*a*) the degree of social deprivation and (*b*) severity of punishment. In both cases there are abrupt "phase transitions" between the two different states, which happen at or before the end of each branch.

small increase in the proportion of criminals. Only at the point where this curve terminates with a sudden upturn *must* there be a shift to the higher branch. Thus, at this point only a slight worsening of deprivation can lead to an abrupt and large increase in crime. Conversely, if we begin on the upper branch (to the right of the graph), even substantial improvements in social conditions (moving to the left) lead to only a small decrease in crime—until the end of the upper branch is reached. Here there is a jump down to the lower curve: improvements that previously seemed ineffectual suddenly bring about a tremendous change for the better.

Ormerod points out that despite the evident simplicity of the model, this result has an important implication that is likely to be general: a particular social policy can have different effects on crime depending on the initial conditions. The liberal view maintains that crime can be reduced by combating deprivation. If a neighborhood is on the high-crime branch and toward the high-deprivation end of the scale, antideprivation measures alone are unlikely (according to the model) to have much effect, and conservatives would then denounce such policies as foolish. But if one were near the end of the upper branch, even small improvements could tip the system down onto the lower branch, creating a dramatic drop in crime.

The same kind of behavior emerges if one considers how, for fixed social conditions, the proportion of criminals changes if the strength of the deterrent is altered. In the model, this amounts to simply changing the probability that susceptible individuals will turn to crime, on the assumption that tougher sentences have a proportionate effect on this probability. Again, a two-branched set of solutions emerges. Increasing the deterrent could have either a small or a large effect depending on where you start from (Figure 13.5*b*). So again, depending on the circumstances, the observations could substantiate arguments either for or against the case for being tough on crime.

There is more to be gleaned from Ormerod's results. I once spent three years generating graphs that look exactly like those in Figure 13.5. But I was not researching crime; I was investigating the conditions under which a liquid and a gas interconvert. As the pressure of the fluid is altered, one finds exactly the same kind of dualism (Figure 13.6). That is to say, there are some pressures at which the fluid can adopt either the liquid or the gaseous state. The two branches show exactly the same gentle slope, ending in a little upturn or downturn, as is generated by Ormerod's model. It is clear that the crime model displays a *first-order phase transition*.

As you may recall, a phase transition happens abruptly as the driving force is changed. For a first-order transition, such as the freezing or evaporation of a liquid, the transition is marked by a sudden jump in some property of the system, such as volume or density. Yet Ormerod's model seems, at face value, to be implying that there are two such jumps, occurring at different values of the "driving force" (the severity of the justice system, or the

13.6   The first-order phase transition between a liquid and gas (evaporation or condensation) can be induced suddenly by only a slight change in pressure. Each state can persist metastably beyond the transition point, until the branch ends at a spinodal point (see page 292).

depth of social deprivation). One jump happens when we reach the end of the upper branch, the other at the end of the lower branch.

For real first-order phase transitions, however, there is only ever one equilibrium state for the system under a given set of conditions, and only one point at which the system jumps from one equilibrium state to the other, as marked on Figure 13.6. This point lies roughly midway along the region of overlap. Beyond this point (to the right of the lower [gas] branch, say) the gas is no longer the most stable state; it is not an equilibrium state. Instead, it represents a *metastable* state—provisionally stable, but only for as long as nothing happens to trigger the formation of the liquid. In theory, a metastable state will always convert to the more stable state if we wait long enough (see page 163). How long we have to wait depends on how near we are to the end of the branch. In practice, metastable states can persist for an awfully long time. For pressures beyond the end of the gas branch, however, a metastable state is no longer tenable. The end of the line, as we saw in the previous chapter, is called a spinodal point.

I am quite sure that all of this pertains to the crime model too. That is to say, the "equilibrium" switching point between the low-crime and high-crime states will lie somewhere near the middle of the region where the two branches overlap. Once we have passed this point *without* the switch having occurred, there is always the potential that it will. In practice, what destroys a metastable state is the phenomenon of *nucleation*. If a sufficiently large region of the more stable state happens to form by chance in the metastable state, it can expand rapidly to engulf the whole system. Thus, for example, in a liquid cooled below freezing point (supercooled), a tiny crystal of the solid will provide the seed from which freezing spreads throughout the liquid (see page 163). By analogy, we would expect, for example, that toward the end of the low-crime branch, a local neighborhood with a high crime rate could provide a nucleus from which criminality spreads throughout the surrounding population. Such behavior makes it even more difficult to discern cause and effect in crime policy. It can be interpreted correctly, however, once the analogy with phase transitions is recognized.

Do crime rates really show such sudden jumps? Malcolm Gladwell's book *The Tipping Point* cites the case history of New York City: a dangerous, fearsome Gotham in the 1980s; a proud, safe, and neighborly place in the late 1990s. The ghettos of Brownsville and East New York were no-go areas after sundown in the early 1990s. But within five years, murders in New York City had declined by about 64 percent and total crimes had fallen by half. The change is commonly attributed to Mayor Rudolph Guiliani's celebrated "zero tolerance" policy, but there is no agreement over precisely what this entailed or how its effects were manifested. Was it all down to better policing? A decrease in drug trading? An increase in the average age of the population? An upturn in the economy? The determined removal of graffiti from subway trains?

Gladwell argues that this abrupt improvement in crime rates, like social phenomena such as the spread of fashions and ideas and the popularity of successful books, should be regarded as a kind of epidemic. In other words, such behavioral trends spread exponentially in a society. This might be true—although, as we shall see in Chapter 16, there is more to the spread of epidemics than biologists and epidemiologists have traditionally realized. But exponential epidemics are not the only scientific model for sudden change. Phase transitions provide another, somewhat more subtle mechanism. A first-order phase transition represents a switch between two alternative global states of a system. Physics shows us that such switches do not always happen under a unique set of conditions (that is, metastable states can persist), and that it is the past history of the system that really matters. Like Gladwell's epidemics, phase transitions are big changes caused by small effects: "All of the possible reasons for why New York's crime rate dropped are changes that happened at the margin; they were incremental changes."[9]

Critical phase transitions, meanwhile, provide another prescription for rapid change: one in which two choices coalesce into one. And spinodal points create yet another scenario, one in which a first-order-like jump in the state of a system is prompted by the disappearance of a metastable state. Gladwell's "tipping point" can therefore come about in several different ways. And the distinctions matter, because they imply different things about how the change unfolds and what might be done to induce or suppress it.

THE TIES THAT BIND

Until relatively recently, marriage in Western society was not so much a choice as a social obligation. George Bernard Shaw once said, "However much we may all suffer through marriage, most of us think so little about it that we regard it as a fixed part of the order of nature, like gravitation."[10] But no longer so. The average age of marrying couples, the proportion of people who never marry, and the number of divorces all increased in Western society through the twentieth century. More and more people seem to be asking themselves: Why marry?

Shaw went on to claim, "The common notion that the existing forms of marriage are not political contrivances, but sacred obligations . . . influences, or is believed to influence, so many votes, that no Government will touch the marriage question if it can possibly help it."[11] Well, times change. Today some governments, alarmed at the declining figures and, in supposed consequence, the diminution of "family values," feel a duty to try to engineer a revitalization of marriage. But how do you do that? It is possible, in effect, to pay people to marry, through the machinery of tax incentives. It might even be possible to engineer a social climate that smiles on families welded by marriage (generally at the cost of implying disapproval of families that are not). Tampering with employment regulations might make families easier to rear (for example, enforcing the right to generous maternity and paternity leave), in the hope that raising a family, or the contemplation of it, increases the likelihood of marriage.

But in the end one will not get far without some understanding of *why* people choose to get married. Needless to say, there is no criterion that fits every couple, but presumably some reasons are more prevalent than others. If so, what are they?

Gary Becker brought an economist's eye to bear on this issue. In the late 1970s he proposed that a cohabiting couple gains in efficiency over single households by specialization, just as in the pin factory that Adam Smith invoked to advertise the benefits of a division of labor. One partner does the domestic duties; the other goes out to earn a wage. In this way the couple *maximize their utility* like dutifully rational market traders.

This may seem an awfully cold-blooded way to prise open the mystery that brings two people together. In Becker's now-classic analysis of the economics of the family (which helped win him a Nobel Prize in 1992), romance and familial love take a backseat while people coolly evaluate what they stand to gain from their spouses (potential or actual), their children, and their parents. "Participants in marriage markets," argues Becker, face a difficult choice because they "have limited information about the utility they can expect with potential mates."[12] People are compelled to marry across boundaries of race, religion, and class when "they do not expect to do better by further search and waiting."[13] Let us be thankful that Shakespeare did not have Romeo and Juliet put it that way.

It is easy to glibly deride Becker's analysis as a case of rationalism gone mad, with family decisions of all sorts hinging on complex differential equations that describe cost-benefit analyses. Applying economic ideas to family life might look like a caricature of scientism, giving rise as it does to statements like this one: "Consequently, with no costs of supervision and no fixed costs of allocating time between different sectors, the output of a multiperson household would depend only on the aggregate inputs of goods and effective time."[14]

But with this no-nonsense language Becker nevertheless does grasp the essence of many issues surrounding family life. Customs and social norms exist, after all, for a reason. They may sometimes be unfair, out of date, or oppressive, but they become established initially because a society finds (or considers) that it needs them. A cool assessment of the "utility" of potential marriage partners is quite consistent with the common practice in Japan of exchanging *curricula vitae* for inspection by the partners' families, or with the use of marriage brokers in some Jewish and Eastern European communities. The marriageable age has often been used by societies as a means to adjust birthrates, as, for example, in the practice in traditional Irish Catholic communities to marry later than is typical in England.

Moreover, Becker's analysis is able to throw considerable light on customs and social traditions whose explanation is by no means obvious. Polygyny—the taking of many wives—is, he argues, more common than polyandry (multiple husbands) in cultures throughout the world because "the marginal contribution of women to output has significantly exceeded that of men."[15] In other words, women are more useful, not just to men but to one another. In some social circumstances a household with several wives makes all the individuals better off than they would be on their own, whereas a woman typically stands to gain less by taking many husbands.

This is not to defend polygyny (or indeed to condemn it); rather, it focuses attention on the socioeconomic conditions that support it. Certain aspects of patriarchy, according to this model, may not be the preconditions but rather the consequences of a particular kind of society.

The same may be true for the inequalities that continue to exist between levels of earnings for working men and women. Men collectively have always earned more than women, and this would be the case today even if they were paid the same for doing the same job, because on average, women tend (or are, perhaps, compelled) to take lower-paid jobs. This disparity may now be narrowing, but it has been huge in the past. The natural assumption is that there has been tremendous discrimination, with men being awarded the higher-paid jobs in preference to women. That, at least, might be the concern of the liberal; the conservative might be tempted to claim that the disparity is simply a matter of biology, with men being better adapted to take on such jobs.

Becker shows that neither need be the case, provided that male-female households seek to maximize their joint utility: to get things done as efficiently as possible, and with the maximum profit. Under such circumstances, a division of labor is always mutually beneficial. The couple are bound to split the duties; and only minor distinctions, such as a slight biological bias toward maternal child care when children are young, can tip the balance. "Even small amounts of market discrimination against women or small biological differences between men and women can cause huge differences in the activities of husbands and wives."[16] We should not expect the effects to follow in the same proportions as their cause.

Thus, says Becker, social inequality can result from purely rational behavior: "An efficient division of labor is perfectly consistent with the exploitation of women by husbands and parents—a 'patrimony' system—that reduces their well-being and their command of their lives."[17] He has been criticized for seeming to imply a kind of determinism—that's just the way things are and we had better put up with it—or worse still, a justification for inequity. But Becker's work does not in fact do this. Rather, it is very much in the spirit of a "physics of society" in asserting that to understand the reasons things are the way they are, we are often ill advised to rely on intuition and preconception. We should instead seek models that illustrate how certain circumstances can arise if certain rules are followed. Far from demonstrating inevitability, such an analysis may then show us what needs to be changed if we wish to arrange things otherwise, and may save us from empirical tinkering motivated by little more than wishful thinking, the consequences of which might well prove the opposite of what we intended.

Some of Becker's conclusions make uncomfortable reading for liberals like me, such as his suggestion that welfare payments can encumber single women with larger child-care responsibilities, because it then becomes in their "rational" interest both to have more children and not to marry. But again, the implication is not that welfare payments are a bad thing but that their impact cannot be assessed in isolation from the wider socioeconomic picture.

The great value of Becker's work is to identify a fallacy in conventional "neoclassical" economic modeling, which regards birth and marriage rates largely as "given." Economists have not traditionally been interested in such things other than as background features of the economic landscape. In contrast, Becker has shown that economics both affects and is affected by these social factors. Inequality in society is not just the result of human greed but "depends on the relation between fertility and family income; on the underinvestment by poorer families in their children's human capital; on the tendency for mating to be determined by education, family background, and other characteristics; on divorce rates and the amount of child support to divorced women; and on any inequity in the distribution of bequests among children."[18]*

On the other hand, his analysis is itself "neoclassical" in its foundations, being rooted in the assumption of rational agents independently maximizing their utility. No one would seek to deny that many decisions about conception, marriage, and divorce fly in the face of what, objectively weighed, would be in a person's "best interests." There is (mercifully) no precise calculus of falling in and out of love. And most significantly, what we decide about our partners or our children depends not just on our relationship with them as individuals but on the patterns that society in general has established. A neoclassical approach can only ever tell a partial truth, because—yet again—it neglects the crucial factor: interaction.

---

*Becker uses the word *fertility*, somewhat unusually but consistently, to denote not the biological capacity to produce children but the actual (often consciously determined) number of children a woman or couple has—their "child productivity," if you will.

THE CONJUGAL CHOICE

When considered from an economic point of view, marriage is destabilized by conditions that make it less "profitable." For example, as more women find well-paid jobs available to them, they find their personal utility compromised rather than enhanced by marriage.

Changes in the gender demographics of employment have indeed been cited as a cause of the decline of marriage. Yet the corollary—that an economic boom with high employment and high wages disrupts marriage by offering more women employment—does not seem to be consistently observed. Some European countries (such as Denmark and the U.K.) have much higher divorce rates than others (such as France and Germany), despite their comparable economies. It appears that social attitudes to marriage also play a large part in determining how likely a couple is to opt for it, and to stick with it.

Thomas Schelling recognized marriage as the prototype of many social situations in which we make interdependent binary choices. Either we marry or we don't. (Or rather, at any one time we are either married or not.) A married person affects the options of nonmarried people by depleting the pool of choices, and also by exerting an influence on social norms. "The social consequences of marriage," said Schelling, "make this activity one of the central phenomena in the landscape of social science."[19]

Paul Ormerod and Michael Campbell saw this as another opportunity for modeling with interacting agents. As with crime, individuals can be regarded as more likely to make a certain choice if their fellow citizens do the same. One can then investigate how the various possible driving forces for marriage alter the proportions of married and unmarried people in an interactive society, while other factors are maintained at a constant level.

In Ormerod and Campbell's model the population is again apportioned into three groups: single, married, and divorced. Being single is, according to this definition, rather like virginity: once you've left this state, there is no going back. But one can be as Liz Taylor–esque as one likes in switching between marriage and divorce. Two general factors are assumed to influence these choices: economic incentives (wage-earning potential, tax breaks, job opportunities, and so on) and social attitudes (public disapproval of unmarried cohabitation, unfashionableness of marriage).

If the strength of social attitudes is weak, then the model produces a proportion of married people that simply increases as the economic incentive to be and stay married increases. But if social attitudes are stronger, the outcome is different (Figure 13.7a). Again we find two branches: a high-marriage and a low-marriage state of the population.

In fact, Ormerod makes the point that the two branches are linked by a continuous curve (shown by the dotted line). This seems to imply that for a range of strengths of the economic incentive, there are *three* possible states—a vertical line in the figure intersects the curve at three points. However, one can show that beyond the "turning points" of the upper and lower curves, the states represented by the dotted curve are neither stable nor metastable, but unstable; they transform instantly into something else. This is exactly what emerges from van der Waals's theory of the liquid-gas phase transition: a single, kinked curve divided into stable, metastable, and unstable portions (Figure 13.7b).

We can therefore expect all the same features in the marriage model as we saw in the crime model: an equilibrium transition between the upper and lower branches, beyond which nucleation of the more stable state within a metastable state can trigger the switch at any time. And there is again a dependence on history: economic incentives might not significantly increase the proportion of married individuals if this proportion is initially low, even

13.7   In a model that describes how the proportion of married people in a population depends on social pressures and economic incentives, identical social and economic conditions can lead to two possible outcomes: a "high marriage" and a "low marriage" society (a). These two states are connected by a continuous looping curve (dotted line), but between the two turning points of the loop, the states that emerge from the model are not stable. This is entirely analogous to the loop that connects liquid and gas states in van der Waals's theory of phase transitions (b). Van der Waals showed that there is no stable state in the dotted portion of the loop.

though a different social system with the same level of incentive might have a high incidence of marriage.

The loop in the curve appears only if the influence of social attitudes is strong enough. A three-dimensional graph that shows the dependence of marriage both on this and on economic factors therefore looks like Figure 13.8a. The lines in Figure 13.8 trace out a *surface* that develops a kink. Ormerod compares it to a folded mattress.

You may not be surprised to learn that this, too, is a familiar image to statistical physicists. Josiah Willard Gibbs owned a plaster cast of this strange wave, made for him by James Clerk Maxwell. It is the curve that shows the first-order liquid-gas transition vanishing into the critical point. The critical point itself corresponds to the place where the kink first starts (or ceases) to appear—where the upper surface is so twisted that it overhangs the lower surface. On Gibb's surface, "strength of social attitudes" is replaced by tem-

13.8   The dependence of the marriage rate on *both* of the factors in the model—social pressures and economic incentives—can be represented as a surface in a three-dimensional plot (*a*). This surface has a fold: outside the fold, a given set of social conditions allows for only one state of the system, while inside the fold there are two possible states. The fold ends at a kind of critical point. Exactly the same behavior is found in the dependence of the density of a fluid (liquid or gas) on its temperature and pressure (*b*).

perature, "economic incentives" by pressure, and "proportion of population married" by density (Figure 13.8b).

In other words, Ormerod and Campbell's marriage model has a critical point. At this point there are no longer two possible states of society, with a high and low proportion of marriage: these states merge into one. So these interacting agents, influenced by the choices made by one another, display the whole gamut of behaviors that characterize the particles of a fluid, influenced by their mutual forces of attraction and repulsion.

When willful human acts such as marriage and crime first entered the roster of social phenomena governed by satistical laws and regularities, the response was a mixture of amazement, delight, and dismay. Gary Becker's attempts to make these things accessible to economic modeling drew similar reactions. We'd be wise to anticipate the same as physics makes its mark in social science. William Newmarch, addressing the Statistical Society of London in 1860, sketched a positivist perspective no doubt buoyed by the first flowering of statistical science. Yet his words remind us that, at the very least, sound policy must, if it is to be just and effective, draw on something more than preconception and intuition:

> The rain and the sun have long passed from under the administration of magicians and fortune-tellers; religion has mostly reduced its pontiffs and priests into simple ministers with very circumscribed functions . . . and now, men are gradually finding out that all attempts at making or administering laws which do not rest upon an accurate view of the social circumstances of the case, are neither more nor less than imposture in one of its most gigantic and perilous forms.[20]

"Crime," added Newmarch, "is no longer to be repressed by mere severity." And indeed we can now see that there are reasons to expect that the effects (on average and not individual behavior) of "mere severity" may be hard to predict or assess if we fail to recognize that the choices we make, about this as about so much else, are consequent on the relations and dealings we have with our fellow people.

MINORITY RULE

Voters, clearly, hope to end up in the majority. Similarly, the kinds of social interactions we have considered regarding crime and marriage tend to lead to normative modes of behavior: if marriage is considered the "normal" thing to do, more people will do it. But there are many situations in which we would prefer to do what everyone else *doesn't*—we want to be in the minority, and the smaller the better.

Driving to work, we hope to choose a route that most others will not take, and thereby to avoid traffic jams. If we are selling our house, we might prefer to wait until it is a seller's market, when there are more prospective buyers than sellers. That way, we can demand a higher price.

The economist Brian Arthur discovered a version of this "minority problem" when he went to study complex systems at the Santa Fe Institute in New Mexico. He dubbed it the El Farol problem, and it was all about how to have a good night out.

El Farol was a bar in Santa Fe, close to where the Institute originally stood. On Thursday nights it offered Irish music—a great attraction to the Irish-born Arthur. But not to him alone. Researchers from the Institute started to flock there in such numbers that the place often became uncomfortably overcrowded. Eventually people began to stay away on Thursday nights—the music was great, but it wasn't worth the crush. And so some Irish nights were relatively quiet, allowing a lucky few to enjoy the bar as they did in the good old days.

And there's the dilemma. Do you risk going to El Farol in the hope that it will be a "good" (that is, uncrowded) night, or do you stay at home? Either way, the minority does best. If those at the bar are the minority, they have a great night out. If the stay-at-homes are the minority, they have a relaxing night in while the rest are jostled and swamped in the overcrowded bar.

How, Arthur asked rhetorically when he gave a talk on the El Farol problem in 1994, would you expect attendance at the bar to vary over time? "Will it converge," he asked, "and if so, to what? Will it become chaotic? How might predictions be arrived at?"[21]

The problem, he pointed out, is that there is no "right" choice, no "deductively rational solution." No individual has any way of knowing what is best, since that person cannot know what others will choose to do. All an individual can do is guess, on the basis of a hunch or intuition. The choice is made on the basis of *beliefs* about the likely actions of others.

To an economist like Arthur, this situation was familiar. Market traders buy and sell on the basis of beliefs about the market: whether the prevailing feeling is optimistic (so that others will buy) or pessimistic (when everyone wants to sell). But whether prices actually rise or fall depends on what everyone decides, largely in ignorance of what everyone else intends to do. Arthur pointed out that while economists have traditionally assumed that traders use *deductive rationality*—that they make decisions on the basis of solutions to well-defined problems—in fact they are generally facing a problem that is ill-defined. *There is no right answer*, except retrospectively, which is not much use. In this situation one can only apply *inductive* reasoning, based on subjectivity and experience.

"As economists, we need to pay great attention to inductive reasoning,"[22] Arthur argued. And he suggested that the El Farol problem provided the perfect model system for studying it. He devised a simple, idealized description of the situation, in which agents use a wide range of different rules of thumb for predicting what the attendance at the bar will be, and thus for deciding whether or not to go there themselves. These rules make use of past attendance figures: one agent's belief might be, for example, that "the attendance will be the same as last week," or that "the attendance will be an average of the past four weeks." Each agent applies several such rules and makes its decision on the basis of the one that experience has shown to be the most accurate.

Arthur found that this model produced constantly fluctuating attendance figures, with no obvious pattern to them. Some nights, only 30 percent of the agents would go to the bar; on other nights, 90 percent. But the attendance averaged about 60 percent, and seldom deviated from this by more than 20 percent. In other words, while the attendance neither settled down to a regular number nor showed any sign of regular rises and falls, it maintained a constant average. Arthur compared the phenomenon to a forest whose edges stay fixed while individual trees grow and die.

Why 60 percent, though? Because this is the number Arthur chose arbitrarily to designate the tolerable limit. If attendance was greater than 60 percent, the bar was overcrowded and it was better to stay at home. Thus the agents automatically "find" the optimal attendance level, on average, even though there is no rule that guarantees this.

In 1997 the physicists Damien Challet and Yi-Cheng Zhang at the University of Fribourg in Switzerland formulated a better-defined model for the El Farol problem, which became known as the minority game: a game that an agent "wins" by ending up in the minority. In Arthur's model, the strategies used by each agent to reach a decision were more or less arbitrary. In the minority game they are more systematically defined. Each agent keeps a record of which of the two available choices (to stay in or to go to the bar, say) produced the majority in the previous "rounds" of the game (that is, on previous nights). This record—which can be written as a string of binary digits, like the 1s and 0s used for computer logic—is then plugged into a particular strategy in order to reach a decision about the next round. For example, a strategy might say that if bar attenders were in the minority over the previous three rounds, the agent should go to El Farol in the next round. Again, each agent has several strategies to choose from, and at any time it uses the strategy that has so far proved the most successful.

Challet and Zhang found that the average attendance was 50 percent: half the agents elect to stay away, on average, and the other half go to the bar. (In this model, the two options have equal weighting: in essence, you just want to end up in the minority, the group containing less than 50 percent of the agents.) So at first glance it looks as though the agents do a pretty good job of organizing themselves. Even though they cannot devise any *collective* plan, the "minority" is as big as it possibly can be, on average, which is to say, essentially the same size as the "majority." But how "efficient" is the game really? Again, the attendance figures fluctuate constantly on either side of the average (Figure 13.9*a*). Each over- or undershoot means that more players *could* have "won" (been in the minority) than was actually the case. The larger the fluctuations, the more "inefficiently" the game is played.

The researchers found that the fluctuations get smaller as the strategies take into account a greater number of past rounds—that is, as the agents' memories improve, or, you might say, as they become smarter. Moreover, the efficiency of the game increases—the size of the fluctuations decreases—if agents are allowed to *evolve* in a Darwinian fashion such that the most successful ones multiply and the worst players (those that end up most often in the majority) die out (Figure 13.9*b*). In this case, the population as a whole is capable of "learning" to play well.

The minority game bears only a superficial, metaphorical resemblance to an economic market. Indeed, one can argue that traders often strive to be the

13.9   In the minority game, the proportion of agents that make each of the two available choices (here, going to El Farol or staying at home) fluctuates around 50 percent (*a*). In an ideal world, the minority would always be as big as possible—that is, as close to 50 percent as it can be. But because of the fluctuations, the minority is often smaller; the game is "inefficient." If, however, agents with poor strategies are weeded out over time in an evolutionary process, the fluctuations get progressively smaller: the efficiency of the game improves (*b*).

majority, as evidenced by the inclination toward herding behavior (see page 216). Nevertheless, it embodies enough features of real markets—competition and "selfish" behavior, the empirical testing and use of a wide range of strategies, the need to make choices on the basis of limited information—to make it highly tempting to apply the minority game to economic phenomena. Challet and Zhang have modified their model so that it mimics more closely the conditions of the marketplace, and have used it to investigate the effects of, for example, "noise traders" (who make decisions based on perceived trends in market indices) and insider trading. They find, for instance, that the range of chosen strategies narrows when there is "exploitable information" (privileged, or "inside" information) available: there then become "best" ways to play the game, which do not exist when everyone has to use the same information.

One of the more remarkable discoveries of the physics of society is that behavior that looks strangely "human" can emerge among agents that are in effect nothing but robotlike automata. We saw in Chapter 6 how such agents can appear to panic when trying to leave a crowded room. Shahar Hod and Ehud Nakar in Israel have demonstrated that the minority game can mimic another human attribute: indecision. They considered a version of the game in which, having weighed up the outcomes of past rounds, all agents use the same calculation to select their next move. But the agents don't necessarily heed this calculation: they do so with a probability that can vary between 1 (the calculation is always heeded) and 0 (the agent always does the opposite of what the calculation suggests). An agent's personal "strategy," therefore, amounts to the probability with which it follows the apparent lessons of past history. In successive rounds, each agent learns from experience to increase or decrease* this probability.

In the usual form of the game, this process leads to very decisive behavior: the agents end up polarized into extreme groups, which invariably choose either to heed or to ignore history. In other words, the strategies become either "always do what the calculation recommends" or "always do the opposite."

But Hod and Nakar found that the situation changes if the payoffs for "good" and "bad" choices are unequal—if, say, it hurts more to lose (to be in the majority) than it pays to win. This might be the case, for example, when selecting the best route to work. A bad choice, which dumps you in a traffic jam and makes you arrive late, could mean that you miss a crucial meeting or even that you get fired, while a good choice means only that the day unfurls as normal. Similarly, traders in a depressed market stand to lose more in a bad deal than they will gain in a good one.

When the outcomes have this kind of inequality—when the stakes are loaded to our disadvantage—the best strategy is to make cautious, safe choices rather than extreme ones. Hod and Nakar found that under such circumstances, agents in the minority game tend to become indecisive: rather than selecting a particular strategy invariably (that is, with probability 1 or 0), they pick one or the other option with equal probability. This generally means that the rewards over time are small, but so are the dangers. Agents who stick to extreme behaviors, on the other hand, lose more than they win on average.

This indecisive, hedging behavior may be best for each individual, but it makes the population as a whole inefficient: the average size of the winning

---

*That an agent should choose to *ignore* the lessons of history might sound odd. But there is no guarantee that the procedure an agent uses to guide future choices by past experiences will prove to be beneficial. If this procedure seems at any moment to be leading to bad recommendations, the agent will attribute less validity to it.

minority is considerably smaller than it could be. It would mean, for example, that the average number of attendees at the El Farol bar is consistently lower than its ideal capacity. Indeed, the population as a whole typically achieves a worse average than it would if each agent made its decision purely at random: attempts to make decisions by inductive reasoning become destructive to the group's overall efficiency.

The minority game is certainly psychologically naïve; in fact, it assumes virtually no individual psychology at all, beyond a crude ability to reason from experience. Yet the range of group behaviors that emerge from it are extraordinary, subtle, and often impossible to predict. We do not need to believe that we really make our decisions this way before we can acknowledge the value of the model's central message, which is surely this: that we should not be too ready to impute complex psychological motives to the decisions made by a group without first appreciating how much complexity can arise even from the most basic, stripped-down description of the process. There is little that is *obvious* about how we make our choices.

FOURTEEN

THE COLONIZATION OF CULTURE

GLOBALIZATION, DIVERSITY, AND
SYNTHETIC SOCIETIES

There are today no longer Frenchmen, Germans, Spaniards, Englishmen . . . ; there are only Europeans.   —Jean-Jacques Rousseau (1772)

They hate being Indians. They want to be Canadians and Americans . . . Nobody ever told us that all this would be coming in with TV. It's like some kind of invasion from outer space or something. First it was the government coming in here, then those oil companies and now it's TV.

—Cindy Gilday, communications director of the Dene Nation, Northwest Territories, Canada (2001)

Too many people still believe Margaret Thatcher's TINA, There Is No Alternative, whereas we should say TATA, There Are Thousands of Alternatives.

—Susan George (2002)

Among my many vivid memories of traveling in Japan is that of Tokyo subway advertisements for lavish Western-style marriages in full white bridal outfits. The modern Japanese convention, money permitting, is to have both a traditional wedding in kimono and a Western-style white wedding. "More cultural imperialism," I muttered at the time.

In the past, the spread of cultural beliefs and values has certainly been bound up with dreams of empire. Today it follows a commercial imperative. The Victorian conviction that "primitive" cultures simply had not yet found the "right" way to behave, to speak, and to worship is, one hopes, dead and buried, but it is being replaced by the multinational conviction that there are parts of the world that have yet to learn the benefits of Coca-Cola and hamburgers.

The transmission of cultural values from one society to another has surely not always been a bad thing. The Moorish architecture of southern Spain is a beautiful monument to the flow of learning from the Islamic world to the early medieval West. From the blight of slavery in the New World arose the sublime consolation of jazz music that enriched American culture. The paintings of van Gogh and Matisse betray the influence of Japanese wood-block prints; modern European dance floors throb to the rhythms of India and Latin America.

When one culture subsumes another, or the two merge into a monoculture, the world loses a degree of diversity and richness of experience. On the other hand, commonality of belief and tradition lowers the potential for conflict. A shared language lubricates communication. But the relative roles of choice and enforcement in cultural transmission are hard to distinguish. Conquering nations often adopt some of the traditions of their subjects; did not even Rome eventually abandon its gods for a Judeo-Christian monotheism? Whenever and however two cultures come together, there will be an exchange of values, of arts, sciences, and technologies, of customs and beliefs, of languages. Within these processes lies the history of the human world, from the siege of Troy to the eastward expansion of the European Union. History can never be solely about the impact of great men and women but must always also examine the interactions among vast groups of people.

Making models of how people pick up criminal habits from their neighbors or succumb to the social influences of friends and family is simply to attempt to describe a tiny microcosm of this ebb and flow of cultural and social values. Some scientists are seeking to explore a wider vista: to uncover the way in which cultural transmission shapes the demographic, political, and linguistic boundaries of our world. It is, needless to say, an awesome objective; but its beginnings can be simple. The questions these researchers ask are broad ones: What determines the diversity of a culture? Why do some minority cultures or traits survive (like the Basque and Welsh languages) while others vanish? Indeed, why do any two cultures not generally converge completely when they interact?

At the beginning of the twenty-first century, such questions have seldom been more urgent. The latest conflicts in the Balkans reflect the persistent lack of convergence in values and beliefs that made an ad hoc nation like Yugoslavia fatally fissiparous. A comparable process proved bloodless in the breakup of Czechoslovakia; but the collapse of the Soviet Union continues to be a painful affair in Chechnya and Georgia. A nation cannot survive without a sense of national identity; and that identity cannot arise without at least some degree of uniformity of culture.

These modern civil wars are at root no different from the one that impelled Hobbes to seek a calculus of social order. They explode out of unresolved conflicts whose fault lines go so deep as to resist decades or even generations of effort to heal them over. For England in the seventeenth century the line was, on the surface, drawn between those who felt that the sovereign should wield the ultimate power and those who would rather see the king as the servant of the people and their parliament. But below this dispute was the old and festering religious divide—the same wound that had bled Europe for a century. It is a wound that claims lives today on the streets of Northern Ireland.

There seems to be nothing inevitable about convergences of cultural, social, or political values. The cohesion of the United Nations has been severely tested by the differing attitudes of its member nations toward the wars in Kosovo and Iraq—if the bombing of Belgrade highlighted these divergences, the conquest of Baghdad by forces led by the United States and Britain stretched them almost to breaking point. Even the unprecedented degree of international unity over the war in Afghanistan in late 2001 may well disappear before a stable nation arises from Kabul's ashes.

Yet as fast as the world map seems to be fragmenting, we also see efforts to weave new unions. The European Union is a uniquely ambitious experiment which searches for shared economic and technical values while attempting to preserve and respect social and cultural differences. In January 2002 it embarked on its grandest experiment, with a near unification of hitherto national currencies. And concerns about the global environment, world trade, and world health have stimulated the emergence of international groups and organizations that share beliefs quite unrelated to matters of nationality.

The stability of such unions may depend in the long term on the extent to which their members can align their views and goals—on their ability to modify behavior through interaction. The resulting push and pull poses difficult questions for liberal advocates of diversity. Many of us may feel that too much homogeneity is a regrettable thing; yet we argue for "universal" human rights and expect nations to conform to certain codes of conduct—a trend that has been dubbed "liberal imperialism." We may rejoice in cultural diversity but lament the segregation that some ethnic minorities believe is necessary to avoid the erosion of their traditions and identity. If there is too sharp a division between cultures, the result is the persecution of minority groups, alienation, and unrest. All these issues point to a burning question: Can distinct cultures and beliefs coexist stably, or must one inevitably engulf the others?

CULTURE SHOCKS

A long list of factors has been advanced by social scientists to explain why some cultural differences persist while others dissolve. There seems to be a natural human tendency to seek group identity, which depends on emphasizing differences with the "outside" and can sometimes encourage the adoption of extreme views. The spread of fads and fashions may, in contrast, gain momentum from a desire to conform. Languages evolve and become differentiated partly by a seemingly inevitable process of random drift resulting from the accumulation of small changes—the equivalent of Darwinian random mutations. Linguistic and technological compatibility greatly facilitates the exchange of ideas, while geographical isolation can inhibit it.

One thing many of these considerations suggest is that the transmission of cultural ideas depends on how much common ground already exists. A shared language is a particularly strong conduit for exchange, although that does not in itself guarantee convergence—one is immediately reminded of George Bernard Shaw's apocryphal remark that Britain and America are "divided by a common language."*

Robert Axelrod has developed an interaction-based model that seeks to sketch in broad outline how cultures and customs are disseminated. The central assumption in this model is that likeness encourages more likeness: Axelrod does not worry about the mechanistic details of *how* cultural exchange happens but merely says that it becomes more likely the more it has

---

*Oscar Wilde perhaps deserves the credit: "We have really everything in common with America nowadays except, of course, language" (*The Canterville Ghost*, 1877).

already happened. There is, in other words, positive feedback in the cultural convergence of the interacting agents.

Axelrod imagines a map divided into a regular grid with an "agent" on every site. These agents are not individual people but subpopulations inhabiting a geographical region—a village perhaps, or a district. Each agent possesses a number of cultural "features": a style of pottery, for example, or a preferred method of agriculture, or a local dialect. There can be any number of such features, and in reality we would expect them to be both rather numerous and difficult to define and disentangle. Nevertheless, it seems reasonable to suppose that in some sense a culture can be identified by a carefully compiled list of such attributes.

To each cultural feature, Axelrod ascribes a certain number of variations. For example, his model might include five different languages: five different "values" of the "language feature." I shall call these different versions of a single feature traits. Axelrod assumes that any two traits of a given feature are equally different. This is clearly a simplification. For example, the technological bases of Japanese and British societies are different, but are clearly more similar than either is to that of Bhutan. Italians and Spaniards can almost understand each other's languages; both, however, are at a considerable and comparable disadvantage with German. But this is, after all, meant to be a simple, minimal model.

Each grid site in the model is labeled with a set of values (denoting different traits) for each cultural feature. Figure 14.1 shows a small part of a grid for a version of the model with five cultural features, each of which may

14.1 In Robert Axelrod's model of cultural dissemination, each site on a regular grid represents a local culture with a specific number of features (here five). Each feature may take one of ten "values" (0–9)—that is, it shows ten different traits. If two adjacent sites share one or more traits, they can converge further by aligning one of the existing dissimilar traits. One such step is shown here for the two sites indicated in gray.

take one of ten values (labeled 0–9; but remember that the numbers have no quantitative significance).

Transmission of cultural values then takes place by iterating the following procedure:

1. Pick a site at random, and select one of its adjoining neighbors at random. (There are four neighbors on a square grid.)

2. The probability of interaction between these two sites depends on their cultural similarity. If they share no features with the same value (that is, no traits), they will not interact. If they share one feature, they interact with a probability of, say, one in ten. If two features are the same, the probability might be one in five.

3. "Interaction" consists of setting one other—randomly selected—feature on the chosen site that is equal to that of the neighboring site. In other words, the cultures become even more alike.

A single step like this is depicted in Figure 14.1. Its effect, obviously, is to make the sites increasingly more similar. The result, then, might appear to be a foregone conclusion: gradual erosion of difference and spread of a monoculture.

Sometimes this is indeed observed. But not always. The model can reach stable states in which culturally different regions persist right up to a stage at which no further change is possible. If, for example, a region of sites of one culture were to appear in the middle of another culture with which it has no traits in common, the two cannot interact—so the "island" culture remains, even if the culture that surrounds it grows to dominate the rest of the grid.

The emergence of two such islands of diversity in a "global" monoculture is shown in Figure 14.2. Here the grid consists of one hundred sites, each of which (as in Figure 14.1) has five cultural features that may each take ten different values. The number of possible states for even this "simple" system is astronomical: 10 raised to the power of 500, which vastly exceeds the number of atoms in the universe. The cultural traits are initially chosen at random, so most adjacent sites share no traits; a few share just one. The similarity between neighbors is indicated by the shading of the lines dividing them: black denotes one or no common traits, white denotes all five values the same.

After thousands of steps the grid has developed several small patches in which the sites all share the same culture. Note that although all the grid

At start   After 20,000 steps   After 40,000 steps

After 80,000 steps   Final state (81,000 steps)

14.2   A simulation of cultural spread on a 10×10 grid. Each grid point is shown in white, and the heavy black and gray lines separate regions whose cultural features differ in at least one respect. The degree of difference is reflected in the shade of the line, with black representing complete difference of all five features. As the simulation advances, differences are eroded until only three cultural "domains" exist: two small ones surrounded by one large one. Further change is then impossible because the wholly different cultures can no longer interact.

sites appear as white in the figure, this does not mean that they all have the *same* culture—each separate white region bordered by black or gray lines represents a different culture. As "time" progresses in its step-by-step manner, some cultures grow while others are enveloped. Eventually a single culture covers most—but not all—of the grid.

What does this exercise tell us? We would, after all, *expect* the grid to develop greater homogeneity purely from the very way the rules are set up. But we see that this need not lead to the complete eradication of cultural diversity. Moreover, we can use the model to deduce something about the dependence of persistent diversity on initial diversity. Suppose, for example, we were to increase the number of different values a cultural feature can take—to allow, for each cultural feature, fifteen different traits instead of ten. What effect does this have on the number of stable regions (the number of "countries," local dialects, or such) in the final, unchanging form of the grid?

Well, what would we expect? It seems reasonable to imagine that by increasing the potential for diversity in this way, we will increase the actual diversity in the final map—that there will be more stable regions. And that's just what we find. With ten values per feature, the average number of stable regions on the 10×10 grid, for many model runs with different starting configurations, is 3.2. With fifteen values per feature, this average rises to 20. If, on the other hand, we reduce the number of traits to five, the average number of stable regions is just 1: we create the conditions for a monoculture.

So far, so predictable. But Alessandro Vespignani of the Abdus Salam International Centre for Theoretical Physics in Trieste, Italy, and his co-workers looked more closely at how the final diversity depends on the number of traits (call it N) in Axelrod's model. They discovered something unexpected: the switch from a monoculture (for small N) to a polyculture (for large N) is not gradual but abrupt. To illustrate this, they measured the size of the largest region as N varies. For small N, this region encompasses more or less the whole grid. At some critical value of N, the largest region becomes dramatically smaller, until it occupies an insignificant proportion of the grid and is really no different from all the other regions (Figure 14.3). This sudden change has the characteristics of a genuine phase transition between a uniform and a fragmented state, rather like the melting transition between an orderly solid and a disorderly liquid. Close to the phase transition, slight differences in the cultural diversity of a territory (that is, in N) can have a substantial effect on the number of different cultures it can support.*

The researchers also found that within the range of N where the rapid change happens—that is, within the transition region—the size distribution of the stable regions follows a power law. This is just the same behavior as is found for the size distribution of different domains (gaslike and liquidlike,

---

*Konstantin Klemm, at the Niels Bohr Institute in Copenhagen, and his co-workers have looked closely at this transition between a polyculture and a monoculture and find that it is modified by randomness. If there is some chance that any cultural trait will change unprovoked—an effect anthropologists call cultural drift—then the polycultural patchwork is at risk of collapsing into a monoculture even for large N. That's to say, the polyculture can become metastable, and a monoculture is the most stable state. So the behavior of the model depends rather subtly on the conditions under which it unfolds.

14.3   The geographical extent of the largest culture in the final, stable state of Axelrod's model depends on how many "values" each cultural feature can take. If this number (N) is small, a monoculture develops: the size of the largest culture is more or less equal to the size of the whole grid. If N is large, even the largest culture remains of insignificant size relative to the whole grid—there is a lot of cultural diversity. The changeover between these two extremes is not gradual but sudden—it has the characteristics of a phase transition. In this example, the phase transition happens at around $N=240$.

say) at a critical point (Chapter 10). Thus Axelrod's model shares several characteristics with simple physical systems of interacting particles. And these characteristics are not *obvious*: we wouldn't be able to predict them just by looking at the rules of the game.

There are more surprises. Suppose we introduce greater diversity in another way: by increasing the number of different cultural *features*. Thus, each grid site will be labeled not with a string of five numbers, but with a string of, say, ten. Will this increase the cultural variation of the final state? Not at all. For a 10×10 grid with ten features, each of which can take ten values, the average number of stable regions is 1. Remember that for five such features, it is 3.2. For fifteen traits per feature, a five-feature model gives an average of 20 regions, whereas a ten-feature model gives only 1.4 on average, and a fifteen-feature model gives just 1.2. Contrary to intuition, increasing the potential diversity in this way actually *reduces* the model's resistance to monocultural takeover.

How can this be? The answer is not hard to find. Remember that adjacent sites can interact only if they share one or more traits. These interactions promote homogenization. The greater the number of *features*, the greater the chance that one of them will match that of a neighbor, and so the greater the potential for interaction. In contrast, as the number of allowed *traits* increases for each feature, the chance of a match is diminished—just as two people are less likely to pick the same number between 1 and 15 than they are between 1 and 5. So increasing the number of versions of each cultural feature diminishes the chance of interaction. Verbal communication can be harder in India, for example, where there are many local languages, than in the United States, where just one predominates.

The actual numbers here have no significance for the real world. But the model, despite its simplicity, offers two general insights. First, it is not obvious exactly what we mean by the "complexity" of a culture. The technologically sophisticated cultures of the West arguably have greater diversity than the cultures of developing countries in terms of number of *traits*—we identify ourselves according to a vast number of highly specialized jobs, the kind of cars we drive, the many types of music we listen to, the wide spread of incomes, and so forth. But it is not clear whether, or how, real-life cultures differ in the number of *features* that characterize them: most features in Western culture, such as music, transport, art, religion, language, and so on, have direct analogues in Bangladesh and Cambodia. If a breadth of traits has different implications from a breadth of features, this distinction matters. Second, Axelrod's model tells us that we cannot necessarily rely on intuition when evaluating how easy it is for a culture to spread.

The model can also help us to explore the influence of the size of the overall territory. What happens if we expand the 10×10 grid to 100×100, or shrink it to 5×5? Intuition suggests that the number of stable regions will increase as the grid expands: if we can fit 3.2 regions into a 10×10 grid, should we not expect a hundred times that number in a 100×100 grid? Again, this commonsense thinking trips us up. In fact, the number of the stable regions decreases as the grid gets bigger. For a 12×12 grid with five features and fifteen traits per feature, there are on average 23 stable regions. For a 50×50 grid the number falls to about 6, and for 100×100 there are just 2. This contrasts, for example, with size effects on biodiversity: island species are less numerous than those on large continents because the number of evolutionary niches is smaller.

We can see in Figure 14.2 how the stable regions emerge from a kind of gradual "freezing out" of the model. The two small cultures that remain in this example are frozen in place by the impossibility of interaction with their neighbors across the border. But what if there were different, nonfrozen regions with which to interact, just beyond the edges of the grid? These might extend their influence into the 10×10 grid, rearranging its component sites and resurrecting the possibility of interaction with the "island" regions. This is just what happens in large grids: interactions go on longer, so there is less likelihood that the map will become frozen into a configuration with many remaining regions.

But the smallest grids also fall prey to monocultures, simply because, as on ocean islands, there is not enough space to accommodate much diversity. There seems to be an *optimal* territory size that will support a large number of stable regions—as though a continent has an ideal number of nation-states it can accommodate.

A map of the world is a peculiar, uneven patchwork. Most African nations are larger than most European nations; is it a coincidence that two of the smallest, Rwanda and Burundi, have recently seen some of the greatest turmoil? (Similarly, the larger nations of Western Europe have enjoyed greater stability than the smaller Eastern European countries.) And why are the West African nations smaller than those in the east and south? While one can offer contingent historical answers to these questions, Axelrod's model seems on the brink of telling us something more general about how geographically distinct regions or continents are likely to draw their national borders, and what consequences this will have.

THE LAND OF SUGAR AND SPICE

We might object that this model is too absurdly simplistic to bear any relation to things that happen in the real world. Think of what it neglects: geographical influences, the effect of technological change, mass media, different organizational and governing systems, tourism . . . One can imagine how to include some of these influences in the model (and this will probably be done in time), but the broader question remains: Given that there will always be some arbitrariness in the assumptions made, why should we have any faith at all in the conclusions reached?

It is a fair objection, and to address it Axelrod has looked at whether the general predictions of the model transfer to other cultural models erected on a different set of assumptions. To this end he has ventured into the world of Sugarscape.

Since its release in the early 1990s, the computer game SimCity has acquired cult status. The aim is to build and maintain an entire city from the ground up. You give it a power system, you feed its inhabitants, you conduct repairs and maintenance. You get to play God, only to discover what a mundane, maddeningly intricate, and yet wholly compulsive business that is.

Sugarscape is like an eighteenth-century SimCity. It was devised by Robert Axtell and Joshua Epstein of the Brookings Institution as a general-purpose testing ground for social theories. Suppose someone claims that this law or that constraint applied to a society will produce a particular result; you can simulate the process in Sugarscape and see if the prediction holds. "Our rules create the cliffs we drive off," say the two researchers. "Computational systems such as Sugarscape can offer 'headlights,' if you will, by permitting us to project, however crudely, the evolutionary consequences of certain rules."[1]

Sugarscape is another grid world—but this grid is imprinted on a doughnut-shaped object, a torus. This seemingly peculiar shape is chosen simply to eliminate boundaries, which makes computation easier and removes "edge" effects.* The grid sites are populated by "agents," but usually sparsely—there is plenty of open space in Sugarscape. The agents have one key impulse: to seek sugar. Sugarcane is distributed in some chosen manner over the grid, and the agents move toward whichever grid point in their field of vision has the most of it. There they chop it down and consume it; eventually, more sugarcane will grow.

The agents also interact with one another—and do they ever interact! Specified rules enable them to fight, trade, collaborate, exchange cultural traits, and have sex (what else do you do with your life, after all?). Cultural interactions are somewhat similar to those in Axelrod's model. The agents have a set of cultural attributes that can become equalized during interaction through a process of random selection. Often agents will form tribal alliances ("nations") in which cultural attributes coalesce according to the rule that the most common attribute is adopted by all.

---

*Sometimes one might wish to retain edge effects in models like this—for example, to gain some understanding of how cultures change at a frontier.

This is an extremely complex model, and it offers a framework for simulating all manner of scenarios. One can observe the effects of a sugar shortage—will it precipitate war or "international" cooperation? Epstein says, "We think of our model as a laboratory for social science . . . We can examine population growth and migration, famine, epidemics, economic development, trade, conflict, and other social issues."[2] The researchers can even examine evolution and how social forces shape mating patterns. Sugarscape is like a fantastic elaboration of chess, where the pieces make their own decisions and have their own desires and fears.

One scenario, for example, concerns the development of trade. Here Epstein and Axtell introduce a second resource, spice. Agents can have different appetites for sugar and spice, so some trade for one commodity and some for the other. They are programmed to obey certain rules for bargaining and exchange, from which prices emerge according to the interplay of supply and demand. If the agents are "humanized" by being given finite life spans and evolving preferences, the markets never reach equilibrium but fluctuate constantly.

Trade does some interesting things to the cultural landscape. It enables more agents to survive—individuals can buy what they need, whereas in the absence of trade they would have starved. But it also skews the distribution of wealth so that the majority of agents are poor while a small number of them accumulate great riches. Here, then, is Pareto's law of wealth (page 246), conjured up from a computer game. Some economists have argued that this emergent power-law distribution represents a kind of immutable property of a trading society. Sugarscape can help explore just how immutable it is in the face of changing social and economic conditions. The model could, for example, be used as a testing ground for ways of redistributing wealth more equitably.

Sugarscape can arguably embody the brutish world that Thomas Hobbes envisaged: it can approximate to his fearsome State of Nature. Axtell and Epstein simulate warfare by allowing one agent to take over the position occupied by another—basically to kill its opponent and reap the spoils. This might happen, for example, if the victorious agent is somehow "bigger" or more powerful than the vanquished agent. Typically that superiority is manifested in Sugarscape by a greater store of sugar, which means that the stronger agent is better fed and provisioned. In Hobbes's world each man has to look out for himself, since each is pitted against all his neighbors. Sugarscape's creators have been more concerned to simulate wars between rival groups, in which entire armies of agents do battle. In such circumstances there is safety in numbers, so warfare is conducted through group cooperation. Depending on the rules of combat, the researchers find they can generate both rapid "blitzkrieg" takeovers of territory and grinding "trench wars" of attrition.

Axtell and Epstein don't claim to have produced a model of today's industrialized society; they suggest instead that this is a new way of doing social science. The behavior that emerges from Sugarscape is sometimes extraordinarily complex, yet the generative rules guiding the behavior of agents are rather simple. Might that, they ask, also be the case in the real world? Might the enormous diversity in behavior that social science seeks to study turn out to be based on simple foundations? "Imagine," they say,

> that we had begun the entire discussion by simply running [the model] which shows a buzz of agents "hiving" the sugar mountains, and that we had then bluntly asked, "What's happening here?" Would you have guessed that the agents are all following [a simple rule]? We do not think we would have been able to divine it. But that really is all that is happening. Isn't it just possible that something comparably simple is "all that is happening" in other complex systems, such as stock markets or political systems?[3]

In 1996, Axelrod and his colleague Michael Cohen at Michigan approached Axtell and Epstein to see whether they could "align" their model of cultural spread with Sugarscape. They wondered whether, if the two quite different models were made as equivalent as possible, they would generate the same results. This entailed, for example, setting the number of cultural features and traits to be the same in the two models, and using grids of the same size and shape. This levels the playing field; it does *not* by any means make the models identical, since the interaction rules differ. Yet the result of the exercise was reassuring: Sugarscape showed the same number of stable regions as Axelrod's model, as the cultural diversity (the number of features and traits) and grid size were varied.

Ambitious things are planned for Sugarscape. In a collaboration called Project 2050, involving the Brookings Institution, the Santa Fe Institute, and the World Resources Institute in Washington, D.C., the model will be used to develop policy recommendations on population growth, use of resources, migration, economic development, and other social issues in order to identify pathways toward sustainable world development. Such difficult matters call for a complex, multifactored model. Once we reach this level of sophistication, any direct analogy with physical models of interacting particles becomes harder to sustain. Nonetheless one can expect—and Axelrod's alignment study supports the notion—that even in the most densely spun artificial society there will remain a core of what we might regard as *inevitable* physics: robust collective modes of behavior, such as phase transitions and power laws, that emerge over a wide range of plausible conditions. Simple models may see these things first and most clearly.

Many social scientists are now pinning their hopes on that being the case. Sugarscape is just one example of a broad range of computer-based model systems that aim to provide an understanding of how social structures, institutions, behaviors, and traditions arise from the bottom up: from a consideration of how individuals interact locally with one another. This is called agent-based modeling, and in earlier chapters we saw it applied to phenomena such as economic trading, the growth of firms, and the movement of pedestrians. It has even been used to predict the progress of rafting trips on the Colorado River through the Grand Canyon National Park, so that more efficient and rational scheduling can be developed.

Thomas Schelling was one of the modern pioneers of this approach, and Herbert Simon believes it may hold the key to a more rigorous science of society. In principle, agent-based modeling should make accessible to rational experiment some of the grandest social and political questions of our time, such as whether the globalization of the economy is likely to lead to greater cultural harmony or to cultural conflict. But some social scientists remain suspicious that any particular agent-based model of a social phenomenon risks coming to conclusions that depend on the underlying assumptions of the model. How do we know whether any one set of rules or assumptions will lead to truly representative behavior and not to an excessively crude caricature of the real situation? In short, such models can hardly be expected to provide a sound basis for policy until we can distinguish what is contingent from what is robust: what a particular model will produce as opposed to what all good models will produce. This is something physics knows about. It is the difference between a model of particles in motion that freeze at 0°C and a knowledge that there are liquid and solid states of matter separated by a phase transition. It is the distinction between a model of society and a physics of society.

FIFTEEN

SMALL WORLDS

NETWORKS THAT BRING US TOGETHER

I consider extremely fruitful this idea that social life should be explained, not by the notions of those who participate in it, but by more profound causes which are unperceived by consciousness, and I think also that these causes are to be sought mainly in the manner according to which the associated individuals are grouped.

—Emile Durkheim (1879)

There are a number of morals to this story. Perhaps the most important is that your friends just aren't normal. No one's friends are.   —Mark Newman (2003)

In 1941 the American movie actor Eddie Albert appeared in *The Wagons Roll at Night*, which starred Humphrey Bogart. Little known today outside the United States, Albert was never a major star, although he was nominated for an Oscar for his role in *Roman Holiday* (1953), alongside Audrey Hepburn and Gregory Peck. To American TV viewers he is perhaps best known for the sitcom *Green Acres*, an episode of which reputedly plays every day somewhere in the world. But Albert was still acting in Hollywood films fifty years after *The Wagons* rolled. In 1989 he appeared in *The Big Picture*, with Kevin Bacon.

Albert is thus the crucial link that gives several famous actors of yesteryear a low Bacon Number. The Bacon Number is the smallest number of movies that link the actor in question with Kevin Bacon, where each movie brings together two of the actors along the path. Albert has a Bacon Number of 1 because he has actually appeared in a movie with Bacon. Bogart has a Bacon Number of 2: he appeared in a movie with Albert, who appeared in a movie with Bacon. James Dean and Ronald Reagan also have a Bacon Number of 2 thanks to Albert, who starred with them both in the TV drama *I'm a*

Fool (1953). Errol Flynn, too, is linked to Bacon via Albert, who shared the screen with Flynn in *The Sun Also Rises* (1957).

The Kevin Bacon game became a favorite among movie buffs and college students in the 1990s. The aim is to find the shortest route to Kevin Bacon—the lowest Bacon Number—for any movie star. But why Bacon? Like Eddie Albert, he became renowned for appearing in a large number of films in which he was *not* the star. In this way he seems to provide a linchpin that binds many bigger and lesser names in a network of movie relationships. Kevin Bacon, it was ironically suggested, was the real center of the Hollywood industry.

The surprising aspect of this game is just how densely connected the network is. Something like 150,000 distributed films have been made since motion pictures began, together featuring about 300,000 different actors as named cast members. And yet nearly all of these can be assigned a Bacon Number (BN) of 3 or less. The last time I checked,* there were 1,686 actors with a BN of 1 (that is, those who have appeared in a film with Bacon), 133,856 with a BN of 2 (those who have appeared with someone who has appeared with Bacon), and a phenomenal 364,066 with a BN of 3. As of October 2003, the average BN for all movie actors (at least, all those listed on the Internet Movie Database at http://www.us.imdb.com) is 2.946.

These numbers certainly seem to support the contention that there is something special about Kevin Bacon—that he connects actors like no other. But is this really so?

The Kevin Bacon game is not a new invention. Mathematicians, not surprisingly, played their own rarefied version first, with so-called Erdös Numbers. Paul Erdös (1913–1996) was an extraordinary, influential, and prolific Hungarian mathematician who wrote hundreds of papers. Because of this unusual productivity, he seems to be the center of the mathematical universe.† Other mathematicians and scientists can be assigned an Erdös Number (EN)

---

*The latest figures can be seen on the Oracle of Bacon Web site at the University of Virginia: http://www.cs.virginia.edu/oracle/.

†Actually, productivity in itself does not guarantee this—the crucial point is that Erdös had many *coauthors* and was thus well-connected in the network. A lone mathematician could churn out a hundred papers a year and still be irrelevant to the network as a whole. The eighteenth-century Swiss mathematician Leonhard Euler was a prime example—he was the most prolific mathematician in history but wrote only by himself.

according to whether they wrote a paper with Erdös himself (EN=1), with a coauthor of Erdös (EN=2), with a coauthor of a coauthor, and so on.

This construction links a large number of not only mathematicians but also physical and social scientists into a vast network of connections. Albert Einstein has an EN of 2. Werner Heisenberg's EN is 4, and that of the German physicist Erwin Schrödinger is a surprisingly large 8.

I have an Erdös Number of similar magnitude.* This does not, of course, mean that I am a scientist comparable in status to Schrödinger, any more than it makes Heisenberg twice as good as him, or Einstein four times better. I have written very few scientific papers, none of them of significance, yet I can trace a link via my coauthors to any of these great names. This is no particular source of pride, since the vast majority of other scientists could do the same. All it is saying is that the network formed by coauthorship in the scientific community is a highly connected one, so that for any scientist an eminent name is only a few steps away.

We all know the phenomenon behind these games. A colleague of mine finds himself talking at a party to a friend of the best man at his sister's wedding. The pianist of a singer I know is an old school friend of my own old school friend's wife. I discover in conversation that someone I've never met before knows my mother. These social networks of friends and acquaintances are continually reminding us that it is a small world.

The highly interconnected nature of social interactions has tantalized social scientists for decades. But by and large they have had little more to go on than anecdotes. In the past few years, however, some physicists have focused their attention on networks that display this "small world" behavior, and have come to understand some of their defining features. These studies reach far beyond the sphere of social dynamics, revealing that the same features can be found in a wide range of networks in systems as diverse as neural wiring in the brain, interdependent metabolic biochemical reactions in the body, and power-line grids. There is, in other words, a kind of universality to certain types of networks. By studying one of them, even on the basis of a simple model, we can gain insights into apparently unrelated systems and processes.

---

*I'm grateful to Mark Newman for establishing that I have an Erdös Number of 8 at most—I might find a shorter link to the great mathematician if I were to search more exhaustively.

We can even figure out whether there is really anything special about Kevin Bacon.

THE SIX DEGREES

The networks formed by friendships are complex and fascinating but extremely hard to pin down in a way that makes them amenable to mathematical or physical analysis. A central difficulty becomes evident each Christmas. Do we know Amy well enough to send her a card? She sent us one last year, but she's not so prominent in our friendship circle as we seem to be in hers. And then there's Roger: we've known him for years, but we don't really keep in touch anymore.

In the movie network, either you've been in a film with Kevin Bacon or you haven't (although one might argue about whether TV films such as *I'm a Fool* really count). Questions of friendship are more ambiguous. In particular, they may not be reciprocal in equal degree: I see Harry as a friend; he sees me as an acquaintance. And if one tries to establish the structure and extent of such a network, as social scientists have done, one soon discovers how bad people are at describing them accurately. We forget to include people, and we are very bad at estimating the size of our friendship circle. And of course the network changes over time as we let old friendships slip away and form new ones.

In the 1970s the social scientist Mark Granovetter at Johns Hopkins University in Baltimore highlighted the importance of different "strengths" of friendship ties. Strong ties are relatively easy to identify and record, but they may not be the most significant in the overall architecture of the network. Many of our close friends know one another, so their own "strong tie" network looks similar to ours. This property binds groups of individuals together in clusters. But Granovetter suggested that it is the weaker "acquaintance" ties reaching between these cliques that bind the global network together (Figure 15.1). He called this characteristic "the strength of weak ties." But these more tenuous links are precisely the ones that are hard to identify reliably.

Sociologists began to take a close interest in social ties and networks earlier still. In the 1950s, the political scientist Ithiel de Sola Pool and the math-

15.1　A friendship network. Clusters of close friends are typically linked by weaker acquaintance ties (dashed lines).

ematician Manfred Kochen at the Massachusetts Institute of Technology wondered how political power gets mobilized in society: How do individuals acquire political power? Friends of the President may whisper in his ear and influence national policy. But do friends of friends also exert an influence? If so, how much? How close do you have to be to power in order to affect it?

Pool and Kochen attempted to formulate a theory of social networks, predicting that people in a population might be more closely connected to one another than they imagine. The two researchers wrote a paper but never got around to publishing it (it appeared in print retrospectively in 1978). The paper was circulated widely, however, and one of the people who saw it was the psychologist Stanley Milgram, down the road from MIT at Harvard University. In 1967, Milgram, inspired by Pool and Kochen, conducted one of the most famous and most elegant studies into the structure of social networks. He sent packages to 196 people in Omaha, Nebraska, with a request that they forward them to the intended recipient, a stockbroker living in Sharon, near Boston.* But he provided no address. The stockbroker was identified only by his name and profession, along with the fact that he lived in the Boston area. The Nebraskans, chosen at random, were asked to send

---

*Milgram apparently chose Omaha because that was the farthest he could imagine from East Coast "civilization."

the packages to friends, acquaintances, or colleagues who they felt might be better placed to channel the package to the right destination—along with a request that they do the same thing.

Someone receiving the package might, for example, send it to a stockbroker he or she knew, in the hope that the stockbroker would have a better idea of where to send it next. No one was to make any systematic attempt to track down the addressee; the recipient was simply to pass the package along to someone else, chosen for his or her possible greater proximity (in geographical, social, or professional terms) to the addressee. At each stage, participants were asked to include in the package the same kind of details about themselves as had been provided about the addressee. In this way, if and when the packages arrived at the right destination, their route could be reconstructed. "When I asked an intelligent friend of mine how many steps he thought it would take," Milgram recalled, "he estimated that it would require one hundred intermediate persons or more to move from Nebraska to Sharon."[1] It didn't. Milgram found that on average just five intermediaries (that is, six separate journeys) were needed before the packages found their way to the addressees.

The implication was that any person in the United States could be linked to any other person chosen at random via an average of five acquaintances. The exact number can be debated—for example, it is unlikely that each intermediary will make the ideal choice of whom to pass the package on to, and the relevance of packages that *never* get delivered owing to apathy has to be factored in. But there is a consensus that between five and seven steps is enough to link everyone—a conclusion supported by Milgram's later experiments, in 1970, with parcels sent from Los Angeles to New York.*

These studies were restricted to the United States, but their message was expanded to the global scale in John Guare's 1990 play *Six Degrees of Separation.*† Guare's character Ouisa says,

---

*Milgram gained greater fame for a notorious experiment in which he showed how terrifyingly easily people can be ordered to cause others pain. Although no one was actually harmed in the tests, the subjects were fooled into thinking otherwise. Such an experiment would probably be forbidden today. Duncan Watts, a physicist-turned-social scientist and an expert on network theory, says that whenever he told other sociologists that he intended to repeat Milgram's experiment, they looked at him with horror and he had to follow up rapidly with the disclaimer, "No, not *that* one."

†Milgram's "six degrees" seems to have been predicted as early as 1929 by the Hungarian poet Frigyes Karinthy, who made the same claim without any scientific justification for it.

Everybody on this planet is separated by only six other people. Six degrees of separation. Between us and everybody else on this planet. The president of the United States. A gondolier in Venice . . . A native in a rain forest. A Tierra del Fuegan. An eskimo . . . It's a profound thought . . . How every person is a new door, opening up into other worlds.[2]

But is it really a profound thought? The world is full of things that surprise us not so much because they are strange or perplexing but because our intuition is askew. It is natural to assume that people who are geographically very distant will also be socially distant. But to a corresponding degree? Is there any reason I should be socially more remote from people in the Shetland Isles than from people in Bromley, a short bus ride from where I live? Will people in Nebraska be proportionately further removed from my social sphere?

These may seem like whimsical, even trivial questions.* But they have ramifications that go far deeper. The people with whom I come into close physical proximity are the pool from which I pick up many of the diseases I shall suffer during my life. More tangible still is the network of sexual and intimate contacts through which, for example, AIDS spreads throughout a population. There is little fuzziness about this network—either you have a sexual encounter that puts you at risk of contracting AIDS or you do not.† Yet when epidemiologists began to try to understand and forecast the expansion of the disease, they soon encountered the practical difficulties of reconstructing such a social network, even though in principle it was relatively clearly defined. In a few cases, crucial connections were nevertheless identified. A single promiscuous Norwegian sailor appears to have been instrumental in some of the earliest cases of AIDS in Europe, having contracted the disease in West Africa in 1960 and then transmitted it to people in Cologne and Reims during his later life as a truck driver in the 1970s. A French-Canadian flight attendant named Gaetan Dugas, later dubbed Patient Zero, was linked to at least forty of the earliest known cases of AIDS in California and New York.

Some movies, books, plays, and music become huge hits not because of

---

*For a healthy dose of whimsical triviality, see the op-ed article in the *New York Times* by Kirby and Sahry entitled "Six Degrees of Monica" (21 Feb. 1998).

†Infection by blood transfusion is harder to trace but is ultimately an equally concrete link in the chain.

major advertising campaigns but by word of mouth. Gratifyingly, the same process can have the reverse effect—a much-hyped movie can flop when word spreads that it's a turkey. In this age of mass communication and global information systems, the process of dispersal of cultural products and ideas becomes very hard to identify precisely. But there is no doubt that social networks play their part, and only a few decades ago person-to-person connections were paramount. For anyone concerned with the dynamics of globalization, networks of encounters and information exchange between people must be a central theme.

BETWEEN ORDER AND CHAOS

Paul Erdös is the ideal subject for the mathematicians' game of connections because of his productivity. But that is not the only thing that makes him appropriate. In the 1950s and 1960s Erdös pioneered the study of networks, and until recently his work provided the starting point for nearly all social-science studies in this area. With his colleague Alfred Rényi (who has an Erdös Number of 1, naturally), he clarified the properties of networks called *random graphs*.

A graph, we learn at school, has a vertical axis, a horizontal axis, and a scattering of data points with a line drawn through them. But Erdös and Rényi were not talking about this kind of graph. For them, a graph is simply a series of points connected by lines (Figure 15.2a). The points were called *vertices* and the lines are *edges*. This abstract entity can represent all manner of things. The vertices could be cities, and the edges the roads that join them: the graph then represents a road transportation network. Or the vertices could be movie actors, and the edges are the connections between pairs of actors who have appeared in a film together. An actor's Bacon Number is the number of separate edges one has to traverse to get to the relevant actor from the vertex representing Kevin Bacon (Figure 15.2b). A graph like this displays the relationships between the entities represented by its vertices.

When the vertices are cities and the edges are roads, it is easy to comprehend the rules of the graph: it is like a map. The distances and compass directions between the vertices representing cities generally reflect those in

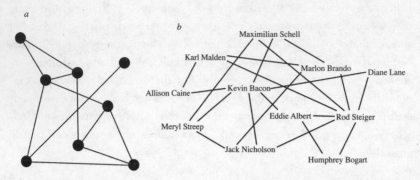

15.2   A typical network graph (*a*): a series of points (vertices) connected by lines (edges). In *b* is shown a small part of the graph that connects movie actors, indicating some of the connections in the vicinity of Kevin Bacon. The respective movies are listed in the endnotes.

geographical reality. But for the movie-actor graph, the rules are not so plain. How close to Kevin Bacon do we place Eddie Albert, and in which direction? Does Jack Nicholson (BN 1: *A Few Good Men*, 1992) get placed at the same distance? We might decide to make all the edges the same length, say, and not to attribute any significance to the directions. But we quickly find that this won't work; we would soon be unable to place two specific costarring actors close enough to link them with a single edge of fixed length, because their links to other parts of the graph keep them apart. In fact, why should we draw the graph in two dimensions on the flat page? Why not a three-dimensional web, like the timber frame of a building? Or, now that mathematicians are comfortable negotiating their way around spaces with more than three dimensions, why not a ten-dimensional web?

The answer to these dilemmas is that it really doesn't matter how we choose to draw the movie-actor graph, as long as the connections between vertices are correct, linking only those actors who have appeared in movies together. To understand the important features of the network, all that matters is this system of connections, which mathematicians call the *topology*. Graphs that look different when drawn on paper can nevertheless be topologically identical. The distances and directions don't *mean* anything. We can call this kind of graph a *relational* graph, because it concerns itself only with showing the relations between its vertices. In contrast, the city network is a *spatial* graph, where distances and positions correlate with something real. Of course, we don't have to make the city graph a spatial one. The iconic map of London's Underground subway lines (Figure 15.3), devised by Harry Beck in 1931, is a wonderful example of a relational graph; but it retains some element of a spatial graph because that makes it easier to draw the links and it helps people to find their bearings. The stations appear roughly in their geographical arrangement, but the distances aren't to scale and the compass directions aren't precise. It can be quicker to walk between the Covent Garden and Holborn stations, one edge apart, than to take the subway; but you wouldn't want to walk from Hatton Cross to Boston Manor, even though they are the same distance apart on Beck's graphlike map.

In the random graphs studied by Erdös and Rényi, a scattering of vertices is interconnected at random. You could make a six-vertex random graph by numbering the vertices and then throwing two dice; an edge is added between the two vertices whose numbers come up. When every vertex is connected by at least one path to any other, the graph is said to be fully connected (Figure 15.4). On such a graph, it is always possible to travel from one vertex to any other. The London Underground is fully connected

15.3 The London Underground is a relational graph, showing how vertices (stations) are connected but giving only approximate indications of their geographical positions and the distances between them. The thick gray line here is the River Thames.

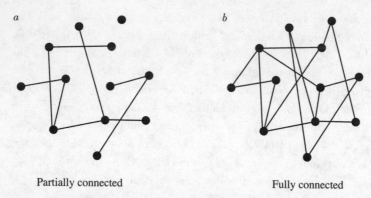

a   *b*

Partially connected   Fully connected

15.4   A random graph becomes fully connected when all the vertices are linked into the network (*b*)—when there are no isolated vertices or clusters, as in *a*.

in this sense. In general there will be several possible routes between destinations, and we'll often be interested in finding the shortest.

Several investigators of social networks, beginning with Anatol Rapoport and his co-workers at the University of Chicago in the 1950s, assumed that these networks have the topology of random graphs. This is almost certainly not the case, but random graphs seemed like a good starting point for such studies because they make only a neutral assumption about the underlying architecture. And besides, Erdös and Rényi had conveniently worked out the mathematical properties of such graphs.

These properties have to be expressed in statistical terms, since the connections are a matter of pure chance. For a large number of vertices—one hundred will do—there is a negligible chance of generating the same graph twice by connecting vertices at random. Thus, just as it makes no sense to ask about individual particle motions in a gas but rather to focus on averages and distributions, so too must we limit our curiosity about random graphs to such quantities as the average number of connections per vertex, and the probability distribution of this number.

Erdös and Rényi showed that the probability distribution is the familiar bell-shaped Gaussian curve. The average number of connections corresponds to the peak of the curve. This average obviously depends on how long we go on adding edges to the graph, but for any specific random graph it has a well-defined value.

Another kind of graph that has been used to study networks lies at the opposite extreme from the random graph. It is the regular grid, an array of identical vertices connected by identical edges (Figure 15.5). There is no need to talk of averages here—every vertex (except those at the edges and corners) has exactly the same number of connections (four in the example shown). These "ordered" graphs are easy to describe mathematically.

Random and ordered graphs have very different properties. To get from one vertex to a distant one on an ordered grid, you have no choice but to proceed in a series of little hops between neighbors. The path length, measured as the number of edges traversed, is long. For a random graph, on the other hand, there is a good chance that a vertex near your starting point will be connected "long distance" to one close to the target vertex. There are lots of shortcuts, in other words. This means that the path length between apparently widely separated vertices may be small.

One way of making this difference precise is to ask, What is the average path length between any two vertices chosen at random? This is called the characteristic path length, and it is a statistical property exactly analogous to the average number of journeys Stanley Milgram's packages had to make to reach their destination. For an ordered graph, the characteristic path length is long, and gets proportionately longer as the number of vertices increases. For a random graph, the characteristic path length is small. Moreover, it doesn't increase much as more vertices are added, because there is a good chance that a random connection will be made between a new, outer vertex and one deep in the middle of the graph.

15.5 On a regular grid, all vertices (excluding those at the edges) have the same number of connections.

Judging from this, you might expect that social networks like the movie-actor graph will be like random graphs. Isn't that the point of the small-world concept—that you're always socially closer to another random individual than you think?

In 1998 two scientists at Cornell University showed that social networks *aren't* like random graphs. Instead they fall into a class of their own, lying somewhere between the perfect disorder of random graphs and the perfect order of regular grids. They are called, suitably enough, small-world networks.

OF CAVEMEN AND CHAT ROOMS

Friendship circles are largely mutual: my friends' close friends tend to be my close friends too. If I know Andy and Betty, Andy and Betty probably know each other too. In other words, my friends and I form a kind of social cluster, albeit one with plenty of links to other clusters. Random networks do not have this clustered structure. In networks with a random topology, the fact of my knowing Andy and Betty has no bearing at all on the probability of their knowing each other.

Ordered grids do have an element of clustering, because each vertex is connected only to its immediate neighbors—there are no big leaps.* Many of the neighbors of one vertex are also neighbors of a neighboring vertex. So there is a good chance that close friends connected by an ordered grid share a mutual close friend.

When Steven Strogatz at Cornell and his graduate student Duncan Watts began to investigate the problem of social networks in the late 1990s, they found a way to represent this clustering phenomenon. They weren't initially interested in human social structures; they were thinking instead about how some animals synchronize their behavior, such as the way a field of crickets will synchronize their chirping. But the question soon led them into the tangled web that social scientists had begun to weave.

Strogatz and Watts imagined two possible extreme societies. The first is

---

*One can create ordered grids with links that reach beyond near neighbors, but that doesn't alter the tendency toward clustering.

fragmented into clusters of people who socialize with one another but with next to no one outside the group. We can think of examples that approximate this situation—for example, businesses that dominate their employees' lives, providing them with almost their sole source of social contacts, as is the case in some Japanese firms. But Strogatz and Watts found a more basic analogy: the world of the cavemen, consisting of sparsely scattered groups, each of which share a cave, do everything together, and shun outsiders. In Thomas Hobbes's time, many peasants in villages were still living in this sort of world. It can be represented as a series of small graphs that are highly connected internally but disconnected from, or only sparsely connected to, one another (Figure 15.6). This is not an ordered graph, but it shares with such graphs the properties of high clustering and long characteristic path length.

The two researchers drew the other extreme not from the past but from the future. Isaac Asimov's 1957 novel *The Naked Sun* portrays a future society in which people interact almost entirely through robots and computers. In Solaria it is as easy, and as likely, for you to forge a relationship with someone on the other side of the world as with your neighbor. These virtual friendships have become so weak and superficial that new ones are established completely independently from those that already exist: everyone in

15.6   In the "caveman world" (*a*), people form social clusters that are closely linked internally but are isolated from one another. Even if this graph is made fully connected by linking the "caves" (*b*), the characteristic path length is long.

society is equally available and accessible. This results in essentially a random-graph network, in which there is no clustering. Some social networks are already a little like this—for example, those that form in Internet chat rooms, where people come and go almost at random because they have so little personal investment in the connections made.

Strogatz and Watts described the differences between the caveman world and Solaria by considering how the likelihood of two people becoming acquainted depends on how many friends they already have in common. In the caveman world, two people who have a mutual friend are almost certain to become friends themselves, because it means they must be in the same cluster. In Solaria, two people who have even a large number of mutual friends have no greater chance of becoming friends with each other than either of them has with any other random person.

Our real world presumably lies somewhere between these two extremes. But where?

The mathematics demands that any graphs one uses to mimic social networks must be fully connected. Otherwise, some vertices remain completely inaccessible to other vertices, and this immediately pushes up the characteristic path length to infinity—which isn't realistic at all. So Strogatz and Watts devised a method for converting a fully connected ordered graph (comparable to a fully connected caveman world) into a fully connected random graph (like Solaria) in steps that always left the intermediate graphs fully connected too.

It is called random rewiring. You start with an ordered grid and select a vertex at random. Then you select one of its connecting edges at random, disconnect it from its destination, and reconnect (rewire) it to another vertex

Ordered network

Random network

Increasing randomness

15.7   Random rewiring of a ring of vertices gradually converts an ordered graph into a random one. Somewhere between these two extremes lies the small world.

in the graph, also selected at random (Figure 15.7). As the rewiring progresses, more and more shortcuts are created that link distant parts of the graph in a single jump, and the graph becomes increasingly random.

The researchers chose to begin with the simplest ordered graph they could think of: a ring of vertices. (A ring is better than a row of vertices, because it eliminates anomalies at the ends.) They began rewiring these graphs and watched what happened to the topologically determined features: the characteristic path length (L) and the amount of clustering, which can be measured as a numerical quantity denoted C.* The amount of rewiring can be quantified by the probability that a vertex chosen at random will have been rewired from its initial configuration. When this probability is zero, the graph is a perfectly ordered (ring) network. When the probability is one, the graph is fully random.

As would be expected, both L and C decrease as the amount of rewiring increases: the ordered grid loses its clustering and acquires more shortcuts. But there are three surprises. First, nearly all the action happens in the first few rewirings. By the time a tenth of the vertices are rewired, the graph typically has properties indistinguishable from those of a random graph. Second, the changeover from ordered-like to randomlike is quite abrupt—Watts compares it to a phase transition in statistical physics. It is as if the ordered, "solidlike" graph melts into the disordered "liquidlike" graph. And finally, L and C do not change together but at different stages of the rewiring process (Figure 15.8).

The last observation seems truly weird. The characteristic path length plummets to a value typical of a random graph while the graph still retains a high degree of clustering (large C). Then, after a little more rewiring, C plummets too. So for a small range of rewiring extent, the graph has small L but large C. The coexistence of these two characteristics is what makes a network a small world. The small world of friendship circles, for example, incorporates a high degree of clustering but possesses numerous shortcuts between clusters, creating the short average path lengths responsible for our

---

*C is defined as the average, over all vertices, of the number of edges connecting a vertex v divided by the total number of possible edges in the neighborhood of v. It is a quantitative answer to the question, Of all the possible ways of connecting vertices in the neighborhood of v, how many are actually realized? If the clustering coefficient C is large, this means that most potential connections within the neighborhood of any vertex are indeed made: there is a lot of clustering. In a friendship network with large C, one's friends are likely to know one another.

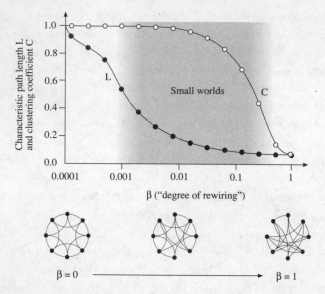

15.8   The characteristic path length L and the clustering coefficient C for a rewired ring graph change rather abruptly as rewiring proceeds. The quantity ß here is essentially a measure of the extent of rewiring. Notice that the horizontal axis has a logarithmic scale, which means that small values of ß are "stretched out." This is so that it is easier to follow the changes in L, which happen mostly in the first few rewiring steps. L and C are both large in the ordered ring graph and small in the random graph. But they do not both change at the same stage. Graphs with large C but small L are small-world graphs.

"six degrees of separation." So Strogatz and Watts dubbed these transitional graphs "small-world graphs."

DO I KNOW YOU?

Do social networks in the real world actually look like the small-world graphs created by random rewiring? In other words, could Strogatz and Watts *prove* that indeed "it's a small world," and thereby perhaps deter us from trotting out this truism every time we discover previously unsuspected connections at parties?

For all its apparent frivolity, the movie-actor network remains one of the best test cases, since the data are unambiguous, abundant, and readily available. It is clear enough from the Kevin Bacon game that there is a short characteristic path length L in this network (the average Bacon Number is low), and it doesn't take much inspection to see that there is a high degree of clustering. Actors of the same nationality tend to form clusters, for instance, which are interlinked by a few cross-cultural linchpins such as Bruce Lee (Hong Kong), Gérard Depardieu (France), and Gong Li (China).

To this extent the movie-actor network has properties compatible with a small world. But what about the network topology? Some kind of comparison can be made by defining a quantity called the vertex contraction parameter, which is a measure of how many shortcuts exist in the network between widely separated vertices. This number can be calculated for any network, and it increases for the ring graphs as they are progressively rewired. Strogatz and Watts constructed a small-world rewiring network whose vertex contraction parameter had the same value as that of the movie-actor network, thereby ensuring that it was somewhat comparable topologically. The question is then how the two networks compare in terms of their characteristic path length L and clustering parameter C.

This comparison is quite good—certainly, the rewiring model can mimic the parameters of the movie-actor network more closely than can either a caveman network or a Solaria network. But we must nevertheless ask whether these are the only alternatives. That is, is the rewiring scheme the *only* way to make a network with small-world properties (low L, high C)? We shall explore this question in the next chapter.

And what of the most burning issue: Is Kevin Bacon really the center of the movie universe? To answer this, one must calculate the average Bacon Number for the entire network and see how it compares with the equivalent measures for other actors: the Elvis Number, the Bogart Number, the Brando Number, and so on. If Kevin Bacon really is the most important linchpin in the network, all other actors will, on average, be closer to him than to anyone else.

It turns out that not only is Kevin Bacon *not* the most important hub of the network, he is not even in the top one thousand (the list of course changes daily as new films are made). Currently up at the top is Rod Steiger (the average Steiger Number is 2.652), followed by Christopher Lee, Dennis Hopper, Donald Pleasence, and Donald Sutherland (who appeared in the movie version of *Six Degrees of Separation*). Marlon Brando is number 202, Frank

Sinatra number 443. By the time we get to Kevin Bacon's level, the differences in the average Actor Number that separate successive actors in the list are tiny, about 0.0001.

So why was Kevin Bacon picked for this game? The answer contains the entire essence of a small world: in such a network, *everyone* appears to be at the center. Some are more "central" than others—but not by very much. Even relatively minor actors like Eddie Albert have a comparable network status to major stars. (Donald Pleasence was a fine actor but hardly a superstar.)

There is something refreshingly egalitarian about this message. If social networks bear any resemblance to the movie-actor graph—and this is a reasonable guess, as we shall see—we need no longer gaze with envy upon those few who seem to represent the hub of our social circles, nor lament our paucity of friends. By and large, it is simply a matter of getting the right perspective on the web that links us all together.

There is a postscript to this story. In 2003, Duncan Watts (now at Columbia University) and his co-workers staged a rerun of Milgram's classic experiment, but using e-mail instead of the postal service. They asked for volunteers to take part in this electronic project and were inundated with offers, finally registering 61,168 individuals in 166 countries. To each of these participants they assigned a target person, to whom an e-mail message was to be relayed. There were eighteen targets in thirteen different countries, ranging from an Ivy League professor in the United States to an inspector of archives in Estonia to a Norwegian military veterinary surgeon. The rules were much the same as Milgram's: participants were to pass on the message to a friend, colleague, or associate whom they considered likely to be "closer" to the designated target.

Well, it was not a good way to deliver the mail. There was an initial drop-out rate of about 63 percent, but that wasn't bad for an e-mail survey, and it meant that 24,163 chains got started. Yet only 384 reached their target. These completed chains contained an average of just four steps, but that number was biased by the fact that the shorter chains were more likely to run to completion. By allowing for the attrition rate, Watts and his colleagues estimated that a typical chain length between any of the participants and their target was five to seven steps—just the same as Milgram had found, although his chains were confined only to the United States.

That bolsters the notion of six degrees of separation, but it does not exactly imply that our small world makes for efficient person-to-person contact. The researchers established that few of the people who dropped out of a chain did so because they could not think of anyone appropriate to pass the message on to. They simply lost interest. What was in it for them?

Yet for one target the attrition rate was significantly lower than for the others: the U.S. professor. Did that mean he was better connected and so more easily found? Probably not, say Watts and his colleagues—they estimated that his average degree of separation was not so different from that of the other targets. But over half of the project's participants were professional college-educated middle-class Americans. Those in the chains directed at the professor probably *believed* that it would be fairly easy to get the message to him, compared with participants faced with relaying a message to, say, a technology consultant in India. This belief gave them that little extra motivation to keep the chain alive.

In other words, say the researchers, the structure of the social network is not the whole story. Whether or not its small-world character will indeed be exploited to make connections depends on the actions and even on the perceptions of the people in it: a timely reminder that social physics cannot afford to do away entirely with individual psychology.

SIXTEEN

WEAVING THE WEB

THE SHAPE OF CYBERSPACE

The Internet Age has been hailed as the end of geography. In fact, the Internet has a geography of its own, a geography made of networks and nodes that process information flows generated and managed from places. The unit is the network, so the architecture and dynamics of multiple networks are the sources of meaning and function for each place. —Manuel Castells (2001)

We are in the presence of a new notion of space, where physical and virtual influence each other, laying the ground for the emergence of new forms of socialization, new life styles, and new forms of social organization. —Gustavo Cardoso (1998)

'Tis true; there's magic in the web of it. —William Shakespeare (1602–1604)

I n May 2000, I received an e-mail announcing its subject with the enticing words I LOVE YOU. Fortunately, by the time I saw it I was aware that a large proportion of the online community had found the same message in their inbox. What lay in store, if I chose to open the message, was not an amorous declaration but a nasty computer virus that would lay waste to my files before posting itself elsewhere.

Computer viruses are not (yet) fatal, but they are a costly bore. They have wreaked havoc in commercial computer networks and have caused individual users untold anguish. The "I Love You" virus immobilized the internal e-mail systems of the U.S. Senate and the British House of Commons and afflicted several other governments and financial institutions. Another virus, dubbed the Love Bug, caused $6.7 billion worth of damage in 2000.

The Slammer computer worm more or less disabled all of South Korea's Internet network in early 2003, and the increasing ingenuity and elusiveness of these virtual parasites must make us fear that one day such an epidemic will assume global proportions, like a modern-day Black Death. The threat highlights the role of the Internet and its companion electronic network, the World Wide Web, in maintaining today's social order. Increasingly, the economy is a wired system, depending on computer-based communications for even the least of its machinations. Many governments share this dependency, so do facets of our social framework ranging from goods distribution systems to schools to policing. "There's a new infrastructure out there," says Larry Irving, head of the U.S. Commerce Department's National Telecommunications and Information Administration, "and we're much more dependent on this infrastructure than we've ever been or ever expected to be."[1]

It is not fanciful to see in the intertwined branches of the Internet and the Web a manifestation of Hobbes's Leviathan: a ruler whose body is composed not of the multitudes of its subjects but of silicon chips, glowing terminal screens, and mile upon mile of copper wire and optical fiber. In his short story "For a Good Purpose," the Italian writer Primo Levi envisaged a similar beast—the automated telephone routing system—acquiring consciousness from the complexity of its connections and engaging in a bout of seemingly random rewiring that, regardless of whether it made for a smaller world, caused chaos among the human population.

How robust is this electronic Leviathan, this Hobbesian Cyber-Commonwealth? "Though nothing can be immortall, which mortals make," conceded Hobbes,

> yet, if men had the use of reason they pretend to, their Common-wealths might be secured, at least, from perishing by internall diseases . . . Amongst the *Infirmities* therefore of a Common-wealth, I will reckon in the first place, those that arise from an Imperfect Institution, and resemble the diseases of a naturall body.[2]

Yes indeed: electronic viruses are just as capable of striking down the Internet as their biological namesakes are of leaving us bedridden. So too is the Internet potential prey to the many other afflictions that flesh is heir to: overload, transmission errors, local breakdowns causing bottlenecks, and attack from outsiders with malicious intent. Hobbes's message for us is that we would do well to build our Leviathans along the lines of a healthy "naturall body." In this chapter we shall discover something quite astonishing about the Internet: it seems to have this "biological" character already, although no one planned it that way. And while this does indeed lend the network a certain robustness, it also exposes an Achilles' heel, a weakness to a particular kind of attack. Identifying such weaknesses may be crucial to making the Internet safe from cyber-terrorism, a danger that grows daily as our reliance on computerized systems increases.

The Internet is just one example how our social structures and institutions are laced with the kind of networks we encountered in the previous chapter, in which countless individual units or agents are linked by interactions into a surprisingly small world. A picture is just beginning to emerge of what these networks really look like and of how they arise unbidden and unplanned among multitudes of agents going about their local business without a thought for the global view.

There is an urgent need to understand this "network society." The global communication and information networks of the Internet and the World Wide Web are shaping novel cultural and institutional structures. Sociologist Manuel Castells believes that the electronic web now being spun over the world—what he calls the Internet Galaxy—provides an entirely new way of conducting business and indulging pleasures. He claims that the decentralized, nonhierarchical "geography" of this network will change the very nature of these activities: "As new technologies of energy generation and distribution made possible the factory and the large corporation as the organizational foundations of industrial society, the Internet is the technological basis for the organizational form of the Information Age: the network."[3]

The growth of this network has been more rapid than anyone could have anticipated, and certainly too fast to allow considered prognostications about its effects or future potential. Widespread use of the Web did not begin until 1995; by the end of that year there were about 16 million users. By early 2001 that figure had grown to more than 400 million; some estimates predict that it will top a billion by 2005, and 2 billion by 2010.

Thanks to the new physics of networks, we can now see that the structure of the Internet reflects the ethos that produced it—as Castells says, "The culture of the producers shaped the medium."[4] That is to say, only a network that grows "organically" according to no master plan, observing principles of free access and meritocratic choice (the best or more useful Web sites get the most connections), will develop the kind of architecture that the Internet and the Web display, with their strengths and pitfalls.

But this culture of "freedom of information" does not guarantee that the network society will be utopian. The Internet is, you might say, free to those who can afford it, well over half of whom live in North America and Europe (and most of the rest in the Asian Pacific Rim). As the power of the Internet grows, those who lack the technological infrastructure, money, or education to access it are increasingly disenfranchised, which creates a "digital divide" not just between developed and developing countries but between different socioeconomic sectors of a single culture. Meanwhile, the far-reaching potential of the Internet is being exploited as avidly by criminals as by business entrepreneurs. Pranksters and malcontents have new opportunities for sabotage; advertisers have a new way to send you their junk.

These considerations make it essential that we understand the virtual anatomy of this information machine. For it and its influence will soon touch us all, in one way or another, as Castells foretells:

> I imagine one could say: "Why don't you leave me alone?! I want no part of your Internet, of your technological civilization. Of your network society! I just want to live my life!" Well, if this is your position, I have bad news for you. If you do not care about the networks, the networks will care about you, anyway. For as long as you want to live in society, at this time and in this place, you will have to deal with the network society."[5]

THE NET IS CAST

"It is no exaggeration," said U.S. judge Stewart Dalzell in 1996,

> to conclude that the Internet has achieved, and continues to achieve, the most participatory marketplace of mass speech that this country—and indeed the world—has yet seen . . . Federalists and anti-Federalists may debate the structure of their government nightly, but these debates occur in newsgroups or chat rooms rather than in pamphlets. Modern-day Luthers still post their theses, but to electronic bulletin boards rather than the door of the Wittenberg Schlosskirche. More mundane . . . dialogue occurs between aspiring artists, or French cooks, or dog lovers, or fly fishermen.[6]

Participatory is the apt adjective. Although the rhetoric of global inclusiveness often overlooks the limited resources of developing countries, there is otherwise nothing elitist about the Internet. Anyone with access to a com-

puter terminal can get an e-mail address, free of charge, and send messages
to others on the far side of the world. Judge Dalzell's comments came in a
summing-up speech in which he ruled attempts to circumscribe the content
of messages transmitted over the Internet as unconstitutional. "The Internet
may fairly be regarded as a never-ending worldwide conversation," he said.
"The Government may not . . . interrupt that conversation."[7]

It is a well-known irony that this open conversation grew out of a network
devised in secrecy by the U.S. military in the 1960s, called ARPANET.
(ARPA, the Advanced Research Projects Agency, was the euphemism for
scientific research funded by the U.S. military.) The Net was originally intended for the transfer of documents; its architects were taken by surprise
when its facility for sending brief messages, initially instigated for their own
convenience during the project's development, became its most attractive
feature. Soon enough the nonmilitary academics and industrial scientists
who'd watched ARPANET grow wanted their own version, and Usenet was
born, sending its packages of coded data down the national and international
telephone lines. On Usenet there emerged the anarchic, anything-goes philosophy that now characterizes the Internet, the meganetwork that grew from
the proliferation of smaller networks with dedicated servers.

The Internet and the World Wide Web—the Net and the Web—are
sometimes perceived as synonymous, but they are not. The Net is a communications network; the Web is an information bank. The man usually
credited with launching the Web is Tim Berners-Lee, an English computer
scientist who was employed by the European particle physics laboratory
CERN in Geneva to help its physicists store and retrieve information.
Berners-Lee's concept is summed up in his description of a program he
devised in 1980 as a kind of personal "memory substitute": "It allowed one
to store snippets of information, and to link related pieces together in any
way. To find information, one progressed via the links from one sheet to
another."[8]

In 1989, Berners-Lee devised a means of implementing this scheme in
an electronic information system of any size: hypertext. In a hypertext document, the words that appear on-screen are interspersed with markers—push
buttons or text that is underlined or differently colored—that allow the user
to leap directly to a different document containing further relevant information. These *hyperlinks* are inserted by whoever creates the document, but

they can link to a page posted by someone else. If the content of the hyperlinked page is later changed or updated, the hyperlink takes the user to the new version.

Berners-Lee's proposal for such a system at CERN was soon seized on by others at the laboratory as the inspiration for a grander vision. They had no wish to limit themselves to a mere institutional web; by 1990 there was talk of a World Wide Web, which would use the global infrastructure established by the Internet to access electronic documents stored on computer systems the world over, as well as articles generated within Internet newsgroups. To navigate around such a complex network, one needed a guide, which came in the form of a *browser*, a program that provided a user-friendly gateway to the universe of resources on the Web. Netscape, now owned by AOL (America Online), and Microsoft's Internet Explorer are the two most commonly used browsers.

It was estimated at the end of 2002 that there were around 3 billion documents then available on the Web. Several million more Web pages are being added every day. With a library that size, one needs an efficient means to get quickly to the information one is seeking. This is the role of *search engines*, programs that can be regarded as robots which trawl through the Web's archives for documents containing keywords or text that match words fed to them as search criteria. Such is the scale and complexity of the Web that today's search engines are able to survey only a small part of it.

The issue of network topology had to be faced by the Internet's inventors from the outset. In large measure this was not so much a mathematical as a technological question. Older readers will remember a time when the orthodox image of a computer system was a central mainframe device linked to a series of terminals that had no autonomy but were mere portals to the central oracle. The concept of a freestanding personal microcomputer did not exist when the Internet's ancestor, ARPANET, was being planned. Thus the "obvious" topology for a computer network in the 1960s was a highly centralized one in which all messages were sent to a central distributing mainframe that passed each of them to the appropriate node (Figure 16.1a).

But Paul Baran, the communications engineer who was set the task of developing a communications network for the U.S. military, realized that this would be a crazy way to build such a system. He knew that a centralized system would be highly vulnerable to attack, since one needed to knock out

16.1   Possible network topologies for the Internet: centralized (*a*), decentralized with local servers (*b*), distributed (*c*), and maximally connected (*d*).

only a single node—the central hub—to disable the whole network. Baran was employed by the RAND Corporation, a U.S. think tank created in 1946 to supply strategic analysis and recommendations to the government for its nuclear defense policy. At the height of the Cold War, many of the military top brass in the United States feared that a nuclear strike from the Soviet Union might knock out the country's communications systems and disrupt retaliatory action. The military needed a network capable of withstanding a nuclear assault.

One possible alternative to the conventional, centralized architecture of a computer system was a decentralized network in which a series of smaller, locally centralized clusters were interconnected by longer links (Figure 16.1*b*). Yet this design still suffered from the defect that any part of the system could be isolated relatively easily by knocking out one of the links, or "edges" (we are clearly talking about graphs again now).

The best design—radical for its time—was a highly interconnected "distributed" network in which each node (vertex) was connected to several others (Figure 16.1*c*). This topology has a high degree of redundancy: there are many possible routes from one vertex to another. So disabling even quite a

high proportion of edges fails to isolate any part of the network. Baran estimated that just three edges per vertex would be enough to guarantee adequate resilience in such a network.

Baran's proposed network languished in bureaucratic purgatory and was
never built, but his ideas were resurrected when the ARPANET scheme was
created. This initiative owed nothing to Cold War fears of nuclear attack; it
aimed merely to increase the efficiency of the computer resources used at
ARPA-funded university sites by interlinking them.

The initial plan was to connect the mainframe computers at each site directly to all the others. But in 1967 the computer scientist Wesley Clark realized that this "maximally connected" network would rapidly get unworkably
complex. With just ten vertices there are already forty-five edges (Figure
16.1d), and the number of edges increases faster than the number of vertices. This would place huge demands on the mainframe computers at each
vertex and be likely to cause "freezing" of the information flow. Clark proposed that instead the mainframes should be linked to a subnetwork of
smaller, interconnected computers, all speaking the same language, that
were dedicated simply to routing the information. Baran's work showed
Clark and others how such a network could be wired up to function efficiently and robustly.

That was about as much topological planning as the Internet ever got. It
has expanded far beyond anything these early pioneers could have envisaged, subject to no regulatory authority or architectural design. The growth,
like that of cities (see Chapter 6), has been organic and uncoordinated. Indeed, some suggest that the Internet is now best regarded as an ecosystem
every bit as complex as those in the natural world. It is no longer possible to
draw an exhaustive map of its connectivity pattern—the structure is too
huge and deeply embedded for it to be traced.

Nevertheless, it seems certain that the Internet does have the highly redundant distributed technology proposed by Baran. The packets of data into
which messages and Web pages are chopped are sent along many different
paths between the source and the destination, perhaps even traversing the
globe in opposite directions before being pasted together at their target point.
This is why the Internet is able to cope with so much traffic at once: if one
pathway is congested or blocked, another usually presents itself.

On the other hand, we know that a map of the Internet would *not* look like
Figure 16.1c. That is more like a regular grid (see page 363), albeit one in

which the number of links per vertex is not uniform but varies between about four and six. As we saw in the previous chapter, there are no shortcuts in such a network: you get from A to B via a series of many short jumps.

So exactly what does the Internet look like? This is a critical question, relating to problems such as how to route mail most efficiently, how robust the system is in the face of local breakdowns, how easily users in one part of the world can reach those elsewhere, and whether the Internet can grow without limit or will reach some saturation point beyond which the network cannot cope. As the economic impact of the network increases—already e-commerce in the United States turns over $100 to $200 billion a year, and similar figures are quoted for the economy-wide savings potentially available from electronic communications—these questions are becoming pressing ones.

The Internet network is a physical entity, like the London Underground. The nodes or vertices are computers, and the links or edges are the transmission lines or satellite channels that run between them. The World Wide Web is a more nebulous realm. Its vertices consist of document pages stored electronically in machines throughout the world, and its edges are the hyperlinks or uniform resource locator (URL) addresses that connect a page directly to another in some other Web site on some other machine. When we follow one of these links, the communication is conducted along the same physical transmission lines that the Internet uses; but the Web itself is not delineated by these lines. One transmission line can in principle connect a million Web pages to a million others in a maze of URL links.

It is not obvious how the Internet or the Web is structured. Could they be random graphs? Are they small-world networks like those we encountered in the previous chapter? Or are they perhaps branching hierarchical trees, like river networks or the passageways of our lungs? The answer is that they appear to be something else entirely.

NO SENSE OF SCALE

Mapping the World Wide Web is like mapping a maze. If you could rise above the maze in a balloon, the task would be easy: you'd simply draw what you saw laid out below you. But there is no balloon that will give us a bird's-eye view of the Web, because it exists only in "cyberspace"—it corresponds to no physical structure we can look at. So we have to do just what we would to map out the maze from the ground: we must enter it and keep track of where it takes us.

What a strange contrivance it is, then, that we have made for ourselves. We have built it, but we cannot easily tell exactly what it is we have built. Rather, we must investigate its shape as though we were blind, feeling the contours little by little. In 1999, Réka Albert, Hawoong Jeong, and Albert-László Barabási of the University of Notre Dame in Indiana confronted this cartographic challenge by sending a robot into the maze of the Web charged with the task of mapping out its convoluted pathways.

The robot was a computer program that was instructed to enter a Web site and follow all the hyperlinks. These took the robot to an assortment of other Web sites, at each of which it would repeat the same process. On each foray the robot kept a record of the number of outgoing hyperlinks from each page it encountered.*

To conduct this search for all the billion or so documents then on the Web would have been far too much for the robot to accomplish. Instead the researchers told their pet program simply to stay within the boundaries of the Web domain specific to the University of Notre Dame (www.nd.edu). This alone encompassed 325,729 HTML documents (HTML, or hypertext markup language, devised by Tim Berners-Lee, is the standard language for writing Web pages), interconnected by almost one and a half million links. This, the researchers hoped, was a data set sufficiently large and representative to stand proxy for the Web as a whole.

Albert and her colleagues found that the probability distributions of both incoming and outgoing links in this graph were power laws (Figure 16.2). That is to say, the probability of a random page on the university Web site having a certain number of links to or from it depended on the number of links according to a power-law relationship. Most pages had few links; a few had many; and each time the number of links was doubled, the number of pages with that many links decreased by a constant factor.

Although we'd intuitively expect fewer pages with many links than with just a few, a power law is by no means the only relationship consistent with this expectation. The result, in other words, is not obvious. Indeed, we might

---

*To construct the network, one also needs to know about the incoming links to each site. The robot could not identify these directly, but they are implicit in where each outgoing link *goes*.

16.2   The probability distribution of incoming hyperlinks to HTML documents on the World Wide Web follows a power law, generating a straight line on this logarithmic plot. The outgoing links have a similar distribution. This implies that the Web is a scale-free network.

even have supposed that the relationship would fit a bell curve, with most pages possessing an average of perhaps three or four links. But the power-law relationship says that there is no such preference—no *scale* to the connectivity of the network. A power law, as we saw earlier, is an indication that the system is *scale free*.

There is nothing preordained in this power law. Everyone who sets up a Web page at the University of Notre Dame, just like anyone who does so on any other Web domain, is completely at liberty to decide how many hyperlinks to make elsewhere. (Of course, they cannot determine how many such incoming links their page will receive.) Yet out of this multiplicity of individual choices comes a mathematical law that holds true, in the case of the Notre Dame data, for pages with between one and more than a thousand links. The network seems to organize itself into this power-law state, just as the sandpile model of self-organized criticality (page 237) generates a critical state with a power-law distribution of avalanches.

The power law tells us at once that the Web is not like the small-world networks conjured up by Steven Strogatz and Duncan Watts. Their "rewiring" graphs *do* have a preferred connectivity: the probability distribution function of the number of edges per vertex rises to a maximum at some particular value and then declines. Neither is the Web network like a random graph, for that displays a different probability distribution too. The probability of finding a vertex with a very high number of connections is far greater for a power-law distribution than it is for either a random graph or a Strogatz-

Watts small-world graph, just as a power law in economic market fluctuations implies an appreciable chance of large events.

Does this mean that the Web is not a small world? Not necessarily; it simply means that it is not the kind of small world Strogatz and Watts devised. Albert and her colleagues tested whether their data set showed the small-world signature: a small average (characteristic) path length between two randomly selected vertices, coupled with a high degree of vertex clustering. It turns out that these two properties combine to give a certain criterion for a small world: as the number of vertices in the network grows, the average path length increases only very slowly.* The researchers found that a graph constructed to have the same power-law distribution of connectivities as they had observed for their section of the Web does indeed show this behavior.

So the World Wide Web *is* a small world—but one with a very specific topology characterized by a power law: by scale-free connectivity. Albert and her colleagues estimate that, if the entire Web has the same structure as the Notre Dame Web domain, then any two of its Web pages are separated by an average of just nineteen links. Because of the Web's small-world structure, this average distance should increase by only two links even if, as predicted, the number of Web pages increases by 1,000 percent in the next few years. All the information we want will remain just a few clicks away.

If that's so, why can it be so damned hard to find? One of the most common complaints from Web users is that a huge amount of useless information stands between you and the item you need. Unfortunately, it is a struggle for search engines to carve a path through the junk because their search criteria are necessarily rather blunt tools. Yes, that vital document might be just nineteen clicks away—but because of the highly interconnected nature of the Web, so are a vast number of other documents, all of which the search engine must trawl through. Moreover, the rapidly changing size, scope, and content of the Web make it hard for search engines to keep pace by indexing new pages and updating or discarding old ones that no longer exist or cease to be relevant. It has been estimated that the best search engines reach only

---

*Mathematically speaking, the characteristic path length depends on the logarithm of the number of vertices.

about a third of the entire indexable Web, and that some of them trawl through only a tenth as much as this.

Lada Adamic, working at Xerox's Research Center in Palo Alto, California, has shown how the small-world character of the Web can be exploited to design better search engines. These take advantage of the high degree of clustering between Web pages devoted to related topics, which distinguishes the topology from that of a random network. An intelligent search engine could use this clustering to limit its sphere of inquiry, after finding the best "hub" sites from which to launch the search. This "smart" scheme can be more efficient than a random walk through the maze of the Web.

Power laws seem to be a recurring leitmotif of the Web. Adamic and her colleague Bernardo Huberman have uncovered this kind of probability distribution in the number of pages per Web site. And they and their collaborators discovered in 1998 that users who surf the Web also obey power-law statistics. Surfing is the common alternative (or often the complement) to using search engines. You find a Web site that looks as though it might contain the information you want, and then you follow the hyperlinks to other pages on the site until you either find what you seek or conclude that it is not there at all.

Most users will happily surf not just from page to page but from site to site. But Huberman, Adamic, and their colleagues considered only the surfing pathways that users take within single sites. They were interested in how "deep" people go—how many clicks they follow, on average, before quitting the site. By looking at various data sets—the behavior of more than 23,000 users registered with the service provider AOL and visitors to the Xerox Web site, for example—they found that the probability distribution function of the number of clicks on a site obeyed a power law, or something very close to it.* This information can assist Web site designers by helping predict the likely number of hits per page.

What does a scale-free network actually look like? In a random graph,

---

*The Xerox team proposed a model of surfing behavior in which visitors to the site execute something like a random walk through its pages until they come across a page whose information value to them lies below some threshold. At this point they quit the site. The model predicts a probability distribution function that has a so-called inverse Gaussian shape. This is not the same as a Gaussian curve (page 61), but is close to a power law, deviating from it gradually as the path length gets longer. The data aren't really good enough to distinguish between this relationship and a true power law.

most vertices have roughly the same number of edges and the network looks rather uniform throughout (Figure 16.3a). In a scale-free network, most vertices have only one or two links, yet a small but significant proportion are highly connected. Thus the structure is very uneven, seemingly dense or pinched in some places but sparse in others (Figure 16.3b). These highly linked nodes provide the shortcuts that make the network a small world.

The Internet, like the Web, also has this scale-free topology with a power-law distribution of connectivity between nodes (Figure 16.4). Herein lies its strength. Albert, Jeong, and Barabási have shown that scale-free networks are much more resilient against random flaws, such as breakdown of a node, than are random networks or the small-world but scale-dominated networks of Strogatz and Watts. (The latter two types of network may both be classed as "exponential," meaning that the probability of highly connected nodes decreases rapidly—exponentially—as the number of links increases.)

Albert and her colleagues found that a scale-free network barely "notices" if up to one in twenty of its nodes is disabled. The characteristic path length is hardly changed at all. In exponential networks, in contrast, a few malfunctioning nodes can increase the path length substantially, making it harder for all the other nodes to communicate. Moreover, exponential networks tend to break into many isolated clusters when the fraction of "dead"

16.3   Random graphs (a) are rather homogeneous, whereas scale-free networks (b) seem to be "pinched" at a few highly connected vertices.

16.4   The structure of a part of the Internet, mapped out by tracing the shortest routes of messages sent from one central computer to thousands of others. Such maps can be seen at www.cybergeography.org/atlas/topology.html.

nodes exceeds about 28 percent. When this happens, the network is totally unable to convey information over any significant distance. A scale-free network, on the other hand, does not "shatter" but "deflates" slowly as nodes are knocked out. A relatively large connected cluster persists even when half the nodes are dead. Because most of the nodes in a scale-free network have only one or two links, breaking links at random mostly just isolates the nodes one at a time (Figure 16.5).

Thus the Internet has precisely the topology it needs to provide a reliable communications network even if some of its nodes are out of action. Such local failures do not arise only from computer malfunction—a node might become temporarily useless because it is jammed by a high volume of information passing through. An effective alternative route can be found quite easily in a scale-free network even if rather a lot of the nodes are simultaneously jammed. (In fact, around 3 percent of data routers in the Internet network are disabled at any time.)

The remarkable thing is that the Internet has grown *unplanned* into this seemingly most robust of conceivable networks. No one designed it this way.

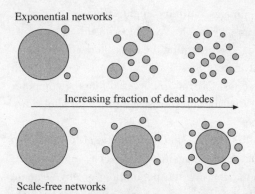

Exponential networks

Increasing fraction of dead nodes

Scale-free networks

16.5   Collapse of networks with dead nodes. Exponential networks, such as the random networks of Erdös and Rényi, break up rapidly as nodes are disabled. In contrast, scale-free networks deflate slowly, retaining a central, highly connected web.

Indeed, if anyone *had* possessed the authority to dictate the topology of the Internet, the chances are that they'd have chosen a far less robust architecture (like Paul Baran's: see Figure 16.1c). The message is clear: sometimes it is best to let technology organize itself. The question remains, however, *why* the Internet has this structure. I shall turn to this at the end of the chapter. First we shall confront a less happy prospect. For despite its robustness, the Internet has a potentially fatal flaw, and it is keeping some people awake at night.

CYBER-ATTACK

It is ironic that the Internet, born out of the desire to provide a communications system robust enough to survive enemy aggression, is now perceived as a vulnerable structure on which the survival of the United States depends. In 1996, President Bill Clinton set up the Commission of Critical Infrastructure Protection to assess ways to protect the electricity, communications, and computer networks from attack.

The form of such attacks is quite different from those feared in the days of the Net's origins. No missiles need be launched, no armies nor air force mobilized. The aggressors need not even leave their homes. They need not be foreign superpowers with immense military might; a single individual armed with conventional computer technology could, with the right know-how, bring a country to its knees. By disrupting the computerized communications of banks and financial institutions, a "cyber-terrorist" could push the American economy into free fall. Misinformation fed into the computer systems of the power and fuel distribution infrastructure could shut down entire states or blow up gas lines.

Terrorism has been defined by the U.S. Federal Bureau of Investigation as "the unlawful use of force or violence against persons or property to intimidate or coerce a government, the civilian population, or any segment thereof, in furtherance of political or social objectives."[9] Cyber-terrorism is an attempt to do that using electronic information networks. It is a real threat, and some politicians are beginning to fear it more than nuclear war. Protection against cyber-terrorism has become a major preoccupation in the United States: the CIA's Information Warfare Center has a staff of more than a thousand, and the FBI, the Secret Service, and the U.S. Air Force all have their own anti-cyber-terrorism squads.

In an age when terrorism has become the specter haunting civilization, this fear is not sheer paranoia. Cyber-terrorism is cheap, it puts the perpetrator in no immediate danger, and it could be devastating. The Pentagon's computer system has already been attacked. In one military exercise, called Eligible Receiver, staff of the National Security Agency used freely available software to gain access to computer systems that would have allowed them to shut down the entire American power grid, and to gain partial control of the U.S. military Pacific Command. Most of the "crackers" eluded attempts to trace them.*

But this kind of network takeover is only one part of the problem, and possibly not the most significant. It would be far simpler, and perhaps just as effective, to use a blunter weapon: merely to stop the network from functioning.

The scale-free small-world topology of the Internet guarantees that attempts to do this by knocking out nodes at random will probably be futile,

---

*A cracker, who aims to use the Internet for disruptive or destructive purposes, is not the same as a hacker, who is primarily interested in developing and improving software according to the unspoken customs of the freewheeling and meritocratic "net culture."

because the network can withstand a considerable amount of such punishment without losing much of its connectivity. But a cyber-terrorist is unlikely to try such a misguided approach. In an exponential network, the existence of a characteristic average scale of connectivity means that none of the nodes is special. In a scale-free network, on the other hand, some nodes are definitely more equal than others. The most highly connected nodes are the crucial shortcuts that keep the average path length low. Remove these nodes and you cut out many links at once, affecting a wide region of the network. If, instead of eliminating nodes at random, we begin with the most highly connected, a comparison of robustness for exponential and scale-free networks reveals a significant difference. In the former case, there is barely any difference in the effects of random and targeted strategies for disabling nodes. In the latter case, the scale-free network breaks up rapidly as the most highly connected nodes are destroyed. Removing one in twenty of the nodes on this basis doubles the characteristic path length—suddenly big detours are necessary to get from place to place.

So a cyber-terrorist who wants to trash the Internet need only to identify a relatively small number of the most highly connected nodes and make those the targets of sabotage. In other words, when an attack is guided by intelligence it can do a disproportionate amount of harm to a scale-free network like the Internet. It is around the linchpin nodes that the most secure safety walls should be raised.

This topological structure makes the Internet more like a "naturall body" than anyone could have guessed. These webs spun by information technology are most often compared to the brain, and perhaps with some justification: Strogatz and Watts have shown that the pattern of connections in the neural network of the parasite nematode worm *Caenorhabditis elegans* displays the characteristic features of a small-world net: large clustering and small characteristic path length between individual neurons. But Albert-László Barabási and his colleagues have uncovered a still more vital network in the cells of a wide range of organisms that supports a striking analogy with the scale-free Internet.

Perhaps the deepest principle of life is metabolism: the conversion of raw materials from the environment into the energy and the molecules that cells need every moment of their existence. Cells need access to a variety of building blocks: our own cells, for example, are supplied with amino acids, sugars, and lipids from food, as well as vitamins and mineral nutrients, water, oxygen, and other essential substances. They use enzymes to rearrange the atoms in these molecules to form new enzymes, nucleic acids, hormones, energy-rich molecules, and so forth. The sequence by which a raw material is converted into a useful molecular form is called a metabolic pathway.

Almost without exception these pathways are not linear but branched and interlinked. A single raw material such as glucose is reconfigured or fragmented in many different ways. The energy-rich molecules made during the breakdown of this sugar are used to power many other metabolic processes. So metabolism defines a large network of chemical reactions in which particular molecular substances can be regarded as nodes, and reactions (usually catalyzed by enzymes) can be considered as the edges that link one node to another (Figure 16.6).

Barabási and his colleagues looked at the metabolic networks of forty-three different organisms ranging from bacteria to plants to "higher" life forms like the nematode worm. They found that the connectivity distribution function in every case was scale free: the probability of a node having a cer-

16.6   A part of the yeast metabolic network. The vertices are molecules either supplied to or manufactured by the network, and the links (edges) are defined by the enzymatic chemical reactions that convert one molecule to another.

tain number of links followed a power law. This implies that there are a few highly connected "hubs" in the network that play a crucial role in holding the web together. Many of the molecules representing these hubs, and their relative importance in the network, are the same for all organisms—a reflection of life's common evolutionary origins.

The scale-free structure of metabolic networks makes sound evolutionary sense, because it makes metabolism relatively insensitive to a small incidence of random failures. If one or two enzymes are defective, perhaps due to a genetic defect, this weakens or even severs the corresponding edges in the graph. While this can have harmful consequences for certain biological functions, it need not, in a scale-free network, break up the whole web and make life impossible. Thus we can regard the scale-free network as an example of good "engineering" wrought by the trial-and-error exploration of options that is natural selection.

On the other hand, the weakness of scale-free networks is their sensitivity to planned attack. If the hub nodes are removed, the network rapidly falls apart. This suggests opportunities for the intelligent design of drugs to combat bacterial infection. Drugs that interfere with the most highly connected molecules will have a disproportionate effect on the viability of the attacked cells. Understanding the network structure would be a first step toward selecting appropriate targets. In this scenario we are the "cyto-terrorists," and our intentions are, from a human perspective, purely benign.

Could the massive breakdowns of power grids in the United States and Italy in 2003 have had anything to do with the topological vulnerabilities of the networks? Barabási thinks so, even though it is not clear that all power grids are scale free. "[The August] blackout has little to do with faulty equipment, negligence or bad design," he says. "The magnitude of the blackout is rooted in an often ignored aspect of our globalized world: vulnerability due to interconnectivity."[10] Barabási points out how systems such as the power grid are prone to cascades of failure where a malfunction at one point shifts power to other lines, leading to an escalation of overloading. "Cascading failures are common in most complex networks," he says. "While celebrating that everybody on earth is only six handshakes from us, we need to accept that so are their problems and vulnerabilities."[11]

SPREADING IT AROUND

Willful destruction and accidental failure of nodes are not the only dangers facing our social networks. Computer viruses represent a more insidious threat to electronic communications than does cyber-terrorism: rather than destroying the network, viruses take advantage of its extreme connectivity to colonize it and wreak havoc at the nodes. Here again, in other words, a beneficial property of the Internet's topology is turned against it.

Computer viruses typically spread by e-mail contact—by the transmission of parcels of information between one computer and another, just as biological viruses are transmitted between individuals as particles in breath or body fluids. Once a link is established between two nodes of the network, they can potentially infect each other.

The study of how diseases spread in a population—the science of epidemiology—has a long history. Computer scientists have borrowed the mathematical models developed by epidemiologists to try to understand the dissemination of computer viruses. One of the standard approaches in epidemiology is to assume that at any given moment an individual can exist in one of two states: healthy and infected. Healthy agents are susceptible to infection from infected agents. An encounter between the two is assumed to lead to infection with a particular probability. At the same time, infected individuals have a chance of becoming healthy (which is to say, susceptible) again. The disease then spreads at a rate equal to the ratio of the probabilities of transmission and of recovery. This is known as the susceptible-infected-susceptible (SIS) model.

Epidemiologists using this model have found that it predicts an epidemic threshold. The disease spreads throughout the population, persistently infecting a constant proportion of people, if the spreading rate is greater than some threshold value; otherwise, it dies out quickly.* Some real epidemics seem to behave this way. The idea of a threshold is central to vaccination programs: as long as a certain proportion of the population is vaccinated, the

---

*The threshold is technically equivalent to a critical point in a nonequilibrium phase transition—like the liquid-gas critical point, except that the states separated by the transition are nonequilibrium states.

spreading rate remains below the threshold value and the disease cannot reach epidemic levels.

Duncan Watts, working with Mark Newman of the Santa Fe Institute in New Mexico, has shown that the small-world networks he devised with Steven Strogatz also possess thresholds for spreading of a "disease" in an SIS-type model. In other words, if social networks are somewhat like these webs, the standard ideas of epidemiology hold true.

But the Internet is not a Strogatz-Watts small world; it is a scale-free small world. Stefan Bornholdt and his co-workers at the University of Kiel have demonstrated that e-mail connections through their university server trace out a scale-free network, suggesting not only that the Internet is *physically* scale free (that is, in terms of user nodes and links between them) but that the "acquaintance network" defined on this electronic web has this same topology.

The physicists Romualdo Pastor-Satorras in Barcelona and Alessandro Vespignani in Trieste have found that this difference completely changes the way in which computer viruses spread themselves through the system. When they used computer simulations to investigate the SIS model on a scale-free network, they discovered that there is no threshold at all. No matter how slow their spreading rate, all viruses can pervade the system, infecting a certain proportion of the nodes. Because infected nodes can be "cured" with antivirus software, the virus does eventually die out. But this process can be very slow. Software to combat a particular virus typically becomes available within days or weeks of the first infection, whereas viruses can survive in the Internet's dense web of connections for many months or even years. The researchers studied the prevalence data of viruses reported between 1996 and 2000 and found that although the survival probability drops sharply in the first few months of a virus's life history, a low level of infection remains for long afterward. (The immensely damaging Love Bug virus, for instance, is still the seventh most prevalent computer virus, even though it is supposed to have been eradicated.)

This behavior is puzzling for conventional epidemiology. The standard models predict that a viral infection should either become an epidemic or die out quickly. Yet the slow, lingering decay, without an epidemic threshold, seems to be a peculiar characteristic of transmission on scale-free networks. The discovery has implications both good and bad. On the downside, even a slow-spreading virus can pervade the system and last for a long time. But it does so only at a very low level—the fraction of infected nodes rapidly becomes very small. This result highlights the fact that network topology matters. For those seeking to combat computer viruses, standard epidemiological wisdom may not be sufficient.

What if real social-contact networks are scale free too? If they are, we may need to rethink our notions about how real diseases spread. As we have seen, social contacts are notoriously hard to map. But a team of sociologists at Stockholm University in Sweden has joined forces with the physicists Gene Stanley and Luis Nuñes Amaral in Boston to uncover the web of sexual contacts in a random sample of almost three thousand Swedes between the ages of eighteen and seventy-four. They found that the distribution in the number of partners over the twelve months before the survey was conducted obeys a power law. In other words, this contact network is scale free.

If this sample is representative of populations elsewhere, there are serious implications for strategies to combat sexually transmitted diseases such as AIDS. It is extremely difficult to eradicate diseases completely in a scale-free network, since they can permeate the population regardless of the spreading rate. Where a vaccine is available, random immunization against the disease cannot contain it even if a large proportion of the population is treated (perhaps at considerable expense).

But the picture is not necessarily so gloomy. As we have seen, the Achilles' heel of scale-free networks is the disproportionate influence of just a few highly connected nodes: the "hubs" that hold it all together. Sever the connections to these hubs and the web quickly falls apart. Pastor-Satorras and Vespignani have shown that in such circumstances, targeting immunization programs at the most promiscuous individuals sharply lowers the network's vulnerability to epidemics of sexually transmitted diseases. (By the same token, an analysis of e-mail networks shows that "immunizing" just 10 percent of the vertices selected on the grounds of their large number of connections can effectively suppress a computer virus outbreak entirely.)

In networks of sexual contacts, it is often far from easy to identify and treat these key individuals. However, Barabási and his student Zoltan Dezso have found that even a rather inefficient targeting campaign has a crucial effect: it raises the epidemic threshold of a scale-free network above zero. In other words, even if only a few of the hub nodes are isolated, there is a chance of eradicating the disease. The more effectively a policy reaches the most highly connected members of the sexual network (in the case of AIDS), the higher the threshold—and the easier it is to stamp out the disease. Faced with limited resources, it is therefore better to combat an infectious disease by making at least some effort to focus control strategies on the "hub" individuals rather than to promote random, blanket immunization and hope for the best.

WORLD OF WEBS

Scale-free networks may be a much more widespread form of small world than the rewiring networks of Strogatz and Watts.* Barabási and Albert revisited the case studies that Strogatz and Watts first considered: the movie-actor network and the western U.S. power grid. For Strogatz and Watts the only indicator of small-world behavior was the coexistence of large clustering of nodes and a small characteristic path length. These criteria do not uniquely specify the topology of connections, however—they apply both to rewiring networks and to scale-free networks. Barabási and Albert found scale-free distributions of connectivity both for movie actors and for power lines. Thus, for example, there is no meaningful "average" connectivity between movie stars; the number of links decreases smoothly (as described by a power law) from Rod Steiger to the most obscure B-movie character actor.

The movie-actor network supplies probably the best-documented evidence we have so far for suspecting that real social networks are scale free—although studies of scientific-collaboration networks (like the "Erdös network"), musical collaborations in the early jazz world, and sexual-contact networks support that contention. On the assumption that this is so, Ricardo Alberich and his co-workers at the University of the Balearic Isles on Mallorca have shown that social networks can provide a kind of watermark to distinguish fact from fiction. Spider-Man's friendship web, they say, just doesn't look real.

Marvel Comics, which began in 1939 under the name of Timely Comics,

---

*In fact it is not clear that *any* of the real small-world networks studied so far is truly like a rewiring network.

describes a bizarre universe of superheroes: Captain America, the Fantastic Four, the X-Men, and thousands more. These characters move in the same world as one another; their stories overlap, so that many may appear together in the same comic. When this happens, say the Spanish researchers, it constitutes a "social connection"—a link between two characters in the Marvel universe network. This network contains about 6,500 characters (nodes) in all, which are featured in around 13,000 comic books. Does it have the same topological features as a real-life web, such as the movie-actor network?

Well, sort of. That's to say, the probability that any given comic book has a certain number of characters in it follows a power law.* To this extent, the Marvel universe network looks realistic. But "it cannot completely hide its artificial origins," the researchers conclude. They find that the *clustering* of the network is much lower than that of scale-free or small-world nets, and indeed scarcely larger than that of a random graph. Clustering describes the extent to which characters form into "friendship circles": in a highly clustered net, remember, two people who have a common friend are more likely to know one another than are two people chosen purely at random.

The small degree of clustering in the Marvel universe reveals its ad hoc, fictional nature, showing that inventing a universe is harder than it seems.

In a sense, this is perhaps not surprising. After all, the Marvel story writers had no notion of what social networks look like, and were not trying consciously to create one that seemed "realistic." They simply put two or more characters together if they imagined the combination would be interesting. Of course, real social networks are also unplanned. It's just that Spider-Man's social network and yours or mine seem to have grown according to different principles. It's not yet clear how those differences arise, but if we can uncover the reasons, we might learn more about how social connections are forged in the real world.

Scale-free networks are now starting to look like such a fundamental aspect of human culture that eyebrows are raised and questions asked when they do *not* appear. The global trade network, the Indian railway system, the trading activity in online auctions such as eBay, the Chinese airline system—all display the familiar power-law distribution (though often with an

---

*Actually the power law obtains only when the number of characters appearing together in a story is more than ten. For smaller groups, the statistics deviate from the power-law distribution.

important proviso, described below). When the emergence of leadership structures is studied using the minority-game model of Challet and Zhang (see page 333), a scale-free hierarchy is generated in which a few key individuals exercise inordinate influence. It seems that once people begin to interact and establish connections, the ubiquitous Gaussian distribution that so dazzled early social statisticians vanishes, and the scale-free distribution emerges in its place.

DO THE RICH ALWAYS GET RICHER?

When a pattern recurs in many different systems that bear no obvious relationship to one another, we must suspect a common causative principle, one which can be couched in the most general terms without reference to the specifics of this or that case. Abrupt phase transitions tend to arise, for example, when competing forces are at play in systems of many "particles": a propensity toward order and a disrupting influence of noise. What, then, are the generic principles that create scale-free networks?

Both the random graphs of Erdös and Rényi and the rewired small worlds of Strogatz and Watts are constructed by purely random processes. One chooses vertices in the graph at random and links them together via an edge. But Barabási and Albert point out that most networks do not grow in this way. Instead there is a bias toward linking a new node to one that is already well connected. New actors are more likely to play supporting roles in films with established stars than they are to be cast in a film full of unknowns. Thus the greater your fame, the greater the likelihood that you will attract new links.

Similarly, a kind of "magnetism of fame" operates in the evolution of the World Wide Web or the network created by scientific citations (another scale-free system). Web pages that already have a high number of incoming hyperlinks become well known and are the natural choice for Web designers who want a pointer from their new page to a source of further information. People cite famous papers because they are the ones everyone else cites, or because they are more likely to have read them. Fame breeds fame, in other words.

One would like to imagine that there is some meritocratic principle un-

derlying all of this: that Web pages receive lots of links, papers get many citations, and actors are frequently cast because they are *good*. It would be
surprising if that were not so. Nonetheless, fame brings increasing returns to
good and bad alike. You can no doubt supply your own examples.

If every new vertex in a growing graph was invariably linked to the most
highly connected vertex, the network would become not scale free but centralized around a single hub. This is not usually what happens when networks grow unplanned. If the graph is large, it will be extremely difficult for
a new vertex to "find" the most highly connected vertex, and the chances are
that some less-than-maximal vertex will be selected instead. In the movie
world this variety of choice is inevitable: even the most famous and well-
connected star can't appear in every new movie. So the formation of new
links to the most highly connected vertices can be only a bias, not a certainty.

Barabási and Albert have shown that this bias is the only ingredient
needed to grow a scale-free network. They imagine a graph that is growing
by the addition of new vertices, each one linked to an existing vertex chosen
at random *but* with a bias that gives highly connected vertices a better
chance of acquiring the new links. The resulting network is scale free. Many
human organizational networks develop on this "rich get richer" principle:
bigger businesses are more likely (but not guaranteed) to get new clients, for
example, partly because they can afford more and better advertising.

If the connectivity of a vertex can be equated in some cases with its
"wealth" (for example, if we are thinking about the trading connections of a
company), this suggests that in a society governed by freedom of choice, and
where one's ability to gather a "market share" is determined by how much
of it one already has, a power-law distribution of inequality is the probable
outcome. We would of course *expect* there to be differences in access to or
command of resources in any free market; but the scale-free growth process
produces greater disparities than a random distribution of wealth. It results
in a significant number of extreme "events": very wealthy individuals or
huge companies. As the social scientist George Kingsley Zipf showed in the
1930s, this power-law behavior is precisely what we see for many size distributions in society, from the size of firms (page 265) to the size of cities and
the distribution of incomes (see page 246).

This is not to say that power-law disparities in a free market are inevi-

16.7   The most highly connected movie actors do not quite follow the power law that characterizes the rest of the network—those with more than three hundred or so collaborations have rather fewer connections than the power law predicts (they plot below the straight line). There is, in other words, an upper limit to the number of collaborations an actor can attain.

table. But it does suggest that if we decide they are undesirable, we shall probably need to restrict some of the freedom with which the market operates.

This type of network growth does not always guarantee quite such a high degree of inequality, however. Gene Stanley and his co-workers at Boston University have looked more closely at the supposedly scale-free networks identified by Barabási and Albert and have discovered that they do after all appear to have a ceiling. For example, the power law in the movie-actor network, which seems inexorably to reward a few privileged individuals with even better connections, begins to falter at the high extreme: the best-connected actors have rather fewer links than the power law would predict (Figure 16.7).*

What truncates the power law? Stanley and his colleagues say that in the real world some factor invariably imposes an upper limit on the number of links a node can acquire. Actors have limited time and finite careers—even the hardest-working of them cannot appear in a thousand films. Old scien-

---

*The statistics of this network are affected, however, by the ambiguities of TV shows. Stanley and his colleagues included them; Barabási and his co-workers did not. A TV show is listed in the actors' database as a single title, even though it might constitute a long-running series with many guest stars who do not actually appear together in a single episode. This can generate some spurious links, and it may affect the extent to which power-law behavior is displayed.

tific papers, even the most seminal, fall victim to neglect—researchers no longer read the old literature, but instead cite a more recent review article or textbook on the subject. Airports cannot handle an infinite amount of traffic, and cost or local demographics sets limits on how much they can expand. If vertices either have a capacity limit or get less able to form new connections as time goes by (a kind of aging), the scale-free structure of the network disappears for the most highly connected nodes.

For some other social networks a power law does not even get established in the first place. Instead the connectivity declines rapidly for the most highly connected links, so that the probability of "superhubs" is vanishingly small. This is true, for example, of the Southern California power grid, and also of the network that links the world's airports by flight paths. Moreover, Stanley and his colleagues found it to be the case for some real social networks: the friendship networks of forty-three Mormons in Utah (where such ties are unusually close and well defined) and of several hundred students at Madison Junior High School in Wisconsin. All these networks seem to have a Gaussian probability distribution: there is an average "scale" to the connectivity. Nevertheless, they are also small-world networks, as revealed by the very slow increase in characteristic path length as the number of vertices increases.

Thus there are several types of small world. The rewiring networks defined by Strogatz and Watts are one example of a "single scale" small world in which there is a preferred average connectivity and a sharp decline in the numbers of highly connected vertices. At the other extreme are the scale-free networks found by Barabási and his team, where greed or promiscuity knows no bounds and even very highly connected vertices are far from rare. In between these two comes a "whole zoo of networks,"[12] in Barabási's words. Whatever its precise topological properties, a small-world network is seldom if ever planned; it simply emerges from the laws that govern the connection of new vertices. Yet Nish Mathias and Venkatesh Gopal in Bangalore, India, have argued that such a network nevertheless has great attractions from an engineering point of view because it typically provides an ideal compromise between two conflicting urges: to maximize the connectivity of vertices and to minimize the length of the "wiring." In general, connecting two nodes in a network has a financial cost proportional to the distance between them. You need twice as much cable to link two power plants one hundred miles apart as you do if they are fifty miles apart. A network communicates most efficiently if every node is connected directly to every other; but that requires an awful lot of wiring. So network builders want fewer and shorter connections, and network users want more and longer ones (and hence many shortcuts). Mathias and Gopal have shown that the first of these options corresponds to a regular grid, and the second to a random network. A small-world network is a way of getting good long-range connections at the cost of little more wiring than is needed for a grid that makes only "local" links.

The physics of networks is a very young science, but already it has revealed some apparently profound rules that apply to our social patterns and institutions. The consequences of network topologies for processes of change such as the spread of disease epidemics or cultural norms are at present only barely sketched. In economics, we have seen that the structure of trading networks has been identified as one of the key questions in understanding how the complex dynamics of the market arise. According to the organizers of a workshop on econophysics at the Santa Fe Institute in 1996, "Markets actually function by means of networks of traders, and what happens in markets may reflect the structure of these networks, which in turn may depend on how the networks emerge."[13]

In a sense we can see network theory as a natural extension of statistical physics: a kind of "sticky" version of many-particle systems in which encounters between individuals lead not simply to collisions but to the formation of permanent, elastic links. The network then becomes a kind of map of the history of the system, just as the expansion of the Internet over time is recorded in the growing web that it weaves. In networks the past matters: a memory of it is frozen in place and shapes the present. These networks provide a graphic image of the extent to which our lives become entwined in ways as innumerable as they are barely fathomable.

SEVENTEEN

ORDER IN EDEN

LEARNING TO COOPERATE

Nor do I know, if men are like sheep, why they need any government: or if they are like wolves, how they can suffer it. —Sir William Temple (1751)

In transgressing the law of Nature, the offender declares himself to live by another rule than that of reason and common equity . . . and so he becomes dangerous to mankind; the tie which is to secure them from injury and violence being slighted and broken by him . . . every man upon this score, by the right he hath to preserve mankind in general, may restrain, or where it is necessary, destroy things noxious to them, and so may bring such evil on any one who hath transgressed that law, as may make him repent the doing of it, and thereby deter him, and, by his example, others from doing the like mischief. —John Locke (1690)

It would be child's play to shell the road behind the enemy's trenches, crowded as it must be with ration wagons and water carts, into a bloodstained wilderness . . . but on the whole there is silence. After all, if you prevent your enemy from drawing his rations, his remedy is simple: he will prevent you from drawing yours.
—Ian Hay (1916)

W hat makes us civilized? The many answers to this question tell us much about those who supply them, but scarcely help us reach a consensus. Most commentators since the Enlightenment have focused on the tension between personal and collective freedoms in a civilized society—a tension that demands compromise. In Sigmund Freud's words,

The liberty of the individual is no gift of civilization. It was greatest before there was any civilization, though then, it is true, it had for the most part no value, since the individual was scarcely in a position to defend it. The development of civilization imposes restrictions on it, and justice demands that no one shall escape those restrictions . . . This replacement of the power of the individual by the power of a community constitutes the decisive step of civilization.[1]

How, though, to circumscribe those powers and those restrictions? Where should the compromise lie? For Hobbes, civilization can emerge from the barbaric State of Nature only if we voluntarily take up the yoke of dictatorship, relinquishing all rights except the right to exist. The social contract of John Locke was altogether more collaborative: citizens relinquish certain freedoms to the State only as long as the State serves their interests. In other words, the State does not have the almost unlimited power that Hobbes would grant it. A populace, said Locke in *Two Treatises of Government* (1690), has the right to rise up and displace a government that breaks the social contract—not capriciously, for that way lies anarchy, but after "a long train of Abuses, Prevarications, and Artifices."[2] Moreover, Locke recommends safeguarding against such abuses of power, to which Hobbes (perhaps strangely, given his pessimistic view of human nature) seems largely indifferent.

At whatever point the balance of power between the state and the people is set, there is a recognition that civil peace and order come at the expense of individual restraint—either self-imposed or state-enforced. All liberal philosophers since the seventeenth century have concurred that this must include, at the very least, the renunciation of attempts to harm others. Karl Popper says,

> Absolute freedom is nonsense . . . We need a society in which the freedom of each person is compatible with the freedom of other persons. The compatibility of my freedom with yours depends on our both renouncing violence towards each other. I won't knock you down, and you won't knock me down.[3]

Freud agreed but acknowledged that, as Hobbes had asserted, this runs contrary to our aggressive and destructive impulses, our "death instinct." As a result, he claimed, mankind's aggression "is introjected, internalized; it is, in point of fact, sent back to where it came from—that is, it is directed towards his own ego."[4] The consequence is a pervasive sense of guilt, which manifests itself as proscriptive religion and the concept of original sin.

Maybe so. Yet the common assumption in such arguments is that by agreeing not to harm one's neighbor, one is thwarting one's best interests—acting contrary to reason, if "reason" is to be equated with a Darwinian survival of the fittest rather than with Locke's divine law or Kant's categorical imperative. Does the cheetah, after all, display altruism toward the gazelle?

In the second half of the twentieth century, all these discussions about liberty, government, and power have been shown to be shortsighted in a profound and astonishing way. Political philosophy has seen humankind as either basically bad and therefore in need of some authority to enforce good behavior, or basically good, thus for the most part guaranteeing that our relations with our neighbors are civilized. But a theory that has its origins in the frivolity of parlor games has demonstrated how good behavior can arise and persist even among enemies, even in the total absence of moral considerations.

Game theory, as it is called, is not generally deemed to be a part of physics at all, but of mathematics (if of a decidedly empirical stamp). Yet it is very decidedly in the spirit of the kind of social physics we have considered so far, in which the behavior of a collection of individuals arises in a nonintuitive but robust manner from the interactions between them. In exploring this theory in this chapter and the next, we shall come across some now familiar phenomena: indifference to details, abrupt changes in patterns of behavior or in statistical populations, sensitivity to fluctuations, and generalizable "laws."

Some of the conclusions of game theory go to the very root of the issue of how we should live and govern our lives. They may challenge our fundamental systems of belief. And since they already *are* taken seriously in the corridors of power, it is important to try to understand as clearly as possible what we may and may not deduce from them, and to what extent they pertain to real human interactions. At one point in its development, game theory seemed to advocate a Cold War policy of belligerent international relations that would have been disastrous had it been rigorously adopted; only later did it emerge why such a conclusion was flawed. Both liberals and conservatives can find support for their beliefs within this discipline, particularly if they select carefully from its findings. More than any other area of mathematical social science, game theory demands that we confront objective results with moral intelligence and exercise great care when we are tempted to couch technical conclusions in anthropomorphic terms. Potentially, it is political dynamite.

IS GOVERNMENT NECESSARY?

In those countries where it has been put into practice, the democratic foundation of John Locke's political philosophy is now so taken for granted as the "right" way to run a society that it is hard to imagine how anyone could have felt otherwise, except out of sheer self-interest. But Locke's *Two Treatises of Government* was written as a counterargument to an alternative that was widely advocated at the time: the notion that kings ruled by divine right. This was asserted in Robert Filmer's *Patriarcha; or The Natural Power of Kings*, published in 1680 but written before the English Civil War, when the political climate was very different. Filmer's sovereign is, if anything, even more fearsome than Hobbes's. At least Hobbes would have him elected from (and by) the masses, whereas for Filmer the king can do as he will with impunity because he is a descendant of Adam and his authority comes from God. This was, naturally enough, music to the ears of Charles I, who knighted Filmer before discovering for himself the fate of kings who become a law unto themselves.

In the seventeenth century Filmer had the weight of history on his side. Most countries and states since antiquity had been monarchies, and the acme of civilization was considered to be embodied in the autocratic ancient Greek and Roman states, some of whose leaders proclaimed their own divinity. Plato acknowledged that monarchy could develop into tyranny under an immoral ruler, but his preferred alternative was an aristocracy—the rule of a few good men. Plato rejected and deplored democracy—literally, "the rule of the people"—because the masses harbored many wicked and immoral people.

Not that it hadn't been tried. Following a succession of tyrannies in the Greek states of the seventh and sixth centuries B.C., Athens established a democratic government in 507. In many respects it was an appealing advertisement for this system of rule: Athenian democracy fostered Sophocles, Euripides, Aristophanes, and Thucydides. This was the society that built the Parthenon and produced the paintings of Apelles, reputedly the greatest of the ancient artists. But the Athenians finally suffered devastating defeat at the hands of Sparta in the Second Peloponnesian War, and their democracy collapsed in 411 B.C.

It is difficult to decide on the best system of government, however, until you have decided why government is necessary in the first place. For Plato and his contemporaries, the need for leadership was taken for granted, and the central question was how to acquire good leaders rather than bad. For Hobbes, that was a secondary matter: almost any leader short of a monster was better than none, for how else could man be prevented from waging war on his neighbor?

Locke was more inclined to see the good in humankind. He too spoke of a State of Nature that existed before nations and governments, but it was a benign Eden: "Men living together according to reason, without a common superior on earth, with authority to judge between them."[5] Here *reason* means adherence to a kind of divine principle that respects such "self-evident" truths as "Thou shalt not kill": "Reason . . . teaches all mankind . . . that being all equal and independent, no one ought to harm another in his life, health, liberty, or possessions."[6] Thus Locke's State of Nature is quite different from Hobbes's vision of every man for himself, and Locke probably had Hobbes in mind when he said,

> And here we have the plain "difference between the state of nature and the state of war," which, however some men have confounded [them], are as far distant, as a state of peace, goodwill, mutual assistance and preservation, and a state of enmity, malice, violence and mutual destruction are from one another.[7]

In his complaint against Hobbes's brutal view of human behavior, Locke was by no means alone. His patron Lord Ashley Cooper, the third Earl of Shaftesbury, echoed him in 1711 by charging that Hobbes "forgot to mention Kindness, Friendship, Sociableness, Love of Company and Converse, Natural affection, or anything of this kind."[8] It is understandable that Locke imagined people to be guided by these things in a State of Nature, since his faith in human nature stems from God's power over men. Hobbes the atheist (by all appearances) took less rose-tinted a view.

Locke's optimistic axioms, then, allow him the luxury of a less oppressive system of rule, so that the social contract between people and government has authority only as long as the leaders serve the common good. And to ensure that the government does not gravitate toward tyrannical oligarchy, Locke insists on checks and balances to power. Leadership is to be divided into legislative, executive, and judicial branches, each with the authority to restrain abuses by the others. In seventeenth-century England these roles were filled by Parliament, the king, and the courts, respectively. In the United States today, their equivalents are Congress, the President, and the Supreme Court.

Despite his generous view of human nature, Locke acknowledges that criminals and miscreants do exist. And his remedy, while men live in a State of Nature, is a biblically ruthless "natural law" which permits that "whoso sheddeth man's blood, by man shall his blood be shed."[9] Thus all is not necessarily so tranquil in Eden, after all.

But when a state exists, such sanctions no longer fall to the individual: natural law gives way to "positive law" decreed and enforced by legislation. Indeed, this is the main purpose of the state: to maintain law, order, and justice—particularly, as far as the materialist Locke is concerned, with regard to one's possessions: "The great and chief end of men uniting into commonwealths, and putting themselves under government, is the preservation of their property."[10]

As long as this is so, all is well. Thus from Locke's faith in human goodness springs the notion that a government should interfere as little as possible in civil life. This is the touchstone of most liberal political philosophy. Immanuel Kant argued similarly against the paternalistic state, even when (perhaps especially when) the state tells us that its dictates are for our own good. Kant's political beliefs were echoed by the Prussian Wilhelm von Humboldt in *Ideas Toward a Definition of the Limits of State Action* (1851), and by John Stuart Mill who, in his essay *On Liberty* (1859), asserted, "The only purpose for which power can be rightfully exercised over any member of a civilized community, against his will, is to prevent harm to others."[11]

Mill even maintained that an individual has the right to harm *himself* so long as this does not injure others. At the root of this libertarian philosophy is the benign social mathematics of utilitarianism, defined thus by Mill:

> The creed which accepts as the foundation of morals, Utility, or the Greatest Happiness Principle, holds that actions are right in proportion as they tend to promote happiness, wrong as they tend to produce the reverse of happiness. By happiness is intended pleasure, and the absence of pain; by unhappiness, pain, and the privation of pleasure.[12]

To some political philosophers, even a minimally interventionist state was too much. Rousseau expounded a romantic anarchism based on the belief that civilization is degenerate and only the primitive "savage" can be noble and good. Man is naturally good, claimed Rousseau in 1754, and it is only institutions that make him bad. Thus the activities and pursuits of civilized society—the arts and sciences, organized agriculture and industry—are corrupting. Rousseau's own version of the "social contract," in his book of that name from 1762, does not reach far beyond Locke's in promoting democracy and debunking the divine right of sovereigns (though even that much was dangerous in prerevolutionary France, and Rousseau was forced to flee to Germany). But his real agenda is more radical:

> Europe is the unhappiest Continent, because it has the most grain and the most iron. To undo the evil, it is only necessary to abandon civilization, for man is naturally good, and savage man, when he has dined, is at peace with all nature and the friend of all his fellow-creatures.[13]

It seems here almost as though Rousseau has gone out of his way to show us the flaw in his dreams, while himself remaining oblivious to it. Lions too will gaze benevolently on their prey, *when they have dined.* But it is the finding of dinner that stokes the fires of conflict and war. When man is hungry, whether he is "savage" or "civilized," he is apt to do whatever he needs in order to get food. And we are many in a world of limited resources. Even in the eighteenth century that was clear, and Malthus's prediction that things could only get worse (see page 56) led Darwin to present a very different picture of the State of Nature: red in tooth and claw, as Tennyson put it.

Darwin's theory of natural selection seemed to cast us back into a Hobbesian world of brutish nastiness, where nature advises us to grab what we can and let the devil take our neighbor. Restraint can be bred into people, but Darwinism makes selfishness the law of the wild.

Darwin himself believed that humankind is exempt from the savage law of nature by virtue of our evolved tendency toward socialization. Indeed, he regarded this as an essential feature of our evolutionary fitness: "The small strength and speed of man, and his want of natural weapons, &c., are more than counterbalanced by his . . . social qualities, which lead him to give and receive aid from his fellow-men."[14] But perhaps predictably, natural selection was soon enough seen by others as a blueprint for justifying aggression rather than cooperation. Kill or be killed: that was the law.

Some sought to escape this deeply unpalatable prospect. In 1902 the Russian prince Peter Kropotkin published *Mutual Aid*, a book intended to show that cooperation, not competition, is innate to humans. He cited examples ranging from the customs of Polynesian islanders to the establishment of medieval guilds. Kropotkin was motivated less by a wish to argue for the best in humankind than by the need to establish human goodness as a foundation for his anarchist political opinions: If people have a natural inclination to cooperate, why should government be necessary? But Kropotkin had only anecdotes to support his case; the Darwinists apparently had a law of nature.

Besides, not even Kropotkin's examples of altruism can evade the unfortunate truth that civilization has its discontents—its freeloaders and its criminals. Not everyone plays the game. Locke's state was designed to apprehend and punish thieves, vagabonds, and murderers. Of course, many people break the law because they are penniless and desperate, or in the throes of violent passion, or mentally disturbed. But others do so simply because they stand to gain and deem the risk of punishment worth taking.

Refusal to cooperate with society need not involve obvious theft or the harming of other individuals. Another person's tax evasion has a negligible effect on me (or anyone else) personally; but for once that purse-lipped catchphrase of the nanny state—"What if everyone did it?"—makes a valid point. The problem is as old as communal living. In the Middle Ages villages had common ground where everyone could graze their livestock. The system depended on restraint. If someone grazed more animals than others did, they gained an advantage at the expense of a slightly faster depletion of the common grazing land. And if one person did it, others would follow. The result was the "tragedy of the commons," which were rapidly grazed bare. Are things any different now, with our seas overfished to near depletion, our rivers polluted, and our skies laden with greenhouse gases?

This is the basic problem for societies based on the principle of unenforced cooperation: they are ripe for exploitation by those who put self before community. So was Kropotkin's dream simply that? Is draconian legislation essential to prevent freeloading?

WHO SHOULD RUN THE WORLD?

There is no field in more urgent need of an answer to the problem of society's "discontents" than that of international relations. For both Hobbes and Locke, a commonwealth of people becomes a single entity, a "body politic"—a Leviathan. And Leviathan was a notoriously belligerent beast. Nations were like men in a State of Nature, since there was no overarching government to contain their actions. Will they then coexist, as Locke put it, in a "state of peace, goodwill, mutual assistance and preservation," or a "state of enmity, malice, violence and mutual destruction?"

Hobbes had no doubts about the answer. For all his attempts at rigorous logic in analyzing the interactions between individuals, when it came to the relations between states he threw up his hands in despair:

> In all times, kings and persons of sovereign authority, because of the interdependency are in continual jealousies, and in the state and posture of gladiators, having their weapons pointing and their eyes fixed on one another; that is, their forts, garrisons, and guns upon the frontiers of their kingdoms; and continual spyes upon their neighbours; which is the posture of war.[15]

The late seventeenth century, the dawn of the supposed Age of Reason, was marked by some of the most violent, petty, brutal, and avaricious decisions ever made by the leaders of European states, and it seems that Hobbes regarded them as beyond hope of rational analysis. Yet how different is his description from the state of the world three hundred years later?

Viewed in this light, the case for world government—favored by early-twentieth-century scientific thinkers such as H. G. Wells and Leo Szilard—looks strong. Even if the majority of nations choose peace unilaterally, what can be done about rogue states in the absence of an enforceable legal framework that binds all? This is, of course, the purpose of the United Nations; but as we have recently witnessed, even Western democracies do not always observe the UN's dictates. The Declaration of Universal Human Rights is not enforced; nations disregard the International Court of Justice whenever it suits them. We cannot pretend there is a world government in anything like the same sense as there are national governments.

Could world government ever be established other than by war? And even if it were possible, would we want it? The idea of a World State dates back at least to Kant and Rousseau; but Kant rejected it, even though it would create a "state of universal peace," because he feared this peace would soon become of a kind enforced by "the most horrible despotism." Rousseau argued for a federal government to free Europe of its perpetual strife, but it was a far cry from today's European Union—capable of compelling every state to join and to do its bidding, an authoritarian vision that even the most fervent modern Europhobe cannot plausibly scare us with.

Robert Kagan of the Carnegie Endowment for International Peace in Washington, D.C., argues that the European Union now has its Kantian perpetual peace, and in a benign form—but that this has been made possible only because the United States has remained in a Hobbesian world where power matters above all else. While Europe advocates a world kept in check by international law and negotiation, argues Kagan, the reality is that at present only U.S. military might maintains order. Europe's "posthistorical paradise" relies on the United States remaining "mired in history."[16] Kagan's analysis is simplistic—he treats nations as though they were just so many Hobbesian brutes ready to wage war purely in pursuit of power, largely ignoring considerations of trade and economics, culture and history. (Some Islamic states are antagonistic toward the United States not because they want to conquer it, for example, but because they dislike its policies in the Middle East.) All the same, Kagan reminds us that the questions of international power politics debated in the Enlightenment are still as pressing, and as unanswered, today—and indeed that the terms of the debate have not shifted very much.

Can nations ever cooperate without a central authority? Why should they do so when powerful nations stand to gain by exploiting small ones? Is it better to act as an aggressive "hawk" or a pacifist "dove"? Or should nations at least strive to *look* like hawks by accumulating massive military forces and nuclear arms? And what is the "right" (Locke would have said "natural") response to aggression? Can there be a just war? Game theory will not alone enable us to resolve these questions, but by stripping them down to their bare bones, it can at least help us formulate them more clearly.

WAR IN THE TRENCHES

We can start to explore such issues amid the horrors of a war that few, if any, would now regard as just, or even sane. The historian Eric Hobsbawm says that "1914 opens the age of massacre. Millions of men faced each other across the sandbagged parapets of the trenches under which they lived like, and with, rats and lice." In this bloody stalemate, France lost one in five men of military age, and half a million Britons under thirty were killed. The Western Front, says Hobsbawm, "naturally helped to brutalize both warfare and politics: if one could be conducted without counting the human or any other costs, why not the other?"[17] The total war of 1914–1918 was arguably the precondition for Hiroshima and the Holocaust, for technologically assisted slaughter of civilians on an immense scale.

Where, in all of this, is there scope to explore questions of cooperation and tolerance?

Actually, the answer is already the stuff of legend. Who does not know of the Christmas Day truces on the Western Front, when enemies greeted one another and played games of soccer between the mud and the barbed wire? And then, according to the legend, they returned to the war.

But in fact many did not. On visiting the trenches in the middle of the fighting, one British officer was outraged at what he found. Not the squalor, the misery, the madness—but the casual attitude the men had to the business of winning the war. The officer confessed to being

> astonished to observe German soldiers walking about within rifle range behind their own line. Our men appeared to take no notice. I privately made up my mind to do away with that sort of thing when we took over; such things should not be allowed. These people evidently did not know there was a war on. Both sides apparently believed in the policy of "live and let live."[18]

Of course, the soldiers knew all too well that there was a war on. They also knew that, as far as the Western Front was concerned, it was unwinnable. They were not acting out of cowardice, nor sloth or despair, nor even out of sporting regard for the enemy. They were simply being rational. They knew what was best for them.

This apparent laxity toward killing the enemy was, at first sight, dangerous as well as distinctly unmilitary. From time to time the Allied generals would order a push, compelling the troops to go over the top and make a head-on assault on the German forces. If they had in the meantime been spending their days chatting, smoking, and ignoring their "duty" to eliminate enemy soldiers, they would have to face greater numbers in these battles and casualties would be heavier. Surely it paid to kill the Germans at every opportunity?

But the point is that the Germans were in exactly the same position. So the choice was simple. Either you fought each other constantly, bombarding the enemy trenches with artillery fire and deploying snipers to pick off anyone foolish enough to show his head above the parapet—while enduring the same treatment from the other side—or you held fire on the tacit understanding that the enemy would do the same, and meanwhile you hoped to be relieved from the Front before the next push came. In the one case you endured fear of sudden death at every moment; in the other you had a quiet life and some hope of going home at the end of it.

Armies have always relied on propaganda to demonize the enemy and instill in the troops a loathing and hatred of their opponents, until the instruction to kill needs no enforcing. Many British soldiers no doubt went to the Front despising "the Boche," and vice versa. But this illusion becomes hard to sustain if you find that the enemy will leave you alone if you leave him alone—all the more so once you start to see how futile the fighting is. Indeed, some troops facing each other on the Western Front made formal truces. Naturally the military commanders regarded such behavior as treasonable and court-martialed the culprits. But the "live and let live" policy that flourished in the trenches did not have to depend on open compacts. It arose by stealth through the complicity of the opposing troops, to the fury and despair of the generals on both sides.

It was not—or it did not begin as—a humanitarian agreement not to kill. Quite the contrary: it was enforced by killing. Both sides came to realize that if they flouted an unwritten agreement by launching a bombardment in the hope of gaining some strategic advantage, they would get as good as they gave. "If the British shelled the Germans," says G. Belton Cobb in *Stand to Arms* (1916), "the Germans replied, and the damage was equal: if the Germans bombed an advanced piece of trench and killed five Englishmen, an answering fusillade killed five Germans."[19]

The fighting was, in other words, conducted on a tit-for-tat basis. Such exchanges are a lethal form of communication. They say, "We will do as we are done by." This is at the same time both a threat and an olive branch, for it also implies that nonagression will be greeted with the same.

Such "live and let live" arrangements may have been triggered by basic necessities. Men must eat, and cannot fight and eat at the same time. Mealtimes thus provide a natural cease-fire. And one does not bombard the enemy supply wagons behind the front line (even though, ballistically speaking, it would not be hard) if one wants to avoid an equal response. In *Goodbye to All That*, Robert Graves recounts how each side would also sometimes hold its fire to allow soldiers to retrieve their dead and injured from No Man's Land.

Why doesn't all war end up this way? The very conditions that made the Western Front so terrible and futile—entrenched forces at deadlock, unable to advance—also promoted the limitation of hostilities. Normally enemy troops encounter each other as part of mobile armies or in sudden attacks by or between guerrilla groups. The same combatants don't enter into sustained confrontation with one another. In these circumstances it pays to kill as many of the enemy as you can. But if you know you are going to be facing the same enemy for a long time, you have a motive to cooperate instead of confront: you can expect the same treatment yourself in the future. By the same token, businesses that deal with one another regularly are unlikely to cheat or default on payment, because they know that the other business will have a chance to do the same to them in the future. If the deal is one of a kind, an unscrupulous company can welsh on it without fear of reprisal.

In addition, it takes time, in the absence of direct contact, to establish the terms of a live-and-let-live policy and to develop sufficient trust that the other side will adhere to it. Deadlocks grant that period of indirect negotiation.

Overt unsanctioned cease-fire in the trench war ran the risk of reprimand or of encouraging the High Command to force a unit onto the offensive with a direct order. So the soldiers at the Front began to develop ways of appearing to be fighting while in fact doing no damage. Each side would recognize this for what it was: a way of sustaining the cease-fire while convincing their officers that conflict continued. These dummy attacks became not just a decoy to deceive the generals but a show of goodwill toward the "enemy." According to Tony Ashworth, a historian who has made a careful study of this live-and-let-live system in the First World War,

In trench war, a structure of ritualized aggression was a ceremony where antagonists participated in regular, reciprocal discharges of missiles, that is, bombs, bullets and so forth, which symbolized and strengthened, at one and the same time, both sentiments of fellow feelings, and beliefs that the enemy was a fellow sufferer.[20]

This ritual element of combat is in fact widespread in warfare and conflict. Among animals it can take the form of leadership contests in which opponents engage in displays that obviate the need for life-threatening combat. Stags brandish their horns, and may proceed to the stage of locking them in a shoving contest, but only rarely does the confrontation progress to a lethal fight, even though both stags are armed with deadly weapons. These rituals are, however, designed to produce a winner. In the trenches they were designed to evade the senseless bloodshed that the generals believed to be necessary.

The live-and-let-live strategy of the trenches reputedly engendered not only tolerance and restraint but mutual goodwill between troops sent out to kill one another. But if this happened, it was a consequence, not a cause, of the cooperative behavior. The soldier shooting into the air did not necessarily have moral compunctions about killing his enemy—he merely knew that by desisting from that, he increased his own chances of survival. Cooperation arose out of self-interest.

We can recognize in a qualitative way how this could have come about. But to make a science of it, we need a tool developed, ironically enough, at a U.S. military think tank.

THE GRAND TOURNAMENT

You are on a train, and as you take your seat you find a wallet. It is fat with money. What do you do?

At face value this is a simple case of binary choice, like those we looked at in Chapter 13. You can either try to return the wallet to its owner—for example, by looking inside for an address or handing it to the train conductor—or you can place it quietly in your pocket.

The behavioral models we have looked at in earlier chapters have tended to assume that agents in a multiplayer scenario respond to the actions of their neighbors in what we might regard as a knee-jerk or at least a some-what mechanical way. That's to say, stimulus A induces response B, either invariably or with a certain probability.

But in a situation like this, choices aren't made so simply. True, some people are invariably honest and some invariably dishonest. But the terrain in between is not negotiated by the random throw of dice. What flashes through our mind, perhaps involuntarily, is the thought: Who would know? And then, maybe: What if *I'd* lost my wallet? How would *I* feel? In such cases, we weigh our options according to some moral code, but that code is bedeviled by *temptation*.

Temptation is arguably the fundamental problem for human societies. It sometimes pays not to be the good, kind, considerate citizen but to rebel, to cheat, to fight, to do the dirty. If my neighbors are all meek, law-abiding people, what is to stop me from appropriating some of their land, or goods, or cattle? A Hobbesian individual in a Hobbesian world is as miserable as everyone else. But a Hobbesian in Eden can run riot, amass a fortune, gorge himself, and fear no reprisal (unless he believes in God). Temptation is a part of the human condition, and that is the problem for all utopias: not everyone is nice, because sometimes crime pays.

It is not obvious how to devise a "particle" that can be led into tempta-tion. But in the 1950s, Merrill Flood and Melvin Dresher at the RAND Cor-poration did more or less just that. They developed a simple mathematical model that incorporated the element of temptation into an interaction be-tween two agents.

The model was presented as a kind of game. Flood and Dresher were ex-ploring the theory of games devised by the mathematical physicist John von Neumann in the 1920s. One of the most formidable mathematicians of the twentieth century, von Neumann helped establish the theoretical basis of the computer and made crucial contributions to the Manhattan Project during the Second World War. He cultivated something of a reputation as a playboy genius, which his passion for gambling and poker did much to enhance. But von Neumann didn't just want to play these games; he wanted to understand them.

For sheer complexity, a mathematician can do no better than to study the game of chess. There is a sense, however, in which poker is much more challenging, for it incorporates the psychological element of bluffing. The question is not, as in chess, what the next best move is, but which move will anticipate, mislead, or disconcert your opponent. The elements of risk and uncertainty in games like poker led von Neumann to see a connection with economics, and in 1944 he set out his ideas in a book coauthored with the economist Oskar Morgenstern, *Theory of Games and Economic Behavior.*

The game devised by Flood and Dresher involved a gamble. It has become known as the Prisoner's Dilemma, and it introduced game theory into sociology, biology, and political science. The game is played by two agents, who are depicted in the explanatory metaphor as prisoners suspected of committing a crime. Each is offered the inducement that if he testifies against the other, thereby securing the other's conviction, he will be set free. If neither agrees to testify, both will receive a sentence—but only a light one, because of the paucity of evidence. If both testify against each other, the sentences will be heavier—but not as heavy as that of the convicted party if only one of them testifies, since the evidence in the former case is equivocal.

The temptation is, of course, to testify against the other prisoner. But if both do so, they both get heavier sentences than if neither testifies. Should each prisoner refuse to testify, hoping that the other will do the same? If they both act "rationally," then both should in fact testify against each other. This choice gives the best outcome for each agent no matter what choice the other prisoner makes. If prisoner 1 testifies and prisoner 2 does not, prisoner 1 gets off the charge—the best of all outcomes for him. If prisoner 2 testifies, on the other hand, then it is still better for prisoner 1 to testify than not to: he gets a lighter sentence in the former case than the latter. So the dilemma doesn't sound like a dilemma at all—it is always better to "rat." The trouble is that since this is equally true for both prisoners, it compels them both to testify and thus to be worse off than if they had cooperated with one another by not testifying.

The essence of the Prisoner's Dilemma can be expressed in terms of a choice either to "cooperate" or to "defect." The best outcome—the maximal payoff—for one agent comes if he defects (testifies) while the other cooperates (that is, refuses to testify—the cooperation here is with the other prisoner, not the authorities). In that case, the other player is the "sucker" and gets the worst outcome. But if the agents play rationally, they get neither this optimal payoff nor the next best thing, which is the payoff from mutual cooperation. Instead, they get the meager rewards of mutual defection, which are a little better than the sucker's payoff.

To recast this dilemma in terms of individuals living in a society, we can regard cooperation as being law-abiding and defection as breaking the law for one's own gain at another's expense. The basic dilemma—that cooperation is good but defection can be even better—was recognized by Rousseau and Spinoza. In his *Discourse on the Origin of Inequality*, Rousseau imagined five men from precivilized times agreeing to cooperate on a stag hunt, on the understanding that each will get a fifth of the spoils. When a hare comes within reach of one of the men, he grabs it—but without his help on the hunt, the stag escapes. The "defector" has the immediate gratification of stewed hare, with none of the difficulties and dangers of catching a stag, but his fellows have nothing.

At face value, the Prisoner's Dilemma seems to confirm Hobbes's pessimism: egoistic individuals guided by logic will always seek to exploit one another. What did it seem to say to the Cold Warriors whom the RAND Corporation addressed? Get the first fist in, for your enemy will try to do the same. Build up your nuclear arsenal with all the resources you can muster, for your enemy is planning to defect, and so you must be ready to do so too. Indeed, you should even consider defecting first—making the first strike. If the other side "cooperates," even to the extent of not launching a strike immediately, you can make them the sucker by doing so at once. You win; they lose.

This does not seem much of a basis for understanding how to construct a world in which people and nations live in harmony. Thankfully, there is more to say about the Prisoner's Dilemma.

The frustrating thing about this game is that the players—the prisoners, if you like—can't communicate. It is obviously in their interests to agree to cooperate rather than to both defect. But since they cannot convey to each other a readiness to do this, they are better off assuming the worst of the other player, which implies that they must defect.

If you play the game more than once, however, there is scope for communication of a kind: even if the players cannot correspond directly, they can signal their intentions by the way they play. If one player reveals a willingness to cooperate by doing so in one round, the other player might decide to reciprocate in the next. The players who, having both begun with ruthless defection, later begin to cooperate find that they achieve better outcomes as a result.* They do not need to experience any sense of guilt or moral obligation in order to switch to cooperation. Pure self-interest is enough to make that the best choice.

This means that the impasse that compels defection from both players in a single round of the Prisoner's Dilemma can be broken simply by playing the game repeatedly. And that is how we commonly encounter comparable situations in real life. If I cheat on my neighbor, he has plenty of opportunity to retaliate. Most businesses deal again and again with the same clients. If two countries share a border, they cannot avoid ongoing political, economic, and social interactions.

So in the repeated, or *iterated*, Prisoner's Dilemma game, the players have the chance to learn from their mistakes and to build a relationship of mutual trust. Cooperation can evolve.

Is this how real people play the game? Psychologists have studied that question extensively in controlled tests and have found that cooperation does develop—but to a degree that varies widely, depending on the character of the players, the nature of the payoffs, and the circumstances of the interaction. One can imagine, for example, that it is easier to defect anonymously than face-to-face.

And let's not forget the element of temptation. If you think you are facing a nice player who will do his best to cooperate, you might be tempted to throw in an occasional defection, thereby boosting your own score at your opponent's expense. If he is forgiving, you might get away with it if you don't try it too often. Sadly, in a cooperative world, defection pays.

This then raises the question, What is the *best* way to play the iterated Prisoner's Dilemma? If you know nothing about your opponent, which strategy should you adopt?

In the late 1970s, Robert Axelrod devised an experiment to try to answer this question. He asked professional game theorists to submit strategies for

---

*In these repeated games a "score" is kept of how well each player did on each round. A player scores highest for defecting when the other cooperates; moderately for mutual cooperation; poorly for mutual defection; and worst of all for cooperating when the other defects (the "sucker's payoff"). One might imagine each player accumulating points or money, for example, rather than prisoners accumulating years of incarceration.

playing the iterated Prisoner's Dilemma and then put each strategy to the test in a round-robin tournament conducted on a computer. The entries played one-on-one against each of the others for many rounds; the winner was the strategy with the highest aggregate score. The fourteen entries came from psychologists, mathematicians, economists, sociologists, and political scientists. Each strategy consisted of a set of rules for determining the choice of cooperation or defection. For example, one might simply choose always to cooperate. (This is obviously a bad choice, since it always comes off worst unless everyone else is also an unconditional cooperator. So no one chose this strategy.) Or one might cooperate on the whole but defect every fourth round.

Many of the submitted strategies were more complex than this. But the tournament was won by the simplest of them all. It was submitted by Anatol Rapaport, a psychologist at the University of Toronto, who called it Tit for Tat (TFT). Its one rule was to begin by cooperating and thenceforth to do whatever its opponent did in the previous round.

Playing against an unconditional cooperator, Tit for Tat cooperates throughout the entire encounter. So both players do equally well. Against an unconditional defector, TFT gets the sucker's payoff in the first round (in which it cooperates), but then it defects consistently, as though determined not to be taken advantage of again. Because of the first round, TFT comes away slightly worse off than the inveterate defector—but only just. Both of them, in any event, do much worse than if they'd cooperated.

By mirroring its opponent, TFT can adapt to whatever the situation calls for. Against cooperators it is nice; against defectors it is tough. When faced with a mixture of cooperation and defection, it gives as good as it gets. So TFT reaps the benefits of cooperation where possible, but cannot be exploited. Neither does it exploit: it never achieves a higher total payoff than its opponent. Some of the other strategies did well against those with a tendency to cooperate; others could hold their own against defectors. But by making the best of both situations, Tit for Tat came out on top in a diverse mixture of strategies. It was a modest, simpleminded victory.

Following the success of his first computerized tournament, Axelrod decided to hold a second, with essentially the same rules as before. News had traveled, and this time there were sixty-two entries from six countries. Again some came from professional scientists and academics, but others were from computer hobbyists, including a ten-year-old boy. All knew the outcome of the first round and so had had a chance to consider the reasons for TFT's success. Anyone could submit any strategy, but only one person chose Tit for Tat: Rapaport. All the others decided they could outdo TFT with something more sophisticated.

They couldn't. Tit for Tat was again the winner.

Does this mean that Tit for Tat is the best way to play the iterated Prisoner's Dilemma? Not exactly. There is, in fact, no best way to play, for it depends whom you're playing. It is very easy to illustrate that this is so. If you are playing against a colony of unconditional cooperators, you will do best to be an unconditional defector—that strategy will fare better than TFT, which would behave like one of the cooperators (except when it plays you). But the message of Axelrod's tournaments seemed to be that if you don't know who you're up against, TFT is the best default strategy.

What, then, makes Tit for Tat so special? For one thing, it is flexible: able to cooperate but not open to exploitation. Cooperation from the other player will immediately elicit cooperation from TFT in the next round. But defection is met unhesitatingly with defection. This sends a clear message: TFT will do as it is done by. It is a strategy from the Old Testament, not the New: an eye for an eye, not turning the other cheek.

This clarity of response is in itself a factor in TFT's favor. In the second tournament, one entry was a strategy designed to try to figure out the rules the other player is using, in order to find some way of exploiting them.* This often happens in real life: one person will check out the other, weighing how much he or she can get away with. If you know beyond doubt that you can never defect without being treated the same way, you have a good incentive to cooperate. If you have reasons to doubt that the retaliation will be relentless, you might be tempted to try your luck. Tit for Tat, in contrast, guilelessly encourages cooperation and discourages defection.

But there is another telling aspect of TFT that contributed to its success: it is never the first to defect. All strategies can be broadly divided into two camps by this criterion: Will they defect first or not? Those that will not are generally called nice strategies. (There is no consensus on what to call the

---

*TFT takes the other player's actions into account too, but only to the extent of mirroring them—it does not seek any deeper understanding of the opponent's strategy.

others, but *nasty* will do.) Axelrod found that nice strategies do consistently better than nasty ones. Indeed, in the first tournament the ranking produced a clear distinction: the eight top-scoring strategies were all nice, and the others, separated from the nice ones by a substantial gap in points scored, were all nasty.

THE SECRET OF COOPERATION

So the Prisoner's Dilemma starts to look less grim when it is iterated: niceness and cooperation fare better than nastiness and exploitation. Even individual selfishness need be no barrier to fair play.

But being cooperative does not in itself guarantee success; Tit for Tat plays a much tougher game than that. Axelrod has identified four characteristics of a successful strategy:

1. Don't be the first to defect (be nice).
2. Always reciprocate.
3. Don't be too clever.
4. Don't be envious.

What does it mean not to be envious, in this context? It means not trying to do better than the other players but simply doing as well as you can for yourself. The Prisoner's Dilemma is not a zero-sum game: someone else's gain does not have to come at the expense of your loss. If you both cooperate, you can both do well (even if not as well as you would if there was an option to exploit the other players). Axelrod says that real players seem to find it hard to relinquish competitiveness and envy. In his tests of the iterated Prisoner's Dilemma with student volunteers, he finds that they tend to measure their performance against that of others, in which case any advantage gained can tempt the others into defecting to try to redress the balance. This can trigger recriminatory outbursts of defection.

The live-and-let-live behavior in the trenches of the First World War can be seen as an example of cooperation arising from a Tit for Tat strategy. Belton Cobb's remark makes it clear that there was no compunction about retaliating in a lethal manner to hostilities from either side—the mutual ceasefire did not depend on good feelings between enemies. (Yet, significantly, it seemed that self-interested cooperation allowed these feelings gradually to develop.) And on the whole both sides followed a "nice" strategy, declining to fire first.

In case we should still wonder whether human sentiments, rather than the mathematical exigencies of game theory, led to this reciprocal cooperation, we might bear in mind that Tit for Tat strategies are also found in the natural world. There is evidence that vampire bats, stickleback fish, monkeys, and even viruses behave according to the rules of TFT. No one can reasonably attribute altruism to viruses; their behavior is purely the result of genetic selection. That is to say, those organisms with a genetic predisposition to show TFT-like behavior gain an evolutionary advantage, and so natural selection works to their benefit—ensuring that this genetic trait becomes more widespread.

This implies that we too may be genetically hardwired to cooperate, perhaps in a TFT-like manner. Indeed it would be astonishing and puzzling if we were not. Edward O. Wilson argues that as civilization evolved, such modes of human behavior will have become converted from instinctive impulses to social norms, then to legal imperatives, and ultimately to moral principles.

One could even argue that the case for a genetic embodiment of the lessons of the Prisoner's Dilemma is supported by the readiness with which we greet their optimistic aspects. We would be sorely dismayed if game theory were *not* capable of producing cooperative behavior; in fact, we might then be tempted to dismiss it as nonsense or as mendacious. We are, it seems, predisposed to look favorably upon altruism and to frown on apparently selfish behavior. That this might be a learned response does not evade the issue; we learn it because those are the cultural norms of our society, and where did *they* come from?

Here, then, is a possible resolution to the divergent views of human nature evinced by Hobbes and Locke, which led them to such differing conclusions about systems of government. People do not, in the absence of a higher authority, necessarily seek to exploit one another in the way Hobbes envisaged. But neither do they desist from it because of a "reason" instilled in them by God. The "reason" can come from nature alone: from the inexorable mathematics of interaction coupled with the winnowing effect of natural selection.

The Prisoner's Dilemma is actually implicit in Hobbes's analysis, since he acknowledges the miseries of mutual defection and argues that men are better off cooperating if this can somehow be arranged: "That a man be willing, when others are so too, as farre-forth, as for Peace, and defence to himselfe he shall think it necessary, to lay down this right to all things; and be contented with so much liberty against other men, as he would allow other men against himselfe."[21]

This connection between *Leviathan*'s State of Nature and the Prisoner's Dilemma was pointed out in 1969 by the political scientist David Gauthier. Without a contract to cooperate, says Hobbes, a man would "expose himself to Prey."[22] But such a contract is liable to dissolve unless there is some authority that can enforce it, since men's appetites will make them liable to defect the moment they see advantage in doing so. Thus, says Gauthier, Hobbes's omnipotent sovereign provides an escape from the Prisoner's Dilemma that men face in the State of Nature, since in a sovereignty defection no longer brings potential rewards but only certain punishment. Even if, as has been argued, it is somewhat misleading to cast Hobbes's scheme in game-theoretical terms when he had no interest in deducing the psychology of people faced with such behavioral dilemmas, it seems clear that Hobbes recognized the underlying problem that arises when antisocial actions offer potential rewards.

But game theory suggests that Hobbes's rather extreme solution—a capitulation of all individual powers and rights beyond self-preservation—may not be necessary. His error, if we may call it that, was to treat people as blind animals who cannot learn from experience—an "experience" that can be handed down from previous generations as a genetic predisposition toward cooperation.

By the same token, we might expect to find other implications of game theory hardwired into human experience. The tendency to form tribal groups increases the likelihood of repeated interactions with other group members and so enables cooperation to develop. Robert Axelrod endorses the notion of prolonged interaction—he calls it "enlarging the shadow of the future"[23]—as a way of promoting and nurturing cooperative behavior. The flip side of this principle is distrust of strangers, since it takes time to establish the mutual trust on which cooperation depends. But this apparent biological predisposition for xenophobia should be moderated by the realization that "nice" strategies do best: even on the first encounter it is preferable to cooperate.

So by arranging to make future exchanges more probable, we can guide two parties toward the benefits of mutual trust. This could entail making a relationship more durable—it was the long-term confrontation of forces at the Western Front that made the tacit cease-fires possible. Or we might increase the rate of interactions: in small communities, the same people deal with one another day after day both socially and economically, so trust is easier to establish than in large cities where interactions are more occasional and impersonal. Increased demographic mobility reduces the durability of interactions and so reduces the incentive for cooperation: transient neighborhoods are rarely cohesive and "neighborly."

It is clearly not news to businesses that their interests are served by developing good long-term relationships with clients. But the way in which such relations can break down gives us some reason to suspect that the reciprocity does indeed stem from a Prisoner's Dilemma–style exchange. A study in 1963 indicated that one of the most common reasons for court action between businesses is the complaint of wrongful termination of a franchise. It is only when relations between businesses are about to cease— when the iterations of the "game" are about to end—that one or both players decides it is worthwhile to start a legal battle and risk bitter recrimination rather than to find a "peaceful" way to resolve differences. In psychological tests, people who play the Prisoner's Dilemma will often sacrifice mutually established cooperation for a few rounds of defection when they know the game is about to finish. In the same way, companies about to go bust are at greatest risk from nonpaying clients, and are themselves more likely to default on debts.

The durability of interactions has implications for modes of government. Karl Popper considers that the most important attribute of a true democracy is not what it does but that it "should keep open the possibility of getting rid of the government without bloodshed, if it should fail to respect its rights and duties, but also if we consider its policy to be bad or wrong."[24] For as Pericles put it in democratic Athens, "Even if only a few of us are capable of devising a policy or putting it into practice, all of us are capable of judging it."[25]

In a democracy, unpopular governments can be removed by elections, which seems incontestably proper. But the termination of any government carries some risks, for a departing government no longer has anything to lose

from acting with blatant self-interest. Bill Clinton's outgoing administration
of 2000 demonstrated this by dispensing political favors that the President
would never have risked in midterm.

To some extent this situation can be remedied by the existence of political parties, which carry long-term accountability for the short-termism of its
members. There is no doubt that in 2002 the British Conservative Party was
still paying the price for its deeply unpopular policies, its arrogance, and its
corruption while in office five years before, even though the perpetrators had
largely disappeared from the political scene (several of them ignominiously).
The U.S. Republican party paid the same long-term price for the Watergate
affair in the 1970s. Thus a political system with a durable party structure
might be expected to be less susceptible to corruption than one with more
ephemeral kinds of political organization. Karl Popper called the party system "horrible," since it makes parliamentarians tend primarily to serve their
party rather than their constituents. "I think," he wrote, "that we should, if
possible, go back to a state where MPs say: I am your representative, I belong to no party."[26] But this could be a recipe for eliminating the accountability that political systems need if they are not to be plagued by abuses of
power.

FOR YOUR OWN GOOD

It would be a foolish evolutionary psychologist who tried to argue that Tit for
Tat is all there is to altruism. For one thing, the self-sacrifice we are capable
of showing toward our own kin has biological roots that owe nothing to game
theory—it seems to be an aspect of the "selfish gene" idea, benefiting individuals who share a close genetic similarity to ourselves.

Moreover, the "behavioral economist" Ernst Fehr and his colleague Simon Gächter in Switzerland have conducted experiments with human subjects which suggest that cooperation can arise in groups even when the
individuals do *not* encounter one another repeatedly. Fehr and Gächter divided 240 students into groups of four, gave each of the students an equal
sum of money, and invited them to invest it (or not) in a group project. The
project produced returns in proportion to the degree of investment. If
all four members invested all their money, they all got a return that ex-

ceeded the investment. So it was in the *group's* interest for everyone to invest everything. But because each member received less than one monetary unit for investing each unit of his or her own, it was in each *individual's* interest not to invest but to freeload, relying on the contributions of the others.

This is analogous to the Prisoner's Dilemma insofar as it presents players with benefits for mutual cooperation but temptations toward individual defection. But Fehr and Gächter mixed up the groups after every round of investment and return, giving them no opportunity to establish mutual trust. They found that cooperation could nevertheless flourish if the rules included some provision for punishing defectors (those who invested little). Players would mete out such punishments even when they were charged a fee for doing so. Without the threat of punishment, cooperation was low; when this threat was introduced, cooperation increased sharply. The researchers call this altruistic punishment, since it is likely to be of no immediate and direct benefit to the punisher—even if punishment reforms the defector, the punisher is unlikely to encounter him or her again. It is nevertheless altruistic because it may benefit those others who will be grouped in later rounds with the player who defected.

This sort of behavior suggests that the possibility of Tit for Tat–style retaliation or punishment may have a part to play in enforcing cooperation in society even when encounters are not repeated. Fehr and Gächter reported that players seemed moved by a sense of injustice: they were simply angry at defectors, and acted on that anger irrespective of whether they stood to gain from it. Moreover, players reported that punishment was a deterrent to their own inclination to defect. It is worth noting, however, that this is precisely the kind of behavior one might expect from players predisposed toward a TFT strategy. By allowing punishment to be meted out after the "game" has been played, the researchers were in effect enacting a kind of two-round game, in which first-round defectors are themselves treated to defection from other players in the second round. The results could imply that we are all imbued with a desire for justice that we will exercise, if necessary, at our own expense.

In all these games, mutual cooperation pays best in the long term. If indeed cooperators are more "successful," we would expect a predisposition toward cooperation to become an irreducible element of our neural circuitry.

Sadly, however, this does not mean that Kropotkin was right to assert that no government is needed because people can be trusted to organize themselves. History shows us what people are capable of, and it does not look much like Eden. Human nature is diverse; it is also mutable, for better or worse. And it is influenced not just by one-on-one interactions, but by the multitudinous society in which each of us is embedded. To deal with that, game theory needs to get more sophisticated.

EIGHTEEN

PAVLOV'S VICTORY

IS RECIPROCITY GOOD FOR US?

The fact is that a man who wants to act virtuously in every way necessarily comes to grief among so many who are not virtuous.　　　　　—Niccolò Machiavelli (1513)

Can one wait with calm confidence for the day when the despotic states that have made wars in the past have been turned, by the social and economic forces of history, into peace-loving democracies? Are the forces of evolution moving fast enough? Are they even moving in the right direction?　　　　　—Kenneth Waltz (1954)

In the interests of peace I am opposed to the so-called peace movement.
　　　　　—Karl Popper (1988)

If the Middle East today bears witness to harsh words and harsh deeds, it has done so before: "You must purge the evil from among you. The rest of the people will hear of this and be afraid, and never again will such an evil thing be done among you. Show no pity: life for life, eye for eye, tooth for tooth, hand for hand, foot for foot."[1]

The uncompromising reciprocity of a Tit for Tat rule may have worked for the children of Moses (although even for them it was not the lawful response to all insults). But can this really form the baseline of a civilized society? Game theory seems best suited to investigating the State of Nature that Hobbes feared as barbaric and Locke idealized as beneficent; it delivers the hopeful message that goodness can arise out of barbarism. But only at the cost, it seems, of uncompromising retaliation to all aggression.

The whole point of government, says Locke, is to eliminate the necessity for every man to be his own judge and enforcer. But does that oblige a government to enforce cooperation among its subjects and its neighbors alike with the same policy of swift and unquestioning retribution? Where in this social calculus might we find room for negotiation, conciliation, mediation, even forgiveness?

To understand the implications of game theory at more than a superficial level, we need to subject Tit for Tat to a rigorous examination. That is my objective in this chapter. We shall circumscribe this strategy's advantages and probe its weaknesses. We shall release it into a community and watch the consequences. We shall ask—as we always must in social physics—not just what the models tell us but what we consider desirable, and whether the two can be reconciled. And so we shall come back to the enduring question: What choices do we have?

ACCIDENTS HAPPEN

After Axelrod's second computer tournament, Tit for Tat looked invulnerable. But it isn't. In the real world it has a fatal flaw: communications are imperfect. Mistakes are made; intentions are misunderstood. In 1983 the Soviet Union shot down a South Korean civilian airplane which had mistakenly strayed into Soviet airspace, in the belief that it was a military craft. All 269 passengers, including several Americans, were killed. A Tit for Tat policy, rigidly applied, would dictate that this error could be avenged only with Russian blood. Fortunately it was not, although the incident did heighten Cold War tensions. The NATO bombing of the Chinese embassy in Belgrade during the attack on Serbian forces in 1998 was at face value another apparent "defection" resulting from a mistake. (There is still debate about whether it was truly unintentional.)

The hair-trigger status of American and Soviet nuclear arsenals during the Cold War highlighted the awful risks of a retaliatory policy in the face of potential mistakes. Taken to its extreme, this creates the scenario gloriously and chillingly lampooned in Stanley Kubrick's film *Dr. Strangelove*, in which a rogue U.S. army general launches a preemptive strike on the Soviet Union. All but one of the B-52 bombers is recalled in time, but the one that cannot be contacted releases its warheads. This triggers global nuclear war even though the Soviets know the bombs were dropped "by mistake," because they have automated their missile system with the Doomsday Machine, which retaliates to any nuclear attack without the option of human intervention. They believed that the absolute certainty of retaliation would enforce cooperation; but the system did not allow for errors.

The problem with mistakes, as far as Tit for Tat is concerned, is not simply that a lone erroneous defection provokes the same in return. Tit for Tat's simplicity means that if this happens between two players who are using the same strategy, they get locked into a cycle of mutual recrimination. One defects by mistake; in the next round it returns to cooperation (since that is what its opponent did in the last round), but the other player returns the defection. This causes the first player to defect in the round after that, and so on: the mistake echoes back and forth for the rest of the match, so that mutual cooperation is never restored. (In the *Strangelove* scenario, of course, a single round of defection from each side is enough to end the game once and for all.)

This sort of behavior arises in many cultures and societies. Axelrod points to the example of family feuds in Albania and the Middle East, which can continue with mutual reprisals for many generations even after the original incident has been long forgotten. Agonizing vicious cycles of mutual slaughter have plagued the Protestant and Catholic communities in Northern Ireland for decades, and currently seem to be destroying all hopes of a peaceful settlement between Israel and Palestine.

Clearly, Tit for Tat does not guarantee a harmonious world.

Nor, amid the mess and confusion of reality, is it always the best strategy. This became apparent when Axelrod's tournament was repeated, this time including the possibility that the players make errors. The players would occasionally choose their response at random rather than according to the rules of their strategy. For an error rate (a "noise" level) of 10 percent—one in ten random choices—TFT is no longer the winner. In fact, TFT then fares even worse when playing against other TFT players than when playing a mixed bag of strategies, since occasional errors create ample scope for fruitless cycles of reprisal.

In such a situation, TFT needs modification if it is to score highly. One alternative, called Generous Tit for Tat (GTFT), lets a certain fraction of defections go unpunished. Another, Contrite Tit for Tat (CTFT), declines to retaliate to a defection that follows a defection of its own—it "accepts" that it got what it deserved. Generous Tit for Tat outperforms all of the entrants in Axelrod's second tournament when there is 1 percent "noise"; CTFT comes sixth. For higher noise levels, CTFT outstrips GTFT.

Tit for Two Tats is a strategy that retaliates only after suffering two consecutive defections: it waits to see whether a defection implies that the other player's intentions are *really* bad, rather than being simply a mistake (that is, noise). Tit for Two Tats was devised by the evolutionary biologist John Maynard Smith, and it was ranked twenty-fourth in Axelrod's second tournament. Maynard Smith did not enter it in the first tournament, but if he had, it would have won, because that first mixture of strategies contained some that impaired TFT's performance by getting locked into mutual retaliation (even without errors). This reinforces the point that there is no best way to play the game.

Another strategy that copes well with noise is less benevolent. Pavlov is a strategy based on pure opportunism; it was named in 1988 (although invented earlier; see below) by David Kraines of Duke University and Vivian Kraines of Meredith College, both in North Carolina. Its philosophy can be summarized as "win-stay, lose-shift." Like TFT, it bases its choice of action on what happened in the previous round. If it did well, it makes the same choice again. If it did poorly, it switches. *Well* here means either the reward for mutual cooperation or the best payoff of all, that for unilateral defection. In short, Pavlov sustains behavior that brings rewards but changes behavior that brings punishment. This recalls the simple, conditioned responses of the Russian physiologist Ivan Petrovich Pavlov's dogs.

With a tough customer like Tit for Tat, Pavlov is quite happy to cooperate. It doesn't fare well with an incorrigible defector—it will try to cooperate every other round. But Pavlov will mercilessly exploit a habitual cooperator once it realizes it can get away with it, whereas TFT would nobly cooperate.

Pavlov performs poorly against the contestants of Axelrod's original tournament—in 1965, Anatol Rapaport gave this strategy the dismissive label of simpleton. And it doesn't do a great deal better even in the presence of noise. But it has the virtue of being able to recover quickly from an isolated error, and if the circumstances are right it can come into its own, as we shall see.

DARWIN'S ALGORITHMS

That history is a sound guide to policy is a cliché, although Friedrich Hegel doubted that nations and governments were so guided. Yet people, businesses, institutions, and even countries surely do sometimes change their behavior in the light of experience—just as the British and German troops whose job it was to eliminate one another on the Western Front ended up participating in unspoken truces of mutual self-preservation. Some law-breakers *can* be reformed. It is this capacity for change that makes inter-national relations both complex and worth arguing over. Some observers believed that Saddam Hussein's government in Iraq might not have been given inevitably to unconditional defection but could have been transformed into a more cooperative regime had it been engaged in dialogue rather than isolated with sanctions and then barraged with bombs.

One of the most interesting and important questions we can ask of the Prisoner's Dilemma is what kind of behavior emerges when the players can *evolve*—when they are allowed to change their strategies. In reality people apply all sorts of moral, ideological, habitual, and whimsical criteria in de-ciding how to behave. But in the spirit of game theory, it is useful to begin by asking what players will do if they are purely pragmatic —that is, merely seeking to optimize their gains. It seems reasonable to assume that players will tend to adopt those strategies that are more successful.

This can be simulated in Axelrod-style tournaments by including an evo-lutionary dimension. At the end of one full round-robin, for example, we might allow players to adopt a new strategy, where the probability of each strategy's being chosen is proportional to that strategy's overall score. In this way, the more successful strategies will multiply while ones that perform poorly will die out. It's not hard to see that this is a Darwinian "survival of the fittest" scenario. It mimics the way in which genetic mutations spread in a population: those carrying a mutation that conveys a reproductive advan-tage generate more offspring, enhancing the prevalence of that "adaptive" mutation.

Martin Nowak at the University of Oxford and Karl Sigmund at the Uni-versity of Vienna conducted just such an experiment in game theory in 1992, with salutary results. They set up a diverse population of strategies, in all of which the choice of whether to cooperate or defect was determined by what the opponent did in the previous round. Some strategies were more in-clined toward defection, others tended to cooperation. Nowak and Sigmund let them all compete against one another, and then altered the proportions of each of them in line with their relative successes.

Naïvely, we might expect that this evolutionary model will be ruled by Tit

18.1   In a mixed population of Prisoner's Dilemma strategies subject to "natural selection" of the most successful, Tit for Tat emerges from a population that has initially become mired in defection. This is accompanied by an abrupt rise in the average payoff for encounters between agents, since TFT allows a greater degree of cooperation.

for Tat, which appears generally to fare best in a mixed population. And this did at first seem to be the outcome. Early in the game the defectors had the upper hand; cooperative strategies died out, and the average payoff of the population fell toward the low payoff gained by mutual defection. But at some point a tiny band of TFT players began to grow rapidly until they dominated the population (Figure 18.1). This takeover was accompanied by an upsurge in cooperation and a rise in the average payoff.

The abruptness of this change is reminiscent of (although not strictly equivalent to) a phase transition. The rise of TFT is a collective effect, a result of many mutual interactions between players. Mutual defection eventually becomes so self-defeating that a small group of TFT players gains more from their mutual cooperation than defectors do from exploiting the TFT players' initial attempts to cooperate. At this point the tables are turned, and it pays to cooperate with the doughty TFT group. Their presence helps seed the spread of cooperation throughout the population.*

But Nowak and Sigmund found that the triumph of TFT is short-lived. Once it has established a culture of cooperation, TFT starts to suffer from its

---

*The crucial role of TFT-type strategies in bringing about this change is emphasized if the evolutionary game is replayed with no TFT players to start with. Then the prognosis is a gloomy one: cooperators die out and we are left with a colony of selfish defectors that plow their Hobbesian furrow to eternity.

Achilles' heel: unforgivingness. These simulations contained an inherent amount of noise in the way the strategies worked, and this meant that TFT was gradually superseded by its more tolerant sibling, Generous Tit for Tat. In the end, only GTFT remained.

Nowak and Sigmund concluded that "Tit for Tat is the pivot, rather than the aim, of an evolution towards cooperation."[2] In other words, it is needed to establish cooperation in a diverse population; but once that has been achieved, "softer" cooperative strategies will take over. In fact, since even GTFT will occasionally get caught up in unproductive recrimination caused by mistakes, a better strategy in a universally cooperative environment is unconditional cooperation: complete forgiveness.

This sounds all very nice and inspirational. But the best strategy of all in a population of unconditional cooperators is unconditional defection: ruthless exploitation of the meek.

Pitted like against like, cooperators do better than defectors; but cooperators are highly vulnerable to rogue defectors. A small band of defectors can wreak havoc in a cooperative culture. Tit for Tat can prevent this from happening, because it treats defectors severely while rewarding cooperation. It can be regarded as the police force of game theory, imposing cooperation with a firm hand. It models an ideal policing strategy in some respects, for (in the absence of noise) it only—and invariably—punishes defection, and never exploits cooperation. The implication seems to be that if we accept some level of defection as inevitable, we have to concede that a society needs at least some TFT players in order to maintain a general culture of cooperation.

Even that, however, may not guarantee a fair society. In 1993, Nowak and Sigmund discovered that TFT's implacable "sense of justice" does not always come out on top. In their earlier evolutionary games, players based their strategies for their next move on their opponent's previous move. But Pavlov, the opportunistic win-stay, lose-shift strategy, does more than this: it takes into account the player's *own* last move too. When the two researchers pitched their earlier strategies against Pavlov, they found that Pavlov's opportunism triumphs. Pavlov does poorly against defectors and lacks TFT's ability to "invade" a defecting population and spread cooperation. But in a (slightly noisy) community imbued with a spirit of cooperation, Pavlov thrives. Nowak and Sigmund found that in such a circumstance Pavlov emerges victorious, even outstripping GTFT. Both these strategies, Pavlov and GTFT, are somewhat tolerant to errors, unlike TFT. But Pavlov has another advantage. If in the model we allow strategies to randomly mutate into new forms, GTFT comes to share some of TFT's transience, becoming "softened" by a gradual drift toward more unconditionally cooperative strategies. Pavlov, however, retains a hard edge. If it discovers by chance that it can get away with unilateral defection, it will continue to do so. So it is a wolf in sheep's clothing: it behaves well while cooperation becomes the norm under the firm authority of TFT, but it remains quite capable of exploiting a cooperative population once the TFT police have been transformed to unconditional co-operators. The motto of a Pavlovian society is no longer "Do as you would be done by," but "Never give a sucker an even break."

The simulations from which Pavlov emerged as victor reveal a fascinating history. Because they involved the interplay of strategies that all based their next move in some way on the previous moves of *both* players, they took a more complex course than the earlier simulations. Most strikingly, there was far less of a sense of inevitability about the changes that took place over time (Figure 18.2). Each simulation "run" produced a different sequence of events. In the sample history shown here, there is an early attempt to establish cooperation: after a turbulent period this fails, and unconditional defectors reign for a long time. Then, after about 92,000 generations, the cooperators gain the upper hand.

This victory is short-lived and soon collapses into defection. When Nowak and Sigmund inspected the breakdown closely, they found that it is

Time (in 100,000 generations)

18.2   The changes in proportions of strategies over many generations of a Prisoner's Dilemma simulation are revealed by the changes in average payoff within the population: higher payoffs reflect a higher proportion of cooperative strategies. In these simulations the strategies evolve according to Darwinian selection, and they can take into account both their own last move and that of their opponent. There are several outbreaks of cooperation, the first two of which eventually collapse. The third is persistent.

caused by the drift of TFT toward GTFT and thence to more forgiving strate-gies, creating a nation of "softies" which is ultimately destroyed by rogue defectors. But this time the defectors are not quite unconditional: the domi-nant strategy is instead one called Grim Trigger, which meets cooperation with cooperation—until it encounters a defection (as, in a noisy game, it in-evitably must). Thereafter, Grim Trigger defects unconditionally. It is rather like *Dr. Strangelove*'s Doomsday Machine.

After about 220,000 generations there is another resurgence of coop-eration, which—after some initial adjustments—proves long-lasting. This switch is again triggered by TFT, but gradually it drifts toward a prominence of GTFT players before ultimately being taken over mostly by Pavlov agents, or close variants of Pavlov. This population is cooperative but potentially opportunistic, and is robust against invasion by defectors. It is not such a bad place to live—but the more virtuous of its citizens are not entirely safe from the threat of exploitation by superficially "nice" Pavlovians.

These simulations display a mixture of chance and certainty. Coopera-tion always wins out if you wait long enough—there is always a happy(ish) ending. And Pavlov is not always the final dominant strategy, although it tri-umphs in about four cases out of five. Most remarkable, though, are the sud-den revolutions that punctuate the course of events: we see "good" and "evil" empires rise and fall, and uprisings that falter and fail. Even in peri-ods of apparent stability (for better or worse), the tally of strategies (and a de-tailed inspection of their characteristics) shows a certain amount of variation and a shifting of norms.

It is hard not to see in all of this an allegory of human history. Marx be-lieved that the socialist revolution was inevitable. Game theory seems to be saying that nothing is so certain, since even if things *are* going to end up a particular way, we can't be sure just where along the evolutionary path we are at the moment. Did the players in the Second Cooperator Uprising (gen-eration 92,000) think that the Age of Perpetual Cooperation, long forecast by the martyred philosophers of the failed First Uprising, had finally arrived? Were the commentators of the Third Cooperator Uprising (generation 220,000) right to conclude that this was the "end of history"?

MAGIC CARPETS

Empires rise and fall not only in time but in space. Rome once ruled from Portugal to the Black Sea, from the Scottish borders to North Africa. Charlemagne's Frankish realm reached deep into Germany, Italy, and the Balkans. The Ottoman Empire overran lands from Transylvania to Egypt. The history of the world is a patchwork of borders, growing and shrinking.

Imperialism seems mercifully to be a thing of the past, but the borders of NATO and of Europe are still liable to shift, and the maps of the Central and Eastern European nations have changed more in the past ten years than in any other decade since the end of the Second World War. Does the Prisoner's Dilemma have anything to tell us about the way national and international boundaries move?

Introducing the element of space into the contests of game theory is no trivial matter, since it places constraints on the interactions that each player can have, and can therefore strongly influence the outcomes. In a tournament in which everyone plays against everyone else, cooperators have the chance to derive mutual benefit from their interactions; but if they are on opposite sides of a map, they can no longer draw strength from the group and may be overwhelmed by defectors. So isolation can militate against cooperation. Israel's geographical location, surrounded by predominantly Islamic states, surely contributes to its self-perception as an embattled nation, and its supporters might argue that this situation prevents the country from being able to adopt the kind of conciliatory policies that European nations can afford to pursue.

On the other hand, players in fixed locations may find a greater incentive to cooperate than do itinerant players, since they are compelled to interact repeatedly with their neighbors rather than moving on after one exchange. One of the difficulties faced by traveling people is that they have little opportunity to establish mutually trusting (and trustworthy) relations with those they encounter on their journeys—the "shadow of the future" is not long enough.

Axelrod began to explore the notion of territoriality in the Prisoner's Dilemma in the 1980s. He considered a checkerboard-like world in which each player occupies one grid space and interacts with the four neighbors with whom it shares an edge. Colonization of the board by successful strategies can take place by means of an "evolutionary" mechanism. In each round of the game, all players play against all four of their neighbors. If one or more neighbors of a particular player gets a better score, the player converts to the strategy that was most successful.

Axelrod was primarily interested in how cooperation can spread in an exploitative society by means of Tit for Tat—or conversely, how a cooperative society can be undermined by defectors. He found that, for certain values of the payoffs, a single defector can spread its baleful influence throughout a community of Tit for Tat players, providing the seed from which defecting strategies gradually expand. Curiously, however, the growing colony of defectors is not so much like a spreading stain as like a snowflake: it sends out branches which split and rejoin to form a complex tapestry (Figure 18.3).

If one considers the full gamut of strategies involved in a tournament like Axelrod's, the number of ways of arranging them into juxtaposed "nations" is immense. But Axelrod tried it anyway, distributing the sixty-three strategies at random on a 14×18 grid so that each strategy was awarded four grid cells. As the contest evolved, all "nasty" strategies (those that defect first)

18.3 The changing shape of a colony of defectors (black squares) growing in a population of TFT players (white squares).

6	6	6	1	44	44	44	44	44	6	6	7	7	6	7	6	6	6
6	1	31	1	1	44	44	44	44	3	6	3	6	6	6	6	6	6
6	6	31	31	1	1	1	1	1	1	3	3	3	52	52	6	6	6
6	1	31	31	31	31	31	31	31	31	31	3	3	6	6	6	6	6
6	9	31	31	31	31	31	31	31	31	31	3	6	6	6	6	6	6
6	31	31	31	31	31	31	31	31	31	31	31	6	6	6	6	6	6
6	31	31	6	6	9	31	31	6	9	41	31	31	6	6	31	31	6
6	31	31	6	6	9	9	9	6	41	41	31	4	31	31	31	31	6
6	31	31	9	9	9	9	9	6	41	41	41	41	31	31	31	31	31
6	31	6	9	9	9	6	6	6	41	41	41	17	31	31	31	31	31
6	6	9	7	9	9	6	6	6	41	41	41	41	31	31	31	31	7
6	6	7	7	7	6	6	6	9	41	41	7	7	7	7	7	7	6
6	6	7	7	7	6	6	6	6	41	6	7	7	7	7	7	7	6
6	6	7	7	6	6	6	44	6	6	6	6	6	7	7	7	6	6

18.4   The territories claimed by different strategies in a Prisoner's Dilemma tournament enacted on a grid. The numbers indicate the ranking of each strategy in Axelrod's original round-robin tournament. All of these strategies are "nice"—they are never the first to defect.

were eliminated and the grid ended up covered with a frozen arrangement of "nice" strategies. Since they always cooperated with their neighbors, none did better than any other, so there was no tendency for further change.

But Tit for Tat did not dominate the board, even though it dominated the round-robin tournament staged between the same selection of strategies. Each distinct initial arrangement of strategies produced a different final configuration, but all those configurations contained large territories that used strategies other than TFT, including some that had hitherto fared rather poorly in the tournament (Figure 18.4). Some "less successful" strategies were able to survive and even to thrive in the spatial contest because they were lucky enough to have neighbors against which they were well matched—it was no longer necessary to do well against all comers, but only against one's neighbors. Tit for Tat is itself not an inherently expansionist strategy: it has no inclination to invade cooperative, nice territories, but only defective ones.

One might interpret this as implying that there is no "absolute" best way to conduct one's affairs with other nations, except for the proviso that it pays to adopt a cooperative demeanor as one's default position. Moreover, the eventual distribution of strategies depends on how they are distributed initially: once again, history matters.

In 1992, Martin Nowak and Robert May at Oxford devised a simpler spatial game to explore the effect of space on cooperation. In this version of the Prisoner's Dilemma there was no allowance even for strategies with the low level of complexity of Tit for Tat; each player was either an uncompromising defector or a persistent cooperator. But again, players could switch from one strategy to the other if it was beneficial to do so. The players were distributed on a square grid, and each interacted with all eight of its neighbors—those with shared corners as well as edges. Each player's behavior in a round copied that of the most successful of these nine players in the previous round: it cooperated or defected depending on which was locally the best option. This is a kind of two-state lattice model of interacting particles, reminiscent of the Ising model used in statistical physics (page 90).

How the game plays out depends on the size of the various payoffs, which may bias the interactions in favor of cooperation of defection. If the bonus for defecting in the face of cooperation is small, cooperation dominates; but this benevolent landscape is threaded with chains of defectors which shift a little from one round to the next (Figure 18.5a). If defection is more lucrative, however, it becomes the most common mode of behavior. Yet small islands of cooperation constantly appear and vanish across the grid (Figure 18.5b). The fraction of cooperators now quickly evolves to a stable average level, irrespective of the starting and configuration—there is a kind of "irrepressible" tendency to cooperate even in a relatively selfish environment. Thus defection and cooperation need not inevitably annihilate each other, but can coexist indefinitely in patterns that are unpredictable in detail but entirely predictable in terms of averages.

The spread of defection and the spread of cooperation are not equivalent. Cooperators do best in dense clusters where they can benefit from the mutual support. Edmund Burke anticipated as much in 1770: "When bad men combine, the good must associate; else they will fall one by one, an unpitied sacrifice in a contemptible struggle."[3] But defectors in the midst of cooperators do best on their own, since they do far better to interact with cooperators than with other defectors. So although defection breeds more defection, the exploiters tend to repel one another, which leads to the formation of the thin threads of defection.

Nowak and May found that a single defector spreads through a cooperative colony in much the same kind of snowflake formation as Axelrod had seen (Figure 18.5c). Again, these ramified "magic carpet" patterns can be regarded as the result of the "repulsion" between defectors, which militates against their forming a dense colony. You might say that each defector prefers to find its own "patch," as isolated as possible from its rivals.

18.5   Spatial games between unconditional cooperators (black) and defectors (gray) show complex behavior that depends on how well it pays to defect against a cooperator. If the pay-off is small, cooperation dominates (*a*). (Here, white squares show sites that have changed from cooperators to defectors in the previous round—that is, those sites where the boundaries are shifting.) For a larger payoff, defectors are more prevalent, but communities of coopera-tors constantly arise and decline (*b*). A single defector can exploit a community of coopera-tors, seeding a community of defectors. Because of their mutual repulsion, the defectors grow not as a solid mass but in a ramified pattern (*c*).

GOVERNED BY REASON?

The Tit for Tat policy and its more generous variants have defined most think-ing about how cooperation evolves. Although TFT can yield to softer "nice" strategies or to the opportunistic Pavlov once cooperation is established, there is still no better way to *initiate* cooperation in a Hobbesian world of exploitation than to meet tit with tat.

On this basis, some argue that firm and immediate reprisals for bad behavior are the only way to make "rogue states" act responsibly: hence the bombings of Serbian Belgrade and the invasions of Afghanistan and Iraq. Karl Popper voiced this conclusion in terms that sound almost shockingly brutal, coming from one who has a reputation as a liberal: "What is happening in Bosnia is proof of the failure, the cowardice, the blindness, of us in the West. It shows we do not want to learn what this century should have taught us: that war is prevented with war."[4]

The idea of a "preventive war" goes back at least as far as Kant, who advocated it in his essay "Perpetual Peace"—not that kings and princes in times past needed any philosophical endorsement to take up arms. It is certainly true that strategies in the Prisoner's Dilemma that make softer or delayed reprisals, such as Tit for Two Tats, fare less well in a diverse population. By itself this seems to argue for air strikes rather than sanctions.

But as we've seen, TFT has its drawbacks too. It is painfully evident in both the Israel-Palestine conflict and the troubles in Northern Ireland how reprisals can hold back progress toward cooperation and peace. They can simply serve to undermine the establishment of trust. There is clearly a need for mechanisms that can "damp out" the echoing cycle of retaliation, if good relations are ever to be resumed between Tit for Tat players who, for whatever reason, have broken rank. The Contrite or Generous Tit for Tat strategies offer some solutions; another is to make use of a Partial Tit for Tat approach, where the reprisal is slightly less severe than the event that provoked it.

Human nature seems likely to complicate attempts to generate cooperation via Tit for Tat exchanges. It would be foolish indeed to ignore the strength of passions or the longevity of resentments when human lives are lost in episodes of "defection," while Tit for Tat remembers nothing beyond the previous round and will "forgive" at the first sign of cooperation. And consider the proposal in the United States that the custodial parent in separated couples who have children might be given the right to deny visits if the other parent does not keep up with child support payments. This scope for retaliation could be regarded as an incentive for the child support payer to cooperate. But quite aside from the fact that this scheme would logically have to allow the converse (payments legally withheld if access to the child was denied), it would be shortsighted not to take into account the sheer irrationality that can arise in partnership breakdown and arguments over child-care responsibilities, which may override the ability of the "players" to decide dispassionately what course of action is most advantageous to them in the long run. More pertinent still is the matter of whether there can possibly be any justification for letting children become bargaining chips in such exchanges.

The threat of a Tit for Tat response has been advanced, both implicitly and explicitly, as a theoretical rationale for the policy of nuclear deterrence. The argument here is that even if the awful prospect of an exchange between nuclear powers does not materialize, the evident capacity and stated readiness of a nation to retaliate to such an attack is an essential element of a peaceful status quo. Popper again supported the notion: "We should have learnt by now that peace on earth needs to be backed up with weapons . . . You could never get peace inside a country by reaching a compromise with the criminals."[5]

This might be valid for war in general; but for nuclear war—and in particular the Cold War concept of mutually assured destruction—the iterative process that is essential for Tit for Tat's superiority (indeed, for its very definition) is not an option. Strategist Hermann Kahn describes with compelling clarity the kind of confused and irrational thinking he often encountered at the RAND Corporation during the early years of the Cold War:

One *Gedanken* [thought] experiment that I have used many times and in many variations over the last twenty-five or thirty years begins with the statement: "Let us assume that the president of the United States has just been informed that a multimegaton bomb has been dropped on New York City. What do you think that he would do?" When this was first asked in the mid-1950s, the usual answer was "Press every button for launching nuclear forces and go home." The dialogue between the audience and myself continued more or less as follows:

KAHN: What happens next?
AUDIENCE: The Soviets do the same!
KAHN: And then what happens?
AUDIENCE: Nothing. Both sides have been destroyed.
KAHN: Why then did the American president do this?

A general rethinking of the issue would follow, and the audience would conclude that perhaps the president should not launch an immediate all-out retaliatory attack.[6]

In other words, according to the political theorist Brian Skyrms, "A strategy that includes a threat that would not be in the agent's interest to carry out were she called upon to do so, and which she would have the option of not carrying out, is a defective strategy."[7]

In any event it would be naïve in the extreme to assume that real players in the international politics game will display the rigorous rationality expected of the idealized players in game theory. Supporters of an "assured Tit for Tat" deterrent, for example, must grapple with the strong possibility (as Popper admitted) that Khrushchev had every intention of using his missiles in 1962, had he been able to get enough of them into Cuba secretly—deterrent or no deterrent. Looked at in this light, Tit for Tat simply creates the climate and the conditions for such a crisis. Moreover, the spectacle of many long-term adherents of the deterrence argument for American nuclear proliferation who are now advocating a missile defense system that undermines the very basis of this argument—what is mutually assured destruction if it is not mutual?—should remind us that the formal logic of game theory is no more than an expendable tool in the face of political ideology and expediency.

Another of the less attractive implications of Tit for Tat is that in effect it compels cooperation from defectors only if they know how implacable it is. This implies that one needs to acquire a reputation for being ready to go on the offensive, which can become manifest as hypersensitivity—refusing to tolerate the slightest perceived threat or insult—or as a bullying propensity to adopt "gunboat" diplomacy. In the 1960s, the United States was prepared to fight a bitter war on the other side of the world primarily to maintain, during the height of the Cold War, its reputation for toughness. This was admitted in a memo sent by John McNaughton, assistant secretary for International Security Affairs, to Secretary of Defense Robert McNamara, outlining the American aims in Vietnam: they were "70 percent—To avoid a humiliating U.S. defeat (to our *reputation* as a guarantor [my italics])," and only "10 percent—To permit the people [of South Vietnam] to enjoy a better, freer way of life."[8]

In other words, the success of Tit for Tat could be regarded as an *incentive* to act belligerently. After all, the disastrous consequences of facing an opponent who does not appreciate one's ruthlessness were brought home in a gloriously sardonic manner in Kubrick's classic study of Cold War diplomacy, when Dr. Strangelove explodes over the phone at his Soviet counterpart: "You fools! A doomsday machine isn't any good if you don't tell anyone you have it!"

Anyone who considers using the Prisoner's Dilemma as a basis for deciding policy should feel duty bound first to enumerate all the factors it neglects. Most obviously, as I have indicated, it takes a highly simplistic view of human nature: the assumption that people act rationally to seek the best for themselves neglects not only irrational passions, the fallibility of our reasoning powers, and sheer foolishness, but also the positive influence of moral codes of conduct. Both experience and evolutionary biology lead us to expect that many people have an innate instinct to cooperate with their fellow beings and do not have to learn that this serves their best interests before they will do it. On the other hand, a few people are probably pathologically inclined to defect within society, sometimes even when they see it does them no good in the long run.

Moreover, the Prisoner's Dilemma provides no scope for negotiation: the "prisoners," remember, are not allowed to collude but must deduce each other's motives only from the way they play the game. Suspicion thrives in such circumstances, and in real life we generally seek to quell it by conducting our transactions more collaboratively.

All the same, there is no denying the strong implication from game theory that an unswervingly retaliatory strategy is the best way to bring about cooperation. Game theorist Karl Sigmund says:

> It would of course be rather silly to attempt to reduce all human interactions to the Iterated Prisoner's Dilemma, or to negate the role of superior authority in civilized communities. But with all due restraint, it is worth pointing out that the brutally simple principle of paying back in kind leads to cooperation in a society of egoists, while the apparently higher summons to dispense with reprisals undermines such cooperation . . . The harsh law of retaliation seems to have been the foundation stone of many, possibly all, stable societies.[9]

At the risk of posing a question that evolutionary psychologists might regard as tautologous, one surely has the obligation then to ask, Is this a *moral* way to behave?

The notion of Tit for Tat sits uncomfortably with liberal sensibilities. "I

came to this project," said Robert Axelrod in 1984, "believing one should be
slow to anger. The results of the Computer Tournament for the Prisoner's
Dilemma demonstrate that it is actually better to respond quickly to a provocation."[10]

Pacifist thinkers from St. Francis to Gandhi, however, have asserted that
to meet violence with violence is self-defeating. That is surely the message
of the New Testament: Love thine enemy, for it is the meek, not the vengeful,
who shall inherit the earth. Many pacifists would argue that nonviolence is a
choice based not on cold logic but on higher moral imperatives, such as the
commandment "Thou shalt not kill." When faced with behavior that can
only be considered as exploitative, even murderously so, this can induce ag-
onies of doubt which, if honestly confronted, belie any suggestion that paci-
fism is a soft option. David Jones, a conscientious objector in the Second
World War, explained the dilemma: "The pain of being a conscientious objector was the increasing knowledge of the enormity of what the Germans
were doing. So that was the really challenging thing, not the war, so much as
how one can justify not somehow trying to do something about it."[11]

Cecil Davies, a kindred spirit in the same conflict, found the question ir-
resolvable:

> Wilfred Owen said he was a conscientious objector with a bad conscience, and I
> think lots of COs often had a bad conscience, and while I still think I was right to do
> what I did when I did, I suppose if I had known about the Holocaust it might have
> been different . . . Life isn't simple.[12]

Indeed not, and the Prisoner's Dilemma should not let us pretend otherwise. But one cannot avoid the conclusion that, *on its own terms*, game theory implies that a retaliatory policy to defection might truly be more "moral"
insofar as it serves the greater good. A Tit for Tat strategy not only protects
oneself from exploitation but also helps safeguard the entire community. For
it is only when TFT is eroded by softer cooperative strategies that the community becomes endangered by exploitative defectors or Pavlovian oppor-
tunists. Unconditional cooperation might seem more noble and kindly, but it
also places the burden of policing on the rest of the community. A utilitarian
would be hard pressed to find an objection to Tit for Tat.

All of this seems like knowledge worth having. But we must beware of how easily it might be twisted or misconstrued. One might naïvely argue that TFT justifies the death penalty for murder—even if we allow that the state, not the individual, should implement the punishment. But the whole point of the iterated Prisoner's Dilemma is that it teaches cooperation through experience and adaptation: defectors convert to nice strategies because they are better off for it. Capital punishment simply eliminates the player, so that no further iterations take place. There is nothing in the Prisoner's Dilemma which suggests that one player learns from the mistakes of *another*—that the death of one defector warns others away from defecting. This could happen, of course, but game theory is silent about it—so the success of TFT is irrelevant to debates about the death penalty.

The same surely applies to nuclear stockpiling during the height of the Cold War. Game theory and the Prisoner's Dilemma were very popular at the Pentagon; but one could hardly draw lessons from Tit for Tat when the game was, by its very nature, one that would be over in a single reciprocal exchange.

John Locke had the wisdom to see that while he advocated the "tooth for a tooth" form of Tit for Tat in his State of Nature, it would not do in a civilized society, and that relinquishing of matters of policing, justice, and punishment to the state both lightens the burden on the individual and lessens the need for savagery in enacting them. In an anarchic state individuals cannot generally protect themselves from assault by locking their would-be assailants away, and so they may be forced to do harm to such "defectors." By implying that punishment is necessary to maintain a peaceable society, the success of Tit for Tat does not say much that will surprise anyone. It cannot tell us a great deal about the form that punishment should take, nor can it prescribe a course that will convert a sinner into a saint.

What the Prisoner's Dilemma does is help us move beyond the pessimism of Hobbes without recourse to the rose-tinted assumptions of Locke. If we know that cooperation is possible even in a world that lacks altruism, we have no reason to despair. Getting there is another matter, on which it is appropriate to let Popper have the last word: "We should cautiously feel the ground ahead of us, as cockroaches do, and try to reach the truth in all modesty."[13]

NINETEEN

TOWARD UTOPIA?

HEAVEN, HELL, AND SOCIAL PLANNING

All successful men have agreed in one thing—they were *causationists*. They believed that things went not by luck, but by law; that there was not a weak or a cracked link in the chain that joins the first and the last of things . . . The most valiant men are the best believers in the tension of the laws. "All the great captains," said Bonaparte, "have performed vast achievements by conforming with the rules of art,— by adjusting efforts to obstacles." —Ralph Waldo Emerson (1860)

. . . it may be that the next great developments in the social sciences will come not from professed social scientists but from people trained in other fields.
—George Lundberg (1939)

We know that going to the Moon was a simple task indeed, compared with some others we have set for ourselves, such as creating a humane society or a peaceful world.
—Herbert Simon (1996)

"We don't impose Heaven on people any more," she said. "We listen to their needs. If they want it, they can have it; if not, not. And then of course they get the sort of Heaven they want."

"And what sort do they want on the whole?"

"Well, they want a continuation of life, that's what we find. But . . . better, needless to say."

"Sex, golf, shopping, dinner, meeting famous people and not feeling bad?" I asked, a bit defensively.

"It varies. But if I were being honest, I'd say that it doesn't vary all that much."
—Julian Barnes (1989)

Utopias come in many forms—as many, perhaps, as the human mind is capable of devising. Some have been socialist and egalitarian, some hierarchical and dictatorial. In some, people commit no crimes; in others all miscreants are exterminated. Some are ruled by women, or have no men at all, or alternatively no women. Sex plays a remarkably large role in many of them. Virtually all defy Voltaire's Dr. Pangloss in deviating markedly from the world in which we live—which is perhaps unsurprising when we remember that one way of understanding the word *utopia* is "any place but here."

Very few utopian visions have actually been realized, and those that have been tended to founder abruptly and often squalidly. The civil war that prompted *Leviathan* also spawned the Diggers, a group of communist radicals led by Gerrard Winstanley, who in 1649 cultivated wasteland to escape the poverty endemic to rural England. Their colony was hounded by landlords and by Cromwell's army; within a year it had collapsed.

The noble experiment of Robert Owen's socialist commune at New Lanark in Scotland in the early nineteenth century enjoyed some success but could not outlive its charismatic founder in the face of ruthless Victorian industrialism. (The Co-operative Societies formed to promote Owen's ideas have proved more robust.) The greater experiment in socialism that began with Russia's October Revolution of 1917 has rather less to be proud of today.

Marx and Engels condemned utopias as foolish fantasies because they ignored the historical inevitability of capitalism's demise and socialism's ascendancy. But the Marxist vision is, in retrospect, no less an expression of an arbitrary world considered desirable by its creators. And it was largely among the European socialists of the 1930s, such as the biologist Julian Huxley, that eugenics gained favor, until it became a part of the terrible utopia outlined in *Mein Kampf*.

That is the problem with utopias: one can never quite be sure how they will turn out. Human nature is the unpredictable element that sends the best laid plans awry, which is no doubt why, as the critic John Carey has put it, "the aim of all utopias, to a greater or lesser extent, is to eliminate real people."[1] Many of them are populated by mild-mannered, passively happy citizens among whom dissent and unrest have been either savagely repressed, banished, or somehow spirited away. Revealingly, the same characteristics appear in many dystopias, such as Aldous Huxley's *Brave New*

World, where bovine passivity is maintained by biological engineering, the provision of free sex and drugs, and mindless leisure activities. One might argue that it is in this sense that traditional microeconomic theory is a utopian vision too, replacing real people with omniscient, rationally maximizing automata so that the market can seek its fictitious equilibrium.

The most striking aspect of utopias through history is that very few of them are at all appealing from today's perspective. Plato's Republic is a bellicose, class-ridden aristocracy in which censorship is the norm, art is discouraged, and children are taken away from their parents at birth. Thomas More's original Utopia might have been enticing to the sixteenth-century commoner, with its absence of poverty and equality of rights; but it comes at the cost of strict regimentation and conformity. The world of William Morris's *News from Nowhere* (that is, news from Utopia) looks pleasant enough, but it is a naïve romantic fantasy whose architect has made little attempt to hide the strings from which it hangs. The America of the year 2000 as idealized in Edward Bellamy's *Looking Backward* (1888) is chillingly plausible: a land of nationalistic conformity enforced by the Religion of Solidarity, run as a vast business syndicate in which all citizens are compelled to enroll in the militaristic "industrial army." Dissidents are sentenced to solitary confinement on a diet of bread and water. Bellamy evidently thought all this sounded like a splendid idea.

And so it goes on, underscoring the old truth that one man's utopia is another's dystopia—for there is too much variety in human nature for it all to fit one mold. "What has always made the state a hell on earth," says the poet Friedrich Hölderlin, "has been precisely that man has tried to make it his heaven."[2]

The notion that we could ever construct a scientific "utopia theory" is, then, doomed to absurdity. Certainly, a "physics of society" can provide nothing of the sort. One does not build an ideal world from scientifically based traffic planning, market analysis, criminology, network design, game theory, and the gamut of other ideas discussed in this book. Concepts and models drawn from physics are almost certainly going to find their way into other areas of social science, but they are not going to provide a comprehensive theory of society, nor are they going to make traditional sociology, economics, or political science redundant. The skill lies in deciding where a mechanistic, quantitative model is appropriate for describing human behavior, and where it is likely to produce nothing but a grotesque caricature. This is a skill that is still being acquired, and it is likely that there will be embarrassments along the way.

But properly and judiciously applied, physical science can furnish some valuable tools in areas such as social, economic, and civic planning, and in international negotiation and legislation. It may help us to avoid bad decisions; if we are lucky, it will give us some foresight. If there are emergent laws of traffic, of pedestrian motions, of network topologies, of urban growth, we need to know them in order to plan effectively. Once we acknowledge the universality displayed in the physical world, it should come as no surprise that the world of human social affairs is not necessarily a tabula rasa, open to all options.

Society is complex but that does not place it beyond our ken. As we have seen, complexity of form and organization can arise from simple underlying principles if they are followed simultaneously by a great many individuals. John Stuart Mill already recognized this in the nineteenth century: "The complexity does not arise from the number of the laws themselves, which is not remarkably great, but from the extraordinary number and variety of the data or elements—of the agents which, in obedience to that small number of laws, cooperate towards the effect."[3]

On the other hand, complexity is not the inevitable result of a multitude of interactions. Perhaps that is one reason to be wary of making *complexity* a buzzword for a new type of science. The real surprise to emerge from the physics of society is that patterns of social behavior can sometimes be so simple, so observant of mathematical "laws." While the steady averages and recurrent error laws of the pioneers in social statistics arouse no fears today that free will has been demolished, the new discoveries of social physics have added further, more profound and less intuitive motifs to the universals of human activity: phase transitions, power laws, self-organizing patterns, collective motions, scale-free networks.

All the same, we should not forget that there is little truly new in a vision that recognizes society and culture as characteristics that emerge from the intercourse of many autonomous individuals, guided only by simple and local concerns. The German sociologist Georg Simmel summed up the philosophy of today's physics of society in 1908: "Society is merely the name for a number of individuals, connected by interaction."[4] Robert E. Park, an American social scientist, refined this notion in 1927 by pointing to the way in which individuals act together: "Institutions and social structures of every sort may be regarded as products of collective action."[5] In a sense, all we are now doing is broadening that idea, highlighting the collectivity that lies behind not just static institutions but also dynamic, changing features of society: traffic jams, economic fluctuations, the evolution of cultural traits and norms.

GOOD FROM BAD

One of the strongest themes to emerge from the physics of society is that individual behavior is often no predictor of its social consequences. Society is not just a scaled-up person: Leviathan is more than the sum of its parts.

Evil flowing unintentionally from acts of benevolence, good ends resulting from evil intentions—these are two of the universal themes of art and literature. *The Wealth of Nations* could serve as their prototype. Acting from pure self-interest, said Adam Smith, traders contrive to furnish goods for all, at an affordable price, leaving no gaps in the market: "It is not from the benevolence of the butcher, the brewer or the baker that we expect our dinner, but from their regard of their own interest. We address ourselves not to their humanity, but to their self-love, and never talk to them of our necessities, but of their advantage."[6] Meanwhile (says Smith), governments that seek to regulate trade in the interests of society as a whole succeed in doing no more than impairing its efficiency.

Smith's beneficent "invisible hand" is prefigured in Montesquieu's remark that actions undertaken for one motive can have quite different effects—for better or worse. Montesquieu, in turn, was influenced by a satirical poem, *Fable of the Bees: or Private Vices, Public Benefits*, written by the Dutch physician Bernard Mandeville (1670–1733) and first published in 1705 as a pamphlet titled *The Grumbling Hive*. It was a mischievous work, but like most satire it aimed to make a serious point. Mandeville suggested that "vices" such as vanity can work to the good of society, in this case by increasing the demand for luxury goods and thus providing employment in their manufacturing:

> To enjoy the World's Conveniences,
> Be famed in War, yet Live in Ease,
> Without great Vice, is a vain

Utopia seated in the Brain.
Bare Virtue can't make nations live in Splendour.[7]

In other words, society positively *needs* a little vice to keep its wheels lubricated. When vice is eliminated by an ethos of self-denial in the beehive that Mandeville described, the bees' society goes into decline, eroded by idleness and poverty. Mandeville was not promoting Hobbesian selfishness, but he despised the hypocrisy that presented virtue as something other than a necessary conceit that made the system work. Let's not kid ourselves, he insisted. He was, naturally enough, vilified as much as Hobbes had been, and his poem was denounced as a public nuisance. Yet Mandeville gradually expanded the work, which began as a divertissement, into a substantial two-volume treatise that provoked serious discussion about its implied moral philosophy.

There is no better example of Mandeville's principle in action than the emergence of altruism from selfishness in game theory. Hard-line evolutionary psychologists would claim that there is no virtue that is not at some level motivated by self-interest—although they are a long way from proving it.

The physics of society is full of actions that generate effects other than what was intended or expected. The rush to escape a crowded room leads to a slower average rate of exit. In a corridor where people move in both directions, an increased randomness in the motion of individuals (greater "noise" or "heating") can cause a counterintuitive blockage ("freezing") by making it harder to form counterflowing lanes. Highway lane rules that try to segregate the fast from the slow can end up reducing the efficiency of traffic flow.

There are few easier targets than governmental, regulatory, and planning decisions that have had the opposite of their intended effects. In many such cases these unwanted outcomes can be put down to a failure to appreciate the interconnected and interactive nature of the system concerned. The increased congestion that can follow from the building of new roads is a classic example. Statistical physics may help liberate planners and policy makers from their propensity for linear thinking and encourage a greater sophistication in their perception of cause and effect.

IS A SCIENTIFIC SOCIOLOGY GOOD FOR US?

When William Petty applied strict mathematical reasoning to social phenomena, particularly in making recommendations on taxation and the government of Ireland, some of his contemporaries found his approach risible. And indeed it sometimes was. Petty's insistence on developing policy based on rationality and quantitative data rather than subjective intuitions and prejudices was pioneering; when things are not done this way today, we are often the worse for it. But he tended to neglect any consideration of what people could, should, or would tolerate, or of the need to adapt policy and legislation to existing custom and situation.

His claim that laboring people could afford to pay more taxes if they skipped their dinner on Friday evenings (thus saving on the cost of procuring it) was of course perfectly true. But only someone deeply lacking in an understanding of the circumstances and psychology of his fellows would venture to make it a serious proposal. In 1729, Jonathan Swift satirized this dry, detached style of the "scientific" political philosophers in his proposal that the poor people of Ireland might make ends meet and at the same time serve the public good by selling some of their children to be cooked and eaten—for "a young healthy Child, well nursed, is, at a Year old, a most delicious, nourishing, and wholesome Food; whether *Stewed, Roasted, Baked,* or *Boiled*; and I make no doubt, that it will equally serve in a *Fricasie,* or a *Ragoust.*"[8]

By bringing mechanistic physics into social science, do we risk becoming like Petty? If people are smoothed down to billiard balls interacting through mathematically defined forces, where do we find room for compassion, charity, for the thousand and one parts of our daily lives that cannot be reduced to numbers but that make our lives worth living?

This cold idealization, the epitome of Hobbesian man, was, in Lewis Mumford's view, a disastrous outcome of the Enlightenment view of society:

> The new order established in the physical sciences was far too limited to describe or interpret social facts . . . Real men and women, real corporations and cities, were treated in law and government as if they were imaginary bodies; whilst presumptuous fictions, like Divine Right, Absolute Rule, the State, Sovereignty, were treated as if they were realities. Freed from his sense of dependence upon corporation and neighborhood, the "emancipated individual" was dissociated and delocalized: an atom of power, ruthlessly seeking whatever power can command.[9]

And yes, the responsibility for this change must ultimately lie with Thomas Hobbes. Whether they agreed with him or not, most later moral philosophers inherited Hobbes's rational approach to some degree. The consequences are most apparent in the most explicitly "scientific" and mathematical of the social sciences: economics. Although Hobbes had rather little to say about how economic markets worked, his supposition that men make their decisions based on a rational assessment of the opportunities for private gain—that people act out of selfishness—underlies all of conventional economic thinking. Hobbes's ruthless world is the natural habitat of the rational maximizer, the so-called *Homo economicus*.

One could argue that people surely *do* act this way, and that this is scarcely Hobbes's fault. Yet it is also fair to wonder whether the prevailing political and economic climate of our times does not actively cultivate, encourage, and even compel this kind of behavior—whether the basic assumptions of economics have not become self-defining and self-sustaining.

After all, it was not always this way. In the late Middle Ages, the idea of a "just price" dominated economic thought. Goods were priced at a level that was congenial both to buyer and seller: the price gave the vendor enough to live on, and allowed the buyer to acquire necessities at inexorbitant rates. The just price was set by guilds, town councils, or other regulators—the economy was extremely interventionist. It was no doubt a system riddled with inefficiencies and abuses, but the principle was based on an idea of social welfare. Merchants accepted at least to some degree that they had social obligations as well as profit motives.

Hobbes, however, wrote at a time when the expansion of international trade had put the economy beyond the control of individual governments or states. At the international markets that flourished in European towns, there was no one to regulate prices; merchants could sell at whatever rate the buyers would accept. In this soil, self-interest could and did blossom. What it needed, and what Adam Smith provided, was a vindicating philosophy.

That prices nevertheless set their own level through the balance of supply and demand was first discussed by the Frenchman Richard Cantillon in his *Essai sur la nature du commerce en général* (1730–1734), to which Smith's *Wealth of Nations* was much indebted. The appeal of the "hidden hand" to merchants was obvious: they wanted to be left free to do whatever they could get away with, unhindered by regulations or price limits. But the attractions of such freedom were less evident to Smith himself. He had no interest in being an apologist to greedy businessmen; indeed, he was striving to establish not an economic theory but a moral and social philosophy. He accepted only reluctantly the idea that men were motivated by self-interest, and deplored the fact that, in consequence, "the labour and time of the poor, are in civilized countries, sacrificed to the maintaining of the rich in ease and luxury."

Yet accept it he did. Why? Because by simplifying economic theory and making it accessible to quantification, *Homo economicus* permits a "scientific" analysis of the market. And that was part of the Enlightenment's program: to bring science into human affairs.* This desire to simplify for the sake of analysis is, as we have seen, why *Homo economicus* has survived for so long. Many economists today still have a Smithian objection to any form of market regulation, clinging determinedly to his notion that all intervention is necessarily harmful to the market and thus —because economic growth has been enthroned as the barometer of a nation's health—harmful to society as a whole. Yet there is not a single economic theory that can show that a totally free market sets the most socially beneficial price for goods, or leads to their optimal distribution.

Indeed, it may in some ways be quite the contrary. The demonstration that new, "interactive" models of trade and economics accentuate inequalities in wealth (page 349) is perfectly in accord with what Montesquieu pointed out in the eighteenth century: that the greatest selfishness is found in "civilized" commercial societies. In contrast, nomadic cultures retain a strong sense of social obligation and make hospitality to strangers an almost sacred duty. Inequality is, in other words, to some degree the signature of commerce. This view echoes Rousseau in implying that the nastiness and brutishness of people, far from being most prevalent in the Hobbesian State of Nature, is partly a product of civic society itself. That too is unquestionably an oversimplification, as the most cursory familiarity with the ruthless social behavior of other animals will indicate. Nevertheless, the idea that "laws" of society follow from fundamentals of basic human nature must surely be modified to embrace the fact that this nature is itself molded by the

---

*Smith was himself in good part a scientist: he studied Newton's *Principia* closely and was a close friend of the chemist Joseph Black and the geologist James Hutton.

customs and norms of the society in which it operates. If we allow that much, it is fair to ask what biases might have been introduced into our cultural structures and institutions by a society shaped by the scientific rationalism of the eighteenth century.

Perhaps this seems a strange thing to say in a book about the value of physics-based models in the social sciences. But it is simply a note of caution. I believe that the current state of conventional economic theory, for example, which continues to accept questionable "first principles" framed in the glow of the Enlightenment's understandable overconfidence, alerts us to the need for such caution.

CHOICE AND CERTAINTY

The issue is really whether we can trust ourselves to distinguish between moral and physical law. What a physics of society cannot do is tell us how we should live our lives, how we should define our individual and collective responsibilities, how to decide what is important. There are few things quite so misguided and dangerous as looking to science for moral guidance. There are no laws of nature that tell us how to behave or how to govern. John Stuart Mill recognized the folly of trying to build a utopia on the basis of "laws" that supposedly determine how things ought to be, rather than how they are:

> A large proportion of those who have laid claim to the character of philosophic politicians have attempted, not to ascertain universal sequences, but to frame universal precepts. They have imagined some form of government, or system of laws, to fit all cases; a pretension well meriting the ridicule with which it is treated by practitioners.[10]

The Scottish philosopher Adam Ferguson, an associate of Adam Smith, made this distinction clear in 1766. While a physical law tells us what *is*, said Ferguson, a moral law dictates what *ought to be*. The latter is a law "in consequence of its rectitude, or the authority from which it proceeds . . . not in consequence of its being the fact."[11]

The problem, of course, is that if there is no objective way of establishing moral laws, one person's law has no stronger claim than another's, so primacy depends on the distribution of power. To Charles I it was a moral law that he should rule absolutely. The attempts of Bentham and Mill to avoid such relativism by using the principle of utilitarianism—the greatest good for the greatest number*—were never quite convincing: social policies cannot simply be determined by arithmetic.

Nonetheless, Ferguson identifies the central ethical axis of a physics of society: the balance between choice and determinism. Ferguson was no less a product of his times than Adam Smith, Auguste Comte, or Condorcet in believing that there *are* "natural" laws that govern society. "Nations," he wrote, "stumble upon establishments, which are indeed the result of human action, but not the execution of any human designs."[12] He did not believe that we must be ruled by these establishments, just that such "laws" as might exist are in the nature of inevitable consequences of certain actions. If we recognize these consequences, we can then decide whether they are to be desired, according to whatever moral precepts we elect to adopt. There is no reason to believe that unwelcome consequences cannot be alleviated or even avoided.

The physics of society supports this idea. Economists who argue that large wealth inequalities are just one of the things we must live with if we want a society that has a better standard of living than the Stone Age are speaking not as scientists but as dogmatists. Better models of the economy can tell us whether, by making this or that change to trading patterns, we can tilt the income distribution toward greater equality. Similarly, we might hope to determine which changes to driving restrictions on a stretch of highway might lessen the tendency for jams to form (jams being undesirable, if not exactly immoral).

Ralph Waldo Emerson, another believer in the idea that natural laws direct some of the affairs of humankind, felt that these laws need not prescribe a particular kind of government. Insofar as our institutions are concerned, we still have choices. "In dealing with the State," he says, "we ought to remember that its institutions are not aboriginal, though they existed before we were born . . . they all are imitable, all alterable; we may make as good; we may make better."[13]

In other words, things don't have to be this way. Karl Marx agreed: "Philosophers have sought to interpret the world. The point, however, is to change it."[14]

Yet the economist Paul Ormerod suggests that Marx had it the wrong way

---

*George Kingsley Zipf has pointed out that "the greatest good for the greatest number" is a meaningless notion, the "fallacy of double optima." Is great good for a few as desirable as moderate good for many?

around: "Politicians have sought to change the world. But the point is to interpret it correctly."[15] In the end, we must do both: understanding and the power to effect change go hand in hand. Politicians are all too ready to seek (or promise) changes they do not know how to bring about, or that might be impossible within the rule systems they have created. (Marx, in contrast, saw only a single, inevitable future—all he wanted was to hasten it.)

The idea of using physics to empower change and to predict its limits has no intrinsic politics. It is not libertarian or repressive, neither left-wing nor right. It might occasionally help us cut through such ideologies. Since most political thinking starts off with a particular worldview and then looks for ways to make the real world conform with it, the suggestion that there may be unavoidable fluctuations and inevitable laws when many individuals get together and interact might be a deeply uncomfortable one for politicians. But if they do not confront it as a real possibility, they risk making choices that are both futile and wasteful.

The historian Richard Olson summarizes the point perfectly in his account of Adam Ferguson's conclusions about how to deal with natural laws of society:

> One way of expressing the relationship between physical and moral laws—i.e., between science and morality—in the formation of society as Ferguson understood the situation is to say that social systems are "softly" deterministic. Left alone, they will inevitably develop along certain lines; but the possibility of changing those lines by conscious and intentional intervention does exist. The whole point of a "social science," then, is to explore the opportunities for and likely consequences of intentional moral action. Without the science, morality is blind; but without the morality, science is useless, pointless, and paralytic.[16]

PLANNING FOR FREEDOM

In antiquity, Ferguson's distinction between "natural" and "moral" was meaningless. Cicero (106–43 B.C.) had a touching faith in the natural morality of the universe. In *De Legibus* (On Laws), he said, "Now we must entirely take leave of our senses, ere we can suppose that law and justice have no foundation in nature, and rely merely on the transient opinions of men."[17]

History lends little support to Cicero's optimism. If we want law and justice, we must build a society that promotes them and protects them from abuses. That is where the state comes in.

To Aristotle, it seemed obvious that humankind had a natural compulsion to form groups: "Man is by nature a political animal,"[18] he asserted. Thomas Aquinas, in his little book *On the Governance of Rulers* (c. 1259), could only agree: "It is natural for man to be a social and political animal, to live in a group, even more so than all the other animals, as the very needs of his nature indicate."[19] But he acknowledged the drawback: "Where there are many men together, and each one is looking after his own interest, the group would be broken up and scattered unless there were also someone to take care of what appertains to the common weal."[20] Thus, a state cannot simply be a group; it must be a group with a ruler. Government, in other words, is inevitable.

Yet this is still a long way from Hobbes's necessary dictatorship. The harshness of Hobbes's solution to government obviously reflects his much more jaundiced view of human nature. Unlike Cicero, he saw no redeeming socializing tendencies in man but only a will to exploit and dominate: "Men have no pleasure, (but on the contrary a great deal of grief) in keeping company, where there is no power able to overawe them all."[21] The twentieth century saw plenty of regimes with the capacity to overawe, and they were not a good advertisement for Hobbes. How do we ensure that society is not just stable but moral?

It is generally taken for granted in the West that the answer, or at least a good part of it, is to make society democratic. But this is a modern view; or at least, it has passed in and out of favor over time. The liberal democracies of Locke and Mill by no means represented the obvious political future of Europe in the 1930s, when it looked vulnerable both to fascism and to a warped and dictatorial socialism. Against such a backdrop we can understand why the Austrian economist Friedrich von Hayek felt compelled to write his famous defense of capitalist freedom, *The Road to Serfdom* (1944). But Hayek's critique of socialism, which he regards as the beginning of an inevitable descent into totalitarianism, is not just a warning to beware of Hitler, Mussolini, and Stalin. It also examines different notions of democracy and addresses the crucial question that all democrats must ask: *How much* should they rule?

Hayek echoes Adam Smith in making the free market the ultimate ar-

biter of political and economic society. Since in Hayek's view money is the
key to freedom (and the extreme naïvety of that attitude must again be seen
in the context of the times), to impose constraints on the ways in which capital can be used is to compromise liberty. Hayek's approval of Smith is apparent as he quotes him thus:

> The statesman who should attempt to direct private people in what manner they
> ought to employ their capitals, would not only load himself with a most unnecessary
> attention, but assume an authority which could safely be trusted, not only to no sin-
> gle person, but to no council or senate whatever, and which would nowhere be so
> dangerous as in the hands of a man who had folly and presumption enough to fancy
> himself fit to exercise it.[22]

Thus "planning," to Hayek, means economic planning, and anything of
that nature smacks of socialism. Far better to entrust the forces of competition to bring about a market that, if not necessarily equitable, is at least
"fair" in the sense that it equalizes opportunity within the boundaries of the
cultural climate. Clearly, this debate about regulation versus market freedom is as relevant today as it was sixty years ago.

Underlying Hayek's advice is a firm belief in self-organizing principles
that maintain a balance. It would be easy to read it as an advocacy of laissez-
faire if Hayek did not take such pains to explain the distinction: "The liberal
argument is in favour of making the best possible use of the forces of competition as a means of coordinating human efforts, not an argument for leaving things just as they are."[23] Thus he feels that the economy needs *tending*
so that these "spontaneous forces of society" can operate to greatest benefit:
"The attitude of the liberal towards society is like that of the gardener who
tends a plant and in order to create the conditions most favorable to its
growth must know as much as possible about its structure and the way it
functions."[24]

Here is the crux of the matter. One can take exception to Hayek's denial
of an absolute system of values—his assertion that each person must be his
or her own judge seems a recipe for disaster (or for a Hitler). One can question his belief that market forces inevitably operate for the "best," and so
should be given free rein. But it is surely incumbent on those who seek to
lead or govern a society that they "know as much as possible about its structure and the way it functions." Hayek is right at least to believe that there

are spontaneous forces of society. In this book we have seen some of the modern efforts to establish what those forces are.

No doubt some of these forces are what we might call exogenous: they arise from outside, from technological change for example (such as mechanization, birth control, information technology), or from the environment (drought and famine, climate change). But many of the forces that act on society are the collective consequence of interactions between one person and another, whether in conducting trade, spreading new fashions, waging war, getting married, or avoiding collisions.

Thomas Schelling points out that these forces are typically a compromise of conflicting wishes—a kind of interplay of attraction and repulsion. What we wish for ourselves may not always be what we wish for others:

> A good part of social organization—of what we call society—consists of institutional arrangements to overcome [the] divergences between perceived individual interest and some larger collective bargain . . . What we are dealing with is the frequent divergence between what people are individually motivated to do and what they might like to accomplish together.[25]

It is particularly (although not uniquely) when we are confronted with such divergences between individual and collective goals that our intuitions about the likely modes of aggregate behavior may be a poor guide, and where instead we need simple yet realistic models in order to predict the outcomes of our actions.

The wider challenge is then to decide whether or not we desire the consequences that these forces bring about. Hayek fails to make a case for *why* a free market is best for society. It might be the most "efficient" in some sense—although new models of the economy cast doubt even on that. But if it permits (indeed, promotes) huge discrepancies in earnings, out of all proportion to efforts or responsibilities, if it allows child pornography (or children) to be traded like bananas, if it places small nations at the mercy of big corporations, then surely we are entitled to ask whether this is what we want. As Schelling points out, "How well each [person] does for himself in adapting to his social environment is not the same thing as how satisfactory a social environment they collectively create for themselves."[26]

That is where the utility of social physics ends. If it becomes a justification for moral choices, it has exceeded its brief. This ought to be obvious, but today science is granted such authority that the small step into morality and ethics can sometimes pass unnoticed. Already there is a tendency to divide science into "good" and "bad," for example by branding it "holistic" or "reductionist." This is disheartening: as the British writer Kenan Malik says, "There are few things more dispiriting than turning science into faith."[27] Hayek himself railed against the inappropriate application of scientific ideas to social problems, and in an age when liberals espoused pseudo-Darwinian arguments for eugenics, he had fair cause for complaint. Knowledge of how things *are* can never supplant the obligation to justify our preferences for how they should be.

And so the physics of society is and can only be a tool, never a moral compass. John Stuart Mill identified with great elegance the way these tools should be employed:

> The aim of practical politics is to surround any given society with the greatest possible number of circumstances of which the tendencies are beneficial, and to remove or counteract, as far as practicable, those of which the tendencies are injurious. A knowledge of the tendencies only, though without the power of accurately predicting their conjunct result, gives us to a considerable extent this power.[28]

It cannot, surely, be anything other than wise to search for these tendencies; indeed, one might regard this as a moral obligation. The American sociologist Franklin Giddings acknowledged in 1924 that this search is the prime objective of the social sciences, and that the goal might be all the more readily achieved if the discipline can indeed evolve into a true science:

> Science is nothing more nor less than getting at the facts, and trying to understand them, and what science does for us is nothing more nor less than helping us to face facts . . . Facing the facts that the social sciences are making known to us, and will make better known, should enable us to diminish human misery and to live more wisely than the human race has lived hitherto.[29]

This, in the end, is all Thomas Hobbes wanted. That his "facts" pointed to a political institution we would today regard as intolerable and more likely to contribute to than to diminish human misery is a warning of how hard it is to escape the tenor of one's times, however "scientific" one strives to be.

Nonetheless, many aspects of Hobbes's methodology and reasoning are remarkably modern, and Hobbes could justifiably claim to be at the head of a long succession of attempts to deduce collective human behavior from individual tendencies.

As you can see, the implications of a physics of society can be formulated very nicely in the words of earlier ages. Indeed, it is often to such words that we must look for a wider perspective on what statistical physicists today are doing, for scientists now are encouraged (with good reason) to interpret their findings with the utmost conservatism and humility. Politicians and policy makers, meanwhile, are tending to withdraw from the visionary political philosophy of their predecessors and to settle for quick solutions to short-term challenges. The most urgent question, to which the physics of society can contribute, is whether we can construct a society blessed with the wisdom and compassion that others, often in harsher or more difficult times, were able to glimpse and to demand.

EPILOGUE: CURTAIN CALL

The performer who provides his own applause is surely either deluded or desperate; but where better than in conclusion to describe the following delightful manifestation of social physics?

In some countries and cultures, particularly in Eastern Europe, the applause with which a gratified audience expresses its appreciation of a performance tends to slew back and forth between random and synchronized. At one moment each person is clapping to his or her own rhythm, and the hundreds of overlapping pulses of sound create a continuous clattering roar like the sound of surf on shingle. But then something remarkable happens: this wash of noise resolves itself into a regular beat, as each pair of hands claps in unison with the others. The synchronization lasts for perhaps a minute or so, then dissolves again into chaos.

There is no one conducting *this* performance, no one to set the pace or to signal when synchrony should begin. It just happens—not once but several times during an ovation. Now, it is no great feat for two or three people to bring their hand claps into step with one another. Indeed, it can be hard for them to avoid doing so, just as two people tend to synchronize their steps when they walk side by side. But the synchronized clapping of an audience of many hundreds is a challenge of another order. That it can crystallize so quickly is surprising enough; but why, once synchronized, does the clapping not stay that way, given that each member of the audience can consciously sustain it with little effort? Why this seesawing alternation between order and chaos?

Tamás Vicsek, Albert-László Barabási, and their co-workers have asked this same question. They made sound recordings of the postperformance applause in several theaters and opera halls in Hungary and Romania, and looked at how the sound volume changed as clapping veered between incoherent and synchronized. They found that although synchronized clapping produces "spikes" of noise that can exceed the sound level during incoherent clapping, nevertheless synchronization decreases the *average* noise intensity in the hall (Figure E.1).

This decrease happens not because the audience members are clapping any less vigorously when they are in step, but because they clap *less often*. People space their claps roughly twice as far apart in the synchronized mode than when clapping freely. This is presumably because it is harder to clap in time with everyone else if the rhythm is too fast—synchronization of a slow hand clap disintegrates if the pace quickens. No one in the audience thinks consciously about this; the measured rhythm of the synchronized clap essentially "selects itself."

The researchers suggest that the difference in average sound volume explains why applause at concerts oscillates between incoherent and synchronized. An appreciative audience wants to make a lot of noise. It also appears to enjoy the communal experience of synchronized clapping, which lends the crowd a single voice. But these two things are in conflict, because synchronization results in a drop in the average noise intensity. No one registers

E.1   Sound levels during audience applause in an Eastern European theater. The switch from random to synchronized clapping is marked by the appearance of sharp, regularly spaced peaks in the noise intensity (top). When this happens, the average noise intensity diminishes (bottom), only to rise again as the synchronized clapping dissolves.

this drop consciously; but the audience, having switched into synchronized mode, nonetheless begins to speed up the rhythm so as to restore the sound volume to its earlier, unsynchronized level. In doing so, they lose the ability to keep their claps in step, and the sychronization dissolves into incoherence. Moments later, the attractive tug of synchrony reasserts itself and the cycle repeats.

A push and a pull; a tension between conflicting desires. This is all it takes to tip our social behavior into complex and often unpredictable patterns, dictated by influences beyond our immediate experience or our ability to control. Regardless of what we believe about the motivations for individual behavior, once we become part of a group we cannot be sure what to expect.

NOTES

INTRODUCTION: POLITICAL ARITHMETICK

1. W. Petty (1690). *Political Arithmetick.* Robert Clavel, London.
2. Ibid., Preface.
3. K. Waltz (1954). *Man, the State, and War*, p. 75. Columbia University Press, New York.

CHAPTER 1: RAISING LEVIATHAN

EPIGRAPHS

Bernard Fontenelle, quoted in A.G.R. Smith (1972). *Science and Society in the 16th and 17th Centuries*, p. 153. Science History Publications, New York.

Bernard Fontenelle (1686). *Conversations on the Plurality of Worlds*, trans. W. Gardiner (1715), pp. 11–12. London.

Robert M. MacIver (1947). *The Web of Government.* Macmillan, New York. Quoted in R. Bier stedt, ed. (1959). *The Making of Society*, p. 493. Random House, New York.

Stephen Cotgrove (1967). *The Science of Society*, p. 181. Allen & Unwin, London.

1. Snorri Sturluson (thirteenth century). *The Prose Edda*, trans. J. I. Young (1964) (with minor changes), p. 86. University of California Press, Berkeley.
2. J. Aubrey (1681). *Brief Lives*, ed. J. Buchanan-Brown (2000), pp. 427–28. Penguin, London.
3. T. Hobbes (1651). *Leviathan*, ed. C. B. Macpherson (1985), pp. 110–11. Penguin, London.
4. T. Hobbes (1651). *Philosophical Rudiments concerning Government and Society*, ch. 1, sec. 7, p. 26. Quoted in *Leviathan* (ed. C. B. Macpherson), p. 28. The *Rudiments* is a translation by Hobbes of his Latin text *De Cive* (1642).
5. Hobbes, *Leviathan*, p. 82.
6. R. Descartes (1641). "Meditations on first philosophy." In *Meditations and Selections from the Principles of Philosophy*, p. 98. Open Court, La Salle, Ill., 1952.
7. Hobbes, *Philosophical Rudiments*, Preface, pp. 10–11.
8. T. Hobbes (1642). *De Cive*, ed. S. Lamprecht (1949), p. 15. Appleton-Century-Crofts, New York.

9. F. Bacon (1627). *New Atlantis*. Quoted in J. Carey, ed. (1999). *The Faber Book of Utopias*, p. 64. Faber & Faber, London.

10. R. M. MacIver (1947). *The Web of Government*. Quoted in Bierstedt, p. 493.

11. L. Mumford (1946). *The Culture of Cities*, p. 93. Secker & Warburg, London.

12. Hobbes, *Leviathan*, p. 151.

13. Ibid., p. 161.

14. Ibid., p. 186.

15. Ibid., p. 190.

16. Job 41:8–34.

17. Hobbes, *Leviathan*, p. 228.

18. F. L. Nussbaum (1965). *The Triumph of Science and Reason 1660–1685*, p. 61. Harper & Row, New York.

19. Hobbes, *Leviathan*, p. 261.

20. Hobbes, *Leviathan*, p. 379.

21. R. Olson (1990). *Science Deified and Science Defied*, vol. 2, p. 41. University of California Press, Berkeley.

22. Quoted in R. Tuck (2002). *Hobbes: A Very Short Introduction*, p. 107. Oxford University Press, Oxford.

23. T. Hobbes (1682). *Behemoth, or The Long Parliament*, ed. F. Tönnies, p. 126. Reprinted by University of Chicago Press, 1990.

23. Hobbes, *Leviathan*, p. 208.

CHAPTER 2: LESSER FORCES

EPIGRAPHS

Piet Hein (1966). "Atomyriades," in *Grooks*, p. 24. MIT Press, Cambridge, Mass.

Albert Einstein (1900), Letter to Mileva Maric, 13 September. In A. Beck & P. Havas, eds. (1987). *The Collected Papers of Albert Einstein*, vol. I, p. 154. Princeton University Press, Princeton, N.J.

James Clerk Maxwell (1856), Inaugural Lecture at Aberdeen, 2 November. In *Notes and Records of the Royal Society of London 28*, 69 (1973).

1. I. Kant, quoted in F. A. Hayek (1962). *The Road to Serfdom*, p. xx. Routledge, London.

2. R. J. Boscovich (1763). *Theoria philosophiae naturalis*. Venice. Trans. as *A Theory of Natural Philosophy* (Open Court, Chicago, 1922). Reprinted by MIT Press, Cambridge, Mass. (1966), p. 141.

3. P.-S. Laplace (1814). *Essai philosophique sur les probabilités*, p. 3. Paris. Reprinted by Gauthiers-Villars, Paris, 1921.

4. S. Carnot (1824). *Réflexions sur la puissance motrice du feu et sur les machines propre à développer cette puissance*. Bachelier, Paris. English trans. by R. H. Thurston in S. Carnot, *Reflections on the Motive Power of Fire and Other Papers on The Second Law of Thermodynamics*, ed. E. Mendoza. Dover, New York, 1960.

5. C. P. Snow (1965). *The Two Cultures*, pp. 14–15. Cambridge University Press.

6. W. Thomson (1852). "On a universal tendency in nature to the dissipation of mechanical energy." *Philosophical Magazine 4*, 304–6.

7. Quoted by J. M. Thomas (1986). *Royal Institution Proceedings*, Royal Institution Inaugural Lecture, p. 3.

8. L. Boltzmann (1872). "Weitere Studien über das Wärmegleichgewicht unter Gasmolekülen." *Sitzungsberichte der Akademie der Wissenschaften*, Wien II, 66, 275–370. In *Wissenschaftliche Abhandlungen*, ed. F. Hasenöhrl, vol. 1, p. 345. J. A. Barth, Leipzig, 1909.

CHAPTER 3: THE LAW OF LARGE NUMBERS

EPIGRAPHS

Wilhelm Wundt (1862). *Beitrage zur Theorie der Sinneswahrnehmung*, p. xxv. Leipzig. John Herschel (1850). "Quetelet on probabilities." *Edinburgh Review 92*, 42.

Herbert Spencer (1880). *The Study of Sociology*, 9th ed., p. 39. Otto Zeller, Osnabrück, 1966.

1. L. Boltzmann, in W. Hoeflechner, ed. (1994). *Ludwig Boltzmann: Leben und Briefe.* Publikationen aus dem Archiv der Universität Graz, Akademische Druck- und Verlaganstalt, Graz. English translation by F. Rohrlich (1992). "A poem by Ludwig Boltzmann," *American Journal of Physics 60*, 972–73.

2. D. L. Goodstein (1975). *States of Matter*, p. 1. Prentice Hall, Englewood Cliffs, N.J.

3. R. Musil (1953–60; written 1930–42). *The Man Without Qualities*, trans. E. Wilkins and E. Kaiser. Secker & Warburg, London.

4. W. Petty (1691). "The political anatomy of Ireland." In C. H. Hull, ed. (1899). *The Economic Writings of Sir William Petty*, vol. 1, p. 129. Cambridge University Press.

5. J. Graunt (1662). *Observations upon the Bills of Mortality.* In C. H. Hull, ed. (1899). *The Economic Writings of Sir William Petty*, vol. 2, p. 334. Cambridge University Press.

6. Quoted in D. J. Boorstin (1985). *The Discoverers*, p. 668. Vintage, New York.

7. T. Short (1767). *A Complete History of the Increase and Decrease of Mankind*, p. i. London.

8. I. Kant (1784). *On History*, ed. L. Beck (1963), p. 11. Bobbs-Merrill, Indianapolis.

9. J.-A.-N. de Condorcet (1785). *Essai sur l'application de l'analyse à la probabilité des décisions rendues à la plurité des voix.* Quoted in J. Carey, ed. (1999). *The Faber Book of Utopias*, pp. 161–62. Faber & Faber, London.

10. J.-A.-N. de Condorcet (1795). *Esquisse d'un Tableau Historique des Progrès de l'Esprit Humain.* Quoted in Carey, p. 165.

11. Condorcet (1795). *Esquisse.* Quoted in R. Bierstedt, ed. (1959). *The Making of Society*, p. 168. Random House, New York.

12. J. T. Desaguliers (1728). *The Newtonian System of the World*, ll. 17–18. Quoted in M. C. Jacob (1981). *The Radical Enlightenment: Pantheists, Freemasons and Republicans*, p. 124. Allen & Unwin, London.

13. D. Hume (1741). "That politics may be reduced to a science." In *Selected Essays*, ed. S. Copley & A. Edgar (1993), pp. 13–24. Oxford University Press.

14. J. Swift (1726). *Gulliver's Travels*, book ii, p. 176. Penguin, Harmondsworth, England, 1985.

15. E. Burke (1790). *Reflections on the Revolution in France*, para. 98. J. Dodsley, London. Republished C. C. O'Brien, ed. (1982). Penguin, Harmondsworth.

16. A. Comte (1830–42). *Cours de philosophie positive*. Quoted in Bierstedt, p. 192.

17. A. Quetelet (1830–31). "Lettre à M. le Bourgmestre, 15 dec. 1831." *Annuaire de l'observatoire de Bruxelles 1*, 285 (1834); "Letter to Bouvard," 5 November 1830, quoted in J. Lottin (1912), *Quetelet: Statisticien et Sociologue*, p. 52. Institut Supérieur de Philosophie, Louvain.

18. A. Quetelet (1840). "Notice scientifique" for his book *Sur l'homme*, *Annuaire de l'Observatoire de Bruxelles 7*, 230.

19. A. Quetelet (1832). "Recherches sur le penchant au crime aux différens ages." *Nouveaux mémoires de l'académie royale des sciences et belles-lettres de Bruxelles 7*, 136.

20. A. Quetelet (1835). *Sur l'homme et le développement de ses facultés, ou essai de physique sociale*, p. 289. Bachelier, Paris.

21. A. Quetelet (1835). "Recherches," p. 6.

22. J. Herschel (1850). "Quetelet on Probabilities." *Edinburgh Reivew 92*, 14.

23. J. S. Mill (1862). *A System of Logic*, in *Collected Works*, vols. 7 and 8 (ed. J. M. Robson) p. 932. University of Toronto Press, 1973.

24. H. T. Buckle (1857–61). *History of Civilization in England*, vol. 2, p. 244. New York, 1913.

25. I. Kant (1784). "Idea of a Universal History from a Cosmopolitan Point of View." See I. Kant, *On History*, p. 11.

26. H. T. Buckle (1857–61). *History of Civilization in England*. See F. Stern (ed.), *The Varieties of History from Voltaire to the Present*, 2nd ed., pp. 121–32. Macmillan, London, 1970.

27. W. Newmarch (1860). "Some Observations on the Present Position of Statistical Inquiry with Suggestions for Improving the Organization and Efficiency of the International Statistical Congress." *Journal of the Statistical Society of London 23*, 362.

28. N. W. Senior (1860). "Opening address of Nassau W. Senior, Esq., . . ." *Journal of the Statistical Society of London 23*, 359.

29. F. K. Hunt (1850). "A Few Facts About Matrimony." In *Household Words 1*, 374.

30. R. W. Emerson (1860). "Fate." In *The Conduct of Life & Other Essays* (1908), p. 159. J. M. Dent & Sons, London.

31. M. Twain (1924). *Autobiography*, vol. 1, ch. 20. Harper & Brothers, New York.

32. F. Nietzsche (1874). *Untimely Meditation. Second Part: Of the Use and Disadvantage of History for Life*. E. W. Fritzsch, Leipzig.

33. Quoted in L. Campbell & W. Garnett (1882). *The Life of James Clerk Maxwell*, pp. 294–95. Macmillan, London.

34. J. C. Maxwell (1873). "Molecules," a lecture. In *The Scientific Papers of James Clerk Maxwell*, ed. W. D. Niven, vol. 2, p. 374. Cambridge University Press, 1890.

35. Quoted in Campbell & Garnett, pp. 438–39.

36. L. Boltzmann (1872). "Weitere Studien über das Wärmegleichgewicht unter Gasmolekülen." In *Wissenschaftliche Abhandlungen*, ed. F. Hasenöhrl, vol. 1, p. 317. J. A. Barth, Leipzig, 1909.

37. P. G. Tait (1886). "On the foundations of the kinetic theory of gases." In *Scientific Papers*, vol. 2, p. 126. Cambridge University Press, 1898–1900.

38. C. S. Peirce (1877). "The fixation of belief." *Popular Science Monthly 12*, 1–15. Reprinted in 1974 in *Collected Papers*, vol. V, p. 226. Harvard University Press, Cambridge, Mass. I am grateful to Vincent Bauchau for bringing this citation to my attention.

39. F. Galton (1875). *English Men of Science: Their Nature and Nurture*. D. Appleton, New York.

40. A.-A. Cournot (1843). *Exposition de la théorie des chances et des probabilités*, p. 181. Hachette, Paris.

41. J. J. Fox (1860). "On the Province of the Statistician." *Journal of the Statistical Society of London 23*, 331.

42. A. Taillandier (1828). Review of "Compte général de l'administration de la justice criminelle en France." *Revue encyclopédique 40*, 612.

43. Editorial "Introduction." *Journal of the Statistical Society of London 1*, 3 (1838).

44. Quoted in M. Diamond & M. Stone (1981). "Nightingale on Quetelet." *Journal of the Royal Statistical Society A 144*, 70.

45. A. de Candolle (1830). "Considérations sur la statistique des délits." *Bibliothèque universelle des sciences, belles-lettres et arts, 104*, 160.

46. I. Kant (1784). *On History*, ed. L. Beck (1963), p. 11. Indianapolis.

47. A. Quetelet (1847). "De l'influence de libre arbitre de l'homme sur les faits sociaux." *Bulletin de la commission centrale de statistique 3*, 142.

48. Quetelet, *Sur l'homme*, p. 9.

49. W. Cyples (1864). "Morality of the doctrine of averages." *Cornhill Magazine 10*, 224.

50. "The Address of the Prince consort on opening as President the Fourth Session of the International Statistical Congress." *Journal of the Statistical Society of London 23*, 280 (1860).

51. F. Dostoevsky, *Letters from the Underworld*, trans. C. J. Hogarth, p. 32. J. M. Dent & Sons, London, 1913.

52. Ibid., pp. 29–30.

53. Ibid., p. 31.

54. L. N. Tolstoy (1869). *War and Peace*, trans. R. Edmunds (1969), vol. 2, p. 1,404. Penguin, Harmondsworth.

55. Ibid., p. 1,426.

56. Ibid., p. 1,440.

57. Quoted in N. Ferguson (1997). *Virtual History*, p. 446. Picador, London.

58. A. Tennyson (1868). "Lucretius," p. 10. Printed for private circulation, Cambridge, Mass.

59. J. C. Maxwell (1873). "Molecules." In W. D. Niven, p. 373.

60. J. C. Maxwell (1867). Letter to P. G. Tait, in C. G. Knott (1911). *Life and Scientific Work of Peter Guthrie Tait*, pp. 213–14. Cambridge University Press.

61. M. Smoluchowski (1918). "Über den Begriff des Zufalls und den Ursprung der Wahrscheinlichkeitsgesetz in der Physik." In *Oeuvres*, vol. 3, p. 87. Kraków, 1924–28.

CHAPTER 4: THE GRAND AH-WHOOM

EPIGRAPHS

Ralph Waldo Emerson (1860). *The Conduct of Life*, p. 210. J. M. Dent, London, 1908.

James Clerk Maxwell (1878). In *The Scientific Papers of James Clerk Maxwell*, ed. W. D. Niven, vol. 2, pp. 715–17. Cambridge University Press, 1890.

Isaac Newton (1704). "Queries." In *Opticks*. Reprinted by Dover, New York, 1952.

1. K. Vonnegut (1965). *Cat's Cradle*, p. 163. Penguin, Harmondsworth.
2. M. Gladwell (2000). *The Tipping Point*. Little, Brown, London.
3. C. Jencks (1995). *The Architecture of the Jumping Universe*. Academy Editions, London.
4. T. S. Kuhn (1962). *The Structure of Scientific Revolutions*. University of Chicago Press.
5. J.F.W. Herschel (1830). *Preliminary Discourse on the Study of Natural Philosophy*, p. 188. Longmans Brown, Green & Longmans, London, 1851.
6. T. Andrews (1869). "On the continuity of the gaseous and liquid states of matter." *Philosophical Transactions of the Royal Society 159*, 575.
7. J. C. Maxwell (1874). "Van der Waals on the continuity of the gaseous and liquid states." *Nature 10*, 477–80, here p. 478.

CHAPTER 5: ON GROWTH AND FORM

EPIGRAPHS

Stephen Hales (1727). *Vegetable Staticks*. W. & J. Innys & T. Woodward, London.

Herbert Spencer (1876). *The Principles of Sociology*. Quoted in R. Bierstedt, ed. (1959). *The Making of Society*, p. 262. Random House, New York.

Georg Christoph Lichtenberg, quoted in F. Cramer (1993). *Chaos and Order*, p. 2. VCH, Weinheim.

1. H. D. Thoreau (1856). *The Journal of Henry David Thoreau*, ed. B. Torrey & F. Allen, vol. 8, pp. 87–88. Houghton Mifflin, Boston, 1906.
2. J. W. Gibbs (1906). "On the equilibrium of heterogeneous substances." In *The Scientific Papers of J. Willard Gibbs*, vol. I, ed. H. A. Bumstead & R. G. Van Name. Longmans, Green, New York.
3. D. W. Thompson (1917). *On Growth and Form*, p. 503. Revised edition published by Cambridge University Press, 1942, and reprinted in 1992 by Dover, New York.
4. Ibid., p. 505.
5. I. Prigogine (1980). *From Being To Becoming*, p. 106. W. H. Freeman, New York.
6. J. L. Borges (1956). "The Garden of Forking Paths." In *Labyrinths*, pp. 44–54. Penguin, Harmondsworth, 1970.
7. Prigogine, p. 106.

CHAPTER 6: THE MARCH OF REASON

EPICRAPHS

Georg Christoph Lichtenberg, quoted in F. Cramer (1993). *Chaos and Order*, p. 114. VCH, Weinheim.

Edward A. Ross (1901). *Social Control*. In R. Bierstedt (1959), *The Making of Society*, p. 336. Random House, New York.

Heinrich von Kleist (1810). *On the Marionette Theatre*. In *Berliner Abendblatter*, 12–15 December. Trans. I. Parry in the *Times Literary Supplement*, 20 October 1978.

1. A. van Leeuwenhoek (1674). In *Antony van Leeuwenhoek and His 'Little Animals.'* Trans. & ed. C. Dobell (1958). Russell & Russell, New York.

2. Hobbes, *Leviathan*, p. 228.

3. J. K. Parrish & L. Edelstein-Keshet (1999). "Complexity, pattern, and evolutionary trade-offs in animal aggregation." *Science 284*, 99–101, here p. 100.

4. Quoted in S. Levy (1992). *Artificial Life*, p. 74. Jonathan Cape, London.

5. Quoted in R. Lewin (1992). *Complexity*, p. 178. Macmillan, New York.

6. Hobbes, *Leviathan*, p. 225.

7. Ibid., pp. 225–26.

8. L. F. Henderson (1971). "The statistics of crowd fluids." *Nature 229*, 381.

9. B. Hillier & J. Hanson (1984). *The Social Logic of Space*, p. 266. Cambridge University Press.

10. Ibid.

11. M. Batty, J. Desyllas and E. Duxbury (2003). "Safety in numbers? Modelling crowds and designing control for the Notting Hill Carnival." *Urban Studies 40*, 1573–90, here p. 1588.

12. Parrish & Edelstein-Keshet, p. 101.

13. P. Ackroyd (2000). *London: The Biography*, p. 2. Chatto & Windus, London.

14. Ibid., p. 103.

15. H. Kett (1787). Quoted in Ackroyd, p. 517.

16. D.D.T. Chen (2000). "The science of smart growth." *Scientific American*, December, pp. 84–92.

17. Ibid.

18. L. Mumford (1938). *The Culture of Cities*, p. 233. Secker & Warburg, London.

19. Ibid., p. 7.

20. H. Spencer (1876). *Principles of Sociology*. Quoted in R. Bierstedt, ed. (1959). *The Making of Society*, p. 262. Random House, New York.

21. Quoted in I. Petersen (1996). "The shapes of cities." *Science News 149*, 6 January, p. 9.

22. H. Simon (1996). *The Sciences of the Artificial*, 3rd ed., pp. 33–34. MIT Press, Cambridge, Mass.

23. Ackroyd, p. 588.

CHAPTER 7: ON THE ROAD

EPIGRAPHS

Thomas C. Schelling (1978). *Micromotives and Macrobehavior*, p. 121. W. W. Norton, New York.

Josiah Willard Gibbs, quoted in A. T. Winfree (1980). *The Geometry of Biological Time* Springer-Verlag, New York.

Rudyard Kipling (1888). L'Envoi to *The Story of the Gadsbys*. A. H. Wheeler & Co., Allahabad.

1. R. Moe (1999). Speech on urban sprawl, Red Hills Spring Event Dinner, Tall Timbers Research Station, Tallahassee, Florida, 24 March.

CHAPTER 8: RHYTHMS OF THE MARKETPLACE

EPIGRAPHS

Samuel Johnson, quoted in R. Heilbroner (2000). *The Worldly Philosophers*, 7th ed., p. 41. Penguin, London.

George Bernard Shaw (attributed).

Johann Wolfgang von Goethe, quoted in R. W. Emerson (1860). The Conduct of Life, p. 197. J. M. Dent, London, 1908.

1. Heilbroner, p. 57.
2. D. Defoe (1706). *Review*, vol. ii, p. 26. Quoted in D. Donoghue (1988). *England, Their England: Commentaries on English Language and Literature*, p. 65. Alfred A. Knopf, New York.
3. Quoted in J.G.A. Pocock (1985), "Josiah Tucker on Burke, Locke, and Price." In *Virtue, Commerce and History: Essays on Political Thought and History, Chiefly in the Eighteenth Century*, pp. 157–91. Cambridge University Press.
4. Emerson, *The Conduct of Life*, p. 202.
5. Ibid.
6. J. Kay (1995). "Cracks in the crystal ball." *Financial Times*, 29 September.
7. I. Fisher (1929). Quoted in R. Heilbroner, p. 251.
8. P. Krugman (1994). *Peddling Prosperity* p. xi. W. W. Norton, New York.
9. Ibid.
10. K. Marx & F. Engels (1848). *The Communist Manifesto*, trans. S. Moore, p. 25. Junius Publications, London, 1996.
11. Ibid., p. 48.
12. P. H. Cootner, ed. (1964). *The Random Character of Stock Market Prices*. MIT Press, Cambridge, Mass.
13. Quoted in G. Stix (1998). "A calculus of risk." *Scientific American*, May, 70–75.

CHAPTER 9: AGENTS OF FORTUNE

EPIGRAPHS

Robert Heilbroner (1999), *The Worldly Philosophers*, 7th ed., pp. 316–17. Penguin, London.
Ralph Waldo Emerson (1860). *The Conduct of Life*, p. 199. J. M. Dent, London, 1908.
W. Brian Arthur, quoted in M. M. Waldrop (1994), *Complexity*, p. 328. Penguin, London.

1. T. Carlyle (1849). "The Nigger question." In *Miscellaneous Essays*, vol. 7, pp. 79–110. Chapman & Hall, London, 1888. I recommend P. Groenewegen's "Thomas Carlyle, 'the dismal science,' and the contemporary political economy of slavery," *History of Economics Review 34*, 74–94 (2001) for a discussion of Carlyle's critique of "economic science" and his views on the economic implications of the emancipation of slaves.

2. F. Y. Edgeworth (1888). "Tests of accurate measurement." In *Papers Relating to Political Economy*, vol. 1, p. 331. Macmillan, London, 1925.

3. F. Y. Edgeworth (1881). *Mathematical Psychics: An Essay on the Application of Mathematics to the Moral Sciences*, pp. 12–13. Kegan Paul & Co., London.

4. Edgeworth, *Mathematical Psychics*, p. 50.

5. Heilbroner, p. 176.

6. J. M. Keynes (1926). "Francis Ysidro Edgeworth," In *Essays in Biography*, p. 224. W. W. Norton, New York, 1963.

7. P. Krugman (1994). *Peddling Prosperity*, p. xi. W. W. Norton, New York.

8. Heilbroner, p. 317.

9. D. Howell (2000). *The Edge of Now*. Macmillan, London.

10. J. Kay (2000). "Economic with the truth." *Prospect*, October, p. 70.

11. J. M. Keynes (1936). *The General Theory of Employment, Interest and Money*, p. 161. Macmillan, London, 1973.

12. G. Soros (1994). Quoted in the editorial introduction to *The Economy as a Complex Evolving System II*, ed. W. B. Arthur, S. N. Durlauf & D. A. Lane, p. 15. Addison-Wesley, Reading, Mass.

13. Kay, p. 70.

14. A. Kirman (1996). "Some observations on interactions in economics." Paper presented at Workshop on Structural Change, Manchester Metropolitan University, U.K., May 20–21, 1996.

15. Ibid.

16. Arthur, Durlauf & Lane, p. 9.

17. C. Davenant (1696). "An Essay on the East-India-Trade," p. 25. London.

18. A. Smith (1759). *The Theory of Moral Sentiments*, p. 82. Ed. D. D. Raphael & A. L. Macfie. Clarendon Press, Oxford, 1976.

19. E. Burke (1797). Letter to Arthur Young, 23 May. In *The Works and Correspondence of the Right Honourable Edmund Burke*, vol. II, p. 398. Francis & John Rivington, London, 1852.

20. Y. Louzoun, S. Solomon, J. Goldenberg & D. Mazursky (2002). "The risk at being unfair: world-size global markets lead to economic instability." Preprint.

21. J. M. Epstein & R. Axtell (1996). *Growing Artificial Societies*, p. 136. MIT Press, Cambridge, Mass.

CHAPTER 10: UNCOMMON PROPORTIONS

EPIGRAPHS

George Kingsley Zipf (1949). *Human Behavior and the Principle of Least Effort*, p. 27. Hafner, New York, 1965.

Herbert Simon (1996). *The Sciences of the Artificial*, 3rd ed., p. 2. MIT Press, Cambridge, Mass.

Philip Anderson (1997). In *The Economy as a Complex Evolving System II*, ed. W. B. Arthur, S. N. Durlauf & D. A. Lane, p. 566. Addison-Wesley, Reading, Mass.

1. L. Laloux, M. Potters, R. Cont, J.-P. Aguilar & J.-P. Bouchard (1999). "Are financial crashes predictable?" *Europhysics Letters 45*, 1–5.
2. P. Bak (1997). *How Nature Works*, pp. 187–88. Oxford University Press.
3. Ibid., p. 191.
4. A. Tennyson (1842). "Locksley Hall," p. 20, line 137, in *Locksley Hall and Other Poems*. Ernest Nister, London.
5. B. Mandelbrot (1983). *The Fractal Geometry of Nature*. Freeman & Co., New York.
6. E. Wolf (1996). "Trends in household wealth during 1989–1992." Submitted to the Department of Labor, New York. And J. Diaz-Gimenez, V. Quadrini & J. V. Rios-Rull (1997). "Dimensions of inequality: facts on the U.S. distributions of earnings, income and wealth." *Quarterly Review of the Federal Reserve Bank of Minneapolis 21*, 3–21.
7. C. Snyder (1940). *Capitalism the Creator*, p. 417. Macmillan, New York.
8. Z.-F. Huang & S. Solomon (2001). "Finite market size as a source of extreme wealth inequality and market instability." *Physica A 294*, 503–13.
9. Zipf, p. 543.
10. Ibid., p. 544.

CHAPTER 11: THE WORK OF MANY HANDS

EPIGRAPHS

Charles Handy (1993). *Understanding Organizations*, 4th ed., p. 313. Penguin, London.

John Burton Sanderson Haldane (1949). "Shapes and weights." In *What Is Life?* p. 186. Alcuin Press, Welwyn Garden City, England.

E. Fritz Schumacher (1974). *Small Is Beautiful*, p. 55. Sphere, London.

1. Quoted in G. Monbiot (2000). *Captive State*, pp. 166–67. Macmillan, London.
2. *A Review of Monopolies and Mergers Policy: A consultative document.* Cmnd 7198, pp. 136–7. HMSO, London, 1978.
3. A. Smith (1776). *An Inquiry into the Nature and Causes of the Wealth of Nations.* Abridged version, ed. L. Dickey (1993), p. 5. Hackett, Indianapolis.
4. Monbiot, p. 179.
5. R. Axtell (2001). "Zipf distribution of U.S. firm sizes." *Science 293*, 1818–1820.
6. J. Sutton (1997). "Gibrat's legacy." *Journal of Economic Literature XXXV*, pp. 42–43.

7. R. Axtell (1999). "The emergence of firms in a population of agents: local increasing re-
    turns, unstable Nash equilibria, and power law size distributions." Center on Social and
    Economic Dynamics, Working Paper no. 3, p. 3. Brookings Institution, Washington, D.C.
8. Sutton, p. 57.
9. J. Kay (2003). "The real economy." *Prospect*, May, p. 28.
10. Axtell (1999), p. 87.

CHAPTER 12: JOIN THE CLUB

EPIGRAPHS

John Stuart Mill (1843). *A System of Logic*, pp. 572–73. Longmans, Green & Co., London,
1884.
Eric Hobsbawm (1994). *Age of Extremes*, p. 147. Abacus, London.
Leo N. Tolstoy (1869). *War and Peace*, trans. R. Edmunds, vol. 2, p. 1426. Penguin, Harmondsworth, 1969.

1. Thucydides, *History of the Peloponnesian War*, trans. B. Jowett, 2nd ed., book III. para.
    11, Oxford University Press, London, 1900.
2. R. Axelrod & D. S. Bennett (1993). "A landscape theory of aggregation." *British Journal
    of Political Science 23*, 211–33, here p. 219.
3. K. Waltz (1979). *Theory of International Politics*, p. 167. Addison-Wesley, Reading,
    Mass.
4. G. H. Snyder (1984). "The security dilemma in alliance politics," *World Politics 36*,
    461–95.
5. Hobsbawm, p. 162.
6. M. Oakeshott (1933). *Experience and Its Modes*, p. 128. Cambridge University Press.
7. Quoted in N. Ferguson (1997). *Virtual History*, p. 1. Picador, London.
8. H. Trevor-Roper (1981). "History and imagination." In *History and Imagination: Eas-
    says in Honour of H. R. Trevor-Roper*, p. 364, ed. V. Pearl, B. Worden & H. Lloyd-Jones.
    Duckworth, London.
9. Ferguson, p. 85.
10. Ibid.

CHAPTER 13: MULTITUDES IN THE VALLEY OF DECISION

EPIGRAPHS

Karl Mannheim (1936). *Ideology and Utopia*. Harcourt, Brace, New York. Quoted in Bierstedt (1959). *The Making of Society*, p. 505.
Joshua Epstein (2001). "Learning to be thoughtless: social norms and individual computation." *Computational Economics 18 (1)*, 9–24.
Malcolm Gladwell (2001). *The Tipping Point*, p. 259. Abacus, London.

1. *The Onion 36 (40)*, 9 November 2000. See www.theonion.com/onion3640/bush_or_
    gore.html.

2. Quoted in M. J. Hinich and M. C. Munger (1997). *Analytical Politics*, p. 3. Cambridge University Press.

3. Epstein, pp. 9–24.

4. T. C. Schelling (1973). *Micromotives and Macrobehavior*, p. 20. W. W. Norton, New York.

5. Ibid., p. 23.

6. P. Ormerod (2002). "Sense on segregation," *Prospect*, February, 12–14.

7. M. Lind (2001). "Are there global political values?" *Prospect*, December 2001, p. 20.

8. F. Dostoevsky (1911). *Crime and Punishment*, p. 64. J. M. Dent & Sons, London.

9. M. Gladwell, p. 8.

10. G. B. Shaw (1908). *Getting Married*, p. 111. In *Selected Passages from the Works of Bernard Shaw*, p. 142. Constable & Co., London, 1912.

11. Shaw, *Getting Married*, p. 130. In *Selected Passages*, p. 145.

12. G. S. Becker (1991). *A Treatise on the Family*, p. 325. Harvard University Press, Cambridge, Mass.

13. Ibid., p. 232.

14. Ibid., p. 32.

15. Ibid., p. 59.

16. Ibid., p. 4.

17. Ibid.

18. Ibid., p. 19.

19. Schelling, p. 36.

20. W. Newmarch (1860). "Some observations on the present position of statistical inquiry with suggestions for improving the organization and efficiency of the International Statistical Congress." *Journal of the Statistical Society of London 23*, 362–63.

21. W. B. Arthur (1994). "Inductive reasoning and bounded rationality." *American Economic Review 84*, 406–11.

22. Ibid.

CHAPTER 14: THE COLONIZATION OF CULTURE

EPIGRAPHS

Jean-Jacques Rousseau (1772). "Considérations sur le Gouvernement de Pologne." In *The Political Writings of Jean Jacques Rousseau*, vol. II, ed. C. E. Vaughan (1915), p. 432. Cambridge University Press.

Cindy Gilday, quoted in J. Mander, "Technologies of globalization." In *The Case Against the Global Economy*, ed. E. Goldsmith & J. Mander (2001), p. 52. Earthscan, London.

Susan George (2002). Quoted in "Globally locally," *New Scientist*, 27 April, p. 44.

1. J. M. Epstein & R. Axtell (1996). *Growing Artificial Societies*, p. 163. MIT Press, Cambridge, Mass.

2. Quoted in I. Peterson (1996). "The gods of Sugarscape." *Science News*, 23 November, p. 332.

3. Epstein & Axtell, p. 52.

CHAPTER 15: SMALL WORLDS

EPIGRAPHS

Emile Durkheim (1879). *Revue Philosophique*, December.

Mark E.J. Newman (2003). "Ego-centered networks and the ripple effect," *Social Networks* 25, 83–95.

1. S. Milgram, quoted in M. Gladwell (2001). *The Tipping Point*, p. 36. Abacus, London.
2. J. Guare (1990). *Six Degrees of Separation: A Play.* Vintage, New York.

Figure 15.2

Kevin Bacon—Eddie Albert: *The Big Picture*
Kevin Bacon—Meryl Streep: *The River Wild*
Kevin Bacon—Jack Nicholson: *A Few Good Men*
Kevin Bacon—Maximilian Schell: *Telling Lies in America*
Kevin Bacon—Allison Caine: *Diner*
Kevin Bacon—Diane Lane: *My Dog Skip*
Rod Steiger—Humphrey Bogart: *The Harder They Fall*
Rod Steiger—Eddie Albert: *The Longest Day*
Rod Steiger—Jack Nicholson: *Mars Attacks!*
Rod Steiger—Diane Lane: *Cattle Annie and Little Britches*
Rod Steiger—Karl Malden: *On the Waterfront*
Rod Steiger—Marlon Brando: *On the Waterfront*
Rod Steiger—Maximilian Schell: *The Chosen*
Marlon Brando—Karl Malden: *On the Waterfront*
Marlon Brando—Maximilian Schell: *The Freshman*
Marlon Brando—Jack Nicholson: *The Missouri Breaks*
Jack Nicholson—Meryl Streep: *Ironweed*
Humphrey Bogart—Eddie Albert: *The Wagons Roll at Night*
Karl Malden—Allison Caine: *Nuts*
Maximilian Schell—Meryl Streep: *Julia*

CHAPTER 16: WEAVING THE WEB

EPIGRAPHS

Manuel Castells (2001). *The Internet Galaxy: Reflections on the Internet, Business, and Society*, p. 207. Oxford University Press, Oxford.

Gustavo Cardoso (1998). *Para uma Sociologia do Ciberespaço: Comunidades Virtuais em Portugues*, p. 116 (trans. M. Castells). Celta Editora, Oeiras, Portugal.

William Shakespeare (1602–1604). *Othello* III, iv, 69.

1. L. Irving (1998). Quoted in E. Wasserman (1998). "Feds take steps against threat of cyber terrorism." CNN, 25 September. See www.cnn.com/TECH/computing/9809/25/cyberterrorism.idg/>.
2. T. Hobbes (1651). *Leviathan*, ed. C. B. Macpherson (1985), p. 353. Penguin, London.

3. Castells, p. 1.

4. Ibid., p. 36.

5. Ibid., p. 282.

6. S. Dalzell, quoted in J. Naughton (1999). *A Brief History of the Future*, pp. 190–91. Weidenfeld & Nicolson, London. See www2.epic.org/cda/cda_dc_opinion.html.

7. Ibid., p. 192.

8. T. Berners-Lee (1990). "Information management: a proposal," CERN Internal Report, May, p. 9. See www.w3.org/History/1989/proposal.html.

9. FBI Definitions, *U.S. Code of Federal Regulations* 28, sec. 0.85.

10. A.-L. Barabási (2003). "We're all on the Grid together." *New York Times*, 16 August.

11. Ibid.

12. A.-L. Barabási (2002). Personal communication.

13. W. B. Arthur, S. N. Durlauf & D. A. Lane (1997). *The Economy as a Complex Evolving System II*, p. 9. Addison-Wesley, Reading, Mass.

CHAPTER 17: ORDER IN EDEN

EPIGRAPHS

Sir William Temple (1751). "An Essay upon the Origin and Nature of Government." *Works* I, p. 99. Quoted in the introduction to T. Hobbes, *Leviathan*, ed. C. B. Macpherson, p. 61.

John Locke (1690). *Essay on Civil Government.* In *Two Treatises of Government*, ed. M. Goldie, p. 118. J. M. Dent, London, 1993.

Ian Hay (1916). *The First Hundred Thousand*, pp. 224–25. William Blackwood, London.

1. S. Freud (1930). *Civilization and Its Discontents*, pp. 32–33. Hogarth Press, London, 1973.

2. J. Locke (1690). Second *Treatise on Government*. In *Two Treatises of Government*, 2nd ed., ed. P Laslett, p. 433. Cambridge University Press, Cambridge, 1970.

3. K. Popper (1997). *The Lesson of This Century*, p. 35. Routledge, London.

4. Freud, p. 60.

5. Locke (ed. Goldie), p. 124.

6. Ibid., p. 117.

7. Ibid., p. 124.

8. A. A. Cooper, 3rd Earl of Shaftesbury (1711). *Characteristicks of Men, Manners, Opinions, Times*, ed. P. Aynes, vol. ii, p. 67 (Wyat, London). Clarendon Press, Oxford, 1999.

9. Genesis 9:6.

10. Locke, p. 178.

11. J. S. Mill (1859). *On Liberty*, ed. G. Himmelfarb, p. 68. Penguin, London, 1985.

12. J. S. Mill (1861). "Utilitarianism." In J. S. Mill and J. Bentham (1987). *Utilitarianism and Other Essays*, ed. A. Ryan, p. 278. Penguin, Harmondsworth, England.

13. J.-J. Rousseau (1754). "Discourse on Inequality." Quoted in B. Russell (1984). *A History of Western Philosophy*, p. 663. Unwin Paperbacks, London.

14. C. Darwin (1871). *The Descent of Man, and Selection in Relation to Sex*, p. 64. John Murray, London, 1874.

15. T. Hobbes (1651). *Leviathan*, ch. 51.
16. R. Kagan (2002). "Power and weakness." *Policy Review 113*, June & July (http://www.policyreview.org).
17. E. Hobsbawm (1994). *Age of Extremes*, pp. 24–26. Abacus, London.
18. G. Dugdale (1932). *Langemarck and Cambrai*, p. 94. Wilding & Son, Shrewsbury.
19. G. B. Cobb (1916). *Stand to Arms*, p. 74. Wells Gardner, Darton, London.
20. T. Ashworth (1980). *Trench Warfare 1914–1918: The Live and Let Live System*, p. 144. Holmes & Meier, New York.
21. Hobbes, p. 190.
22. Ibid.
23. R. Axelrod (1984). *The Evolution of Cooperation*, p. 124. Basic Books, New York.
24. Popper, p. 70.
25. Pericles' funeral oration is recounted by Thucydides in *History of the Peloponnesian War*, bk. II. This paraphrase is given by Popper, p. 72.
26. Popper, pp. 36–37.

CHAPTER 18: PAVLOV'S VICTORY

EPIGRAPHS

Niccolò Machiavelli (1513). *The Prince*, trans. G. Bull, ch. XV, p. 91. Penguin, London, 1981.

Kenneth N. Waltz (1954). *Man, the State, and War*, p. 108. Columbia University Press, New York.

Karl Popper (1988). "Reflections on the theory and practice of the democratic state," talk in Munich, 9 June. In *The Lesson of This Century*, p. 79. Routledge, London, 1997.

1. Deuteronomy 19:19–21.
2. M. A. Nowak & K. Sigmund (1992). "Tit for tat in heterogeneous populations." *Nature* 355, 250–253.
3. E. Burke (1770). "Thoughts on the Causes of the Present Discontents." Quoted in M. J. Hinich & M. C. Munger (1997), *Analytical Politics*, p. 136. Cambridge University Press.
4. Popper (1997). *The Lesson of This Century*, p. 49. Routledge, London.
5. Ibid., p. 55.
6. H. Kahn (1984). *Thinking About the Unthinkable in the 1980s*, p. 59. Simon & Schuster, New York.
7. B. Skyrms (1996). *Evolution of the Social Contract*, p. 24. Cambridge University Press.
8. *Pentagon Papers*, ed. N. Sheehan & E. W. Kenworthy (1971), p. 432. Times Books, New York.
9. K. Sigmund (1995). *Games of Life*, p. 191. Penguin, London.
10. R. Axelrod (1984). *The Evolution of Cooperation*, p. 184. Basic Books, New York.
11. Quoted in F. Goodall (1997). *A Question of Conscience*, p. 198. Sutton Publishing Group, Stroud, U.K.
12. Quoted in Goodall, p. 199.
13. Popper, p. 91.

CHAPTER 19: TOWARD UTOPIA?

EPIGRAPHS

Ralph Waldo Emerson (1860). *The Conduct of Life*, p. 176. J. M. Dent, London, 1908.

George Lundberg (1939). *Foundations of Sociology*. Macmillan, London. In R. Bierstedt, ed. (1959). *The Making of Society*, p. 518. Random House, New York.

Herbert Simon (1996). *The Sciences of the Artificial*, 3rd ed., p. 139. MIT Press, Cambridge, Mass.

Julian Barnes (1989). *A History of the World in 10½ Chapters*. Jonathan Cape, London.

1. J. Carey (1999). *The Faber Book of Utopias*, p. xii. Faber and Faber, London.

2. Quoted in F. A. Hayek (1962). *The Road to Serfdom*, p. 18. Routledge, London.

3. J. S. Mill (1843). *A System of Logic*, book 6. In Bierstedt, p. 205.

4. G. Simmel (1908). *The Sociology of Georg Simmel*, ed. and trans. K. H. Wolff. Free Press, Glencoe, Ill., 1950.

5. R. E. Park (1927). "Human nature and collective behavior." *American Journal of Sociology XXXII*, no. 5, 733.

6. A. Smith (1776). *An Inquiry into the Nature and Causes of the Wealth of Nations*, ed. L. Dickey, p. 11. Hackett, Indianapolis, 1993.

7. B. Mandeville (1705). *Fable of the Bees*. Quoted in B. Willey (1940), *The Eighteenth Century Background*, p. 96. Columbia University Press, New York.

8. J. Swift (1729). *A Modest Proposal for Preventing the Children of Poor People in Ireland from Being a Burden to Their Parents or Country and for Making Them Beneficial to the Public*. In *Selected Prose Works of Jonathan Swift*, ed. J. Hayward (1949), p. 430. Cresset Press, London.

9. L. Mumford (1938). *The Culture of Cities*, p. 93. Secker & Warburg, London.

10. Mill, *Logic*. In Bierstedt, p. 203.

11. A. Ferguson (1766). *Institutes of Moral Philosophy*. Quoted in G. Bryson (1945), *Man and Society*, p. 35. Princeton University Press, Princeton, N.J.

12. A. Ferguson (1767). *An Essay on the History of Civil Society*, ed. D. Forbes, p. 122. Edinburgh University Press, 1966.

13. R. W. Emerson (1844). "On Politics." In *The Essays of Ralph Waldo Emerson*, p. 325. Grant Richards, London, 1903.

14. K. Marx (1845). *Theses on Feuerbach*, no. 11. Published and edited by F. Engels, as an appendix to F. Engels (1888), *Ludwig Feuerbach und der Ausgang der Klassischen Deutschen Philosophie*. J.H.W. Dietz, Stuttgart.

15. P. Ormerod (1998). *Butterfly Economics*, p. 182. Faber & Faber, London.

16. R. Olson (1990). *Science Deified and Science Defied*, vol. 2, pp. 210–11. University of California Press, Berkeley, Calif.

17. Cicero, *De Legibus*. In Bierstedt, p. 41.

18. Aristotle, *Politics*, trans. W. D. Ross. In Bierstedt, p. 24.

19. St. Thomas Aquinas (c. 1259). *On the Governance of Rulers*, trans. G. B. Phelan. In Bierstedt, p. 52.

20. Ibid., p. 53.
21. Hobbes, *Leviathan*, p. 185.
22. Smith, *Wealth of Nations*, p. 130.
23. F. von Hayek (1944). *The Road to Serfdom*, p. 27. Routledge, London, 1962.
24. Ibid., p. 14.
25. T. C. Schelling (1978). *Micromotives and Macrobehavior*, pp. 127–28. W. W. Norton, New York.
26. Ibid., p. 19.
27. K. Malik (2000). "Natural science." *Prospect*, August/September, p. 37.
28. Mill, *Logic*. In Bierstedt, p. 209.
29. F. Giddings (1924). *The Scientific Study of Human Society*. In Bierstedt, p. 362.

BIBLIOGRAPHY

Abul-Magd, A. Y. (2002). Wealth distribution in an ancient Egyptian society. *Physical Review E 66*, 057104.

Adamic, L. (1999). The Small World Web. Preprint, Xerox Palo Alto Research Center, www.hpl.hp.com/shl/papers/smallworld/smallworldpaper.html.

Alberich, R., J. Miro-Julia & F. Rosselló (2002). Marvel Universe looks almost like a real social network. Preprint, arxiv.org/abs/cond-mat/0202174.

Albert, R. & A.-L. Barabási (2000). Topology of evolving networks: local events and universality. *Physical Review Letters 85*, 5234–5237.

Albert, R. & A.-L. Barabási (2002). Statistical mechanics of complex networks. *Reviews of Modern Physics 74*, 47–97.

Albert, R., H. Jeong & A.-L. Barabási (1999). Diameter of the World-Wide Web. *Nature 401*, 130–131.

Albert, R., H. Jeong & A.-L. Barabási (2000). Error and attack tolerance of complex networks. *Nature 406*, 378–382.

Allingham, M. (2002). *Choice Theory: A Very Short Introduction.* Oxford University Press.

Allsop, V. (1995). *Understanding Economics.* Routledge, London.

Amaral, L.A.N., A. Scala, M. Barthélémy & H. E. Stanley (2000). Classes of small-world networks. *Proceedings of the National Academy of Sciences USA 97*, 11,149–11,152.

Anderson, P. W., K. J. Arrow & D. Pines, eds. (1988). *The Economy as an Evolving Complex System.* Addison-Wesley, Redwood City, Calif.

Andersson, C., A. Hellervik, K. Lindgren, A. Hagson & J. Tornberg (2003). The urban economy as a scale-free network. Preprint, arxiv:cond-mat/0303535.

Anghel, M., Z. Toroczkai, K. E. Bassler & G. Korniss (2003). Competition in social networks: emergence of a scale-free leadership structure and collective efficiency. Preprint, arxiv.org/abs/cond-mat/0307740.

Arthur, W. B. (1989). Competing technologies, increasing returns, and lock-in by historical events. *Economic Journal 99*, 116–131.

Arthur, W. B. (1994). Inductive reasoning and bounded rationality (the El Farol problem). *American Economic Review 84*, 406–411.

Arthur, W. B. (1999). Complexity and the economy. *Science 284*, 107–109.

Arthur, W. B., S. N. Durlauf & D. A. Lane (1997). *The Economy as a Complex Evolving System II*. Addison-Wesley, Reading, Mass.

Aubrey, J. *Brief Lives*. Ed. J. Buchanan-Brown, 2000. Penguin, London.

Avery, J. (1997). *Progress, Poverty and Population*. Frank Cass, London.

Axelrod, R. (1984). *The Evolution of Cooperation*. Basic Books, New York.

Axelrod, R. (1997). The dissemination of culture. *Journal of Conflict Resolution 41*, 203–226.

Axelrod, R. (1997). Advancing the art of simulation in the social sciences. In R. Conte, R. Hegselmann & P. Terna, eds., *Simulating Social Phenomena*, pp. 21–40. Springer, Berlin.

Axelrod, R. (2000). On six advances in cooperation theory. *Analyse & Kritik 22*, 130–151.

Axelrod, R. & D. S. Bennett (1993). A landscape theory of aggregation. *British Journal of Political Science 23*, 211–233.

Axelrod, R., W. Mitchell, R. E. Thomas, D. S. Bennett & E. Bruderer (1995). Coalition formation in standard-setting alliances. *Management Science 41*, 1493–1508.

Axtell, R. L. (2001). Zipf distribution of U.S. firm sizes. *Science 293*, 1818–1820.

Axtell, R. (1999). The emergence of firms in a population of agents: local increasing returns, unstable Nash equilibria, and power law size distributions. Preprint, working paper no. 3, Brookings Institution, Washington, D.C.

Axtell, R., R. Axelrod, J. M. Epstein & M. D. Cohen (1996). Aligning simulation models: a case study and results. *Computational and Mathematical Organization Theory 1*, 123–141.

Bak, P. (1997). *How Nature Works*. Oxford University Press.

Bak, P., K. Chen & M. Creutz (1989). Self-organized criticality in the "Game of Life." *Nature 342*, 780–781.

Ball, P. (1998). *The Self-Made Tapestry*. Oxford University Press.

Ball, P. (2000). Jams tomorrow. *New Scientist*, 15 January, 34–38.

Ball, P. (2002). The physical modeling of society: a historical perspective. *Physica A 314*, 1–14.

Bankes, S. C. (2002). Agent-based modeling: a revolution? *Proceedings of the National Academy of Sciences USA 99*, 7199–7200.

Barabási, A.-L. (2001). The physics of the Web. *Physics World*, July, 33–38.

Barabási, A.-L. (2002). *Linked*. Perseus, Cambridge, Mass.

Barabási, A.-L. & R. Albert (1999). Emergence of scaling in random networks. *Science 286*, 509–512.

Barabási, A.-L., R. Albert & H. Jeong (1999). Mean-field theory for scale-free random networks. *Physica A 272*, 173–187.

Batten, D. F. (2000). *Discovering Artificial Economics*. Westview Press, Boulder, Colo.

Batty, M. (1997). Predicting where we walk. *Nature 388*, 19–20.

Batty, M. (2003). Agent-based pedestrian modelling. In P. Longley & M. Batty, eds., *Advanced Spatial Analysis*, chapter 5. ESRI Press, London.

Batty, M. (2003). The emergence of cities: Complexity and urban dynamics. Preprint.

Batty, M., J. Desyllas & E. Duxbury (2003). Safety in numbers? Modelling crowds and designing control for the Notting Hill Carnival. *Urban Studies 40*, 1573–1590.

Batty, M. & P. Longley (1994). *Fractal Cities.* Academic Press, London.

Becker, G. S. (1968). Crime and punishment: an economic approach. *Journal of Political Economy 76(2)*, 443–478.

Becker, G. S. (1991). *A Treatise on the Family.* Harvard University Press, Cambridge, Mass.

Bentley, W. A. & W. J. Humphreys (1931). *Snow Crystals.* McGraw-Hill Book Co., New York. Reprinted in 1962 by Dover, New York.

Bernardes, A. T., D. Stauffer & J. Kertész (2001). Election results and the Sznajd model on Barabasi network. Preprint, arxiv.org/abs/cond-mat/0111147.

Berry, B.J.L., L. D. Kiel & E. Elliott (2002). Adaptive agents, intelligence, and emergent human organization: capturing complexity through agent-based modeling. *Proceedings of the National Academy of Sciences USA 99*, 7187–7188.

Bicchieri, C., R. Jeffrey & B. Skyrms (1999). *The Logic of Strategy.* Oxford University Press.

Bierstedt, R., ed. (1959). *The Making of Society.* Random House, New York.

Boorstin, D. J. (1985). *The Discoverers.* Vintage Books, New York.

Borland, L. (2002). Option pricing formulas based on a non-gaussian stock price model. *Physical Review Letters 89*, 098701.

Bouchard, J.-P., P. Cizeau, L. Laloux & M. Potters (1999). Mutual attractions: physics and finance. *Physics World*, January, 25–29.

Bouchard, J.-P. & M. Mézard (2000). Wealth condensation in a simple model of economy. *Physica A 282*, 536–545.

Brams, S. J. (1975). *Game Theory and Politics.* Free Press, New York.

Brams, S. J. (1978). *The Presidential Election Game.* Yale University Press, New Haven, Conn.

Brown, L. R. (2001). *Eco-Economy.* Earthscan, London.

Buchanan, M. (2000). That's the way the money goes. *New Scientist*, 19 August, 22–26.

Buchanan, M. (2000). *Ubiquity.* Weidenfeld & Nicolson, London.

Buchanan, M. (2002). *Small World.* Weidenfeld & Nicolson, London.

Burda, Z., et al. (2002). Wealth condensation in Pareto macroeconomies. *Physical Review E 65*, 026102.

Camazine, S., J.-L. Deneubourg, N. R. Franks, J. Sneyd, G. Theraulaz & E. Bonabeau (2001). *Self-Organization in Biological Systems.* Princeton University Press, Princeton, N.J.

Campbell, M., & P. Ormerod (1997). Social interaction and the dynamics of crime. Preprint.

Capocci, A., G. Caldarelli, R. Marchetti & L. Pietronero (2001). Growing dynamics of Internet providers. *Physical Review E 64*, 035105.

Carvalho, R. & A. Penn (2003). Scaling and universality in the micro-structure of urban space. Preprint, arxiv.org/abs/cond-mat/0305164.

Castellano, C., M. Marsili & A. Vespignani (2000). Nonequilibrium phase transition in a model for social influence. *Physical Review Letters 85*, 3536–3539.

Castells, M. (2001). *The Internet Galaxy: Reflections on the Internet, Business, and Society.* Oxford University Press.

Cercignani, C. (1998). *Ludwig Boltzmann: The Man Who Trusted Atoms.* Oxford University Press.

Challet, D., M. Marsili & G. Ottino (2003). Shedding light on El Farol. Preprint, arxiv.org/abs/cond-mat/0306445.

Challet, D., M. Marsili & Y.-C. Zhang (1999). Modeling market mechanism with minority game. Preprint, arxiv.org/abs/cond-mat/9909265.

Challet, D. & Y.-C. Zhang (1997). Emergence of cooperation and organization in an evolutionary game. *Physica A 246*, 407–418.

Chen, D.D.T. (2000). Smart growth. *Scientific American*, December, 85–91.

Chowell, G., J. M. Hyman, S. Eubank & C. Castillo-Chavez (2003). Scaling laws for the movement of people between locations in a large city. *Physical Review E 68*, 066102.

Cohen, D. (2002). All the world's a net. *New Scientist*, 13 April, 24–29.

Cohen, R., K. Erez, D. ben-Avraham & S. Havlin (2000). Resilience of the Internet to random breakdowns. *Physical Review Letters 85*, 4626–4628.

Cohen, M. C., R. L. Riolo & R. Axelrod (2000). The role of social structure in the maintenance of cooperative regimes. *Rationality and Society 13*, 5–32.

Collins, J. J. & C. C. Chow (1998). It's a small world. *Nature 393*, 409–410.

Cont, R. (1999). Modelling economic randomness: statistical mechanics of market phenomena. In M. T. Batchelor & L. T. Wille, eds., *Statistical Physics in the Eve of the 21st Century: the James B. McGuire Festschrift*, 47–64, World Scientific, Singapore.

Cont, R. (1999). Statistical properties of financial time series. Lecture presented at the Symposium on Mathematical Finance, Fudan University, Shanghai, 10–24 August.

Cont, R. & J.-P. Bouchard (2000). Herd behavior and aggregate fluctuations in financial markets. *Macroeconomic Dynamics 4*, 170–196.

Cont, R., M. Potters & J.-P. Bouchard (1997). Scaling in stock market data: stable laws and beyond. In B. Dubrulle, F. Graner & D. Sornette, eds., *Scale Invariance and Beyond, Proceedings of the CNRS Workshop on Scale Invariance, Les Houches*. Springer-Verlag, Berlin.

Conte, R. (2002). Agent-based modeling for understanding social intelligence. *Proceedings of the National Academy of Sciences USA 99*, 7189–7190.

Costa Filho, R. N., M. P. Almeida, J. S. Andrade Jr. & J. E. Moreira (1999). Scaling behavior in a proportional voting process. *Physical Review E 60*, 1067–1068.

Cramer, F. (1993). *Chaos and Order*. VCH, Weinheim.

Csányi, G. & B. Szendröi (2003). Structure of a large social network. Preprint, arxiv.org/abs/cond-mat/0305580.

Cuniberti, G., A. Valleriani & J. L. Vega (2001). Effects of regulation on a self-organized market. Preprint, arxiv.org/abs/cond-mat/0108533.

Czirók, A., A.-L. Barabási & T. Vicsek (1999). Collective motion of self-propelled particles: kinetic phase transition in one dimension. *Physical Review Letters 82*, 209–212.

Czirók, A., M. Vicsek & T. Vicsek (1999). Collective motions of organisms in three dimensions. *Physica A 264*, 299–304.

Czirók, A. & T. Vicsek (1999). Collective motion. In D. Reguera, M. Rubi & J. Vilar, eds., *Statistical Mechanics of Biocomplexity, Lecture Notes in Physics 527*, 152–164. Springer-Verlag, Berlin.

Czirók, A. & T. Vicsek (2000). Collective behavior of interacting self-propelled particles. *Physica A 281*, 17–29.

David, P. A. (1993). Path-dependence and predictability in dynamical systems with local network externalities: a paradigm for historical economics. In *Technology and Wealth of Nations: The Dynamics of Constructed Advantage*, ed. D. Foray & C. Freeman. Pinter, London.

Davidsen, J., H. Ebel & S. Bornholdt (2002). Emergence of a small world from local interactions: modeling acquaintance networks. *Physical Review Letters 88*, 128701.

Davidson, C. (2001). E-mmune from attack. *New Scientist*, 31 March, pp. 35–37.

De Fabritiis, G., F. Pammolli & M. Riccaboni (2003). On size and growth of business firms. *Physica A 324*, 38–44.

Dezso, Z. & A.-L. Barabási (2002). Can we stop the AIDS epidemic? *Physical Review E 65*, 055103(R).

Dodds, P. S., R. Muhamad & D. J. Watts (2003). An experimental study of search in global social networks. *Science 301*, 827–829.

Dorogovtsev, S. N. & J. F. F. Mendes (2000). Evolution of networks with aging of sites. *Physical Review E 62*, 1842–1845.

Ebel, H., L.-I. Mielsch & S. Bornholdt (2002). Scale-free topology of email networks. Preprint, arxiv.org/abs/cond-mat/0201476.

Eguíluz, V. M. & K. Klemm (2002). Epidemic threshold in structured scale-free networks. *Physical Review Letters 89*, 108701.

Eguíluz, V. M. & M. G. Zimmermann (2000). Transmission of information and herd behavior: an application to financial markets. *Physical Review Letters 85*, 5659–5662.

Emerson, R. W. (1860). The Conduct of Life. In *The Conduct of Life and Other Essays*. J. M. Dent, London, 1908.

Epstein, J. M. (2001). Learning to be thoughtless: social norms and individual computation. *Computational Economics 18 (1)*, 9–24.

Epstein, J. M. & R. Axtell (1996). *Growing Artificial Societies: Social Science From the Bottom Up*. Brookings Institution Press, Washington, D.C.

Eriksen, K. A., I. Simonsen, S. Maslov & K. Sneppen (2003). Modularity and extreme edges of the Internet. *Physical Review Letters 90*, 148701.

Farkas, I., D. Helbing & T. Vicsek (2002). Mexican waves in an excitable medium. *Nature 419*, 131–132.

Farmer, J. D. (1999). Physicists attempt to scale the ivory towers of finance. *Computing in Science and Engineering* Nov./Dec., 26–39.

Fehr, E. & S. Gächter (2002). Altruistic punishment in humans. *Nature 415*, 137–140.

Ferguson, N., ed. (1997). *Virtual History*. Picador, London.

Florian, R. & S. Galam (2000). Optimizing conflicts in the formation of strategic alliances. *European Physical Journal B16*, 189–194.

Föllmer, H. (1974). Random economies with many interacting agents. *Journal of Mathematical Economics 1*, 51–62.

Frank, T. (2002). Talking bull. *Guardian* weekend supplement, 17 August, 21.

Freud, S. (1973). *Civilization and Its Discontents*, trans. J. Riviere, ed. J. Strachey. Hogarth Press, London.

Gauthier, D. (1969). *The Logic of Leviathan*. Clarendon Press, Oxford.

Giardina, I. & J.-P. Bouchard (2002). Bubbles, crashes and intermittency in agent based market models. Preprint, arxiv.org/abs/cond-mat/0206222.

Gimblett, R., et al. (2000). An intelligent agent based model for simulating and evaluating river trip scenarios along the Colorado River in Grand Canyon National Park. In *Integrating GIS and Agent Based Modeling Techniques for Understanding Social and Ecological Processes*, ed. H. R. Gimblett, 245–275. Oxford University Press.

Gladwell, M. (2001). *The Tipping Point*. Abacus, London.

Glance, N. S. & B. A. Huberman (1994). The dynamics of social dilemmas. *Scientific American*, March, 76–81.

Gleiser, P. M. & L. Danon (2003). Community structure in jazz. Preprint, arxiv.org/abs/cond-mat/0307434.

Goldsmith, E. & J. Mander, eds. (2001). *The Case Against the Global Economy and for a Turn Towards Localization*. Earthscan Publications, London.

Gopikrishnan, P., V. Plerou, L. A. Nuñes Amaral, M. Meyer & H. E. Stanley (1999). Scaling of the distribution of fluctuations of financial market indices. *Physical Review E 60*, 5305–5316.

Granovetter, M. (1973). The strength of weak ties. *American Journal of Sociology 78*, 1360–1380.

Granovetter, M. (2003). Ignorance, knowledge, and outcomes in a small world. *Science 301*, 73–74.

Guare, J. (1990). *Six Degrees of Separation: A Play*. Vintage, New York.

Handy, C. (1993). *Understanding Organizations*, 4th ed. Penguin, London.

Harman, P. M. (1998). *The Natural Philosophy of James Clerk Maxwell*. Cambridge University Press.

Hayek, F. A. (1962). *The Road to Serfdom*. Routledge, London.

Hayes, B. (2002). Statistics of deadly quarrels. *American Scientist*, January–February.

Helbing, D. (1994). A mathematical model for the behavior of individuals in a social field. *Journal of Mathematical Sociology 19*, 189–219.

Helbing, D. (1996). A stochastic behavioral model and a "microscopic" foundation of evolutionary game theory. *Theory and Decision 40*, 149–179.

Helbing, D. (2001). Traffic and related self-driven many-particle systems. *Reviews of Modern Physics 73*, 1067–1141.

Helbing, D., I. Farkas & T. Vicsek (2000). Simulating dynamical features of escape panic. *Nature 407*, 487–490.

Helbing, D., A. Hennecke & M. Treiber (1999). Phase diagram of traffic states in the presence of inhomogeneities. *Physical Review Letters 82*, 4360–4363.

Helbing, D. & B. A. Huberman (1998). Coherent moving states in highway traffic. *Nature 396*, 738–740.

Helbing, D., J. Keltsch & P. Molnar (1997). Modelling the evolution of human trail systems. *Nature 388*, 47–49.

Helbing, D. & P. Molnar (1995). Social force model for pedestrian dynamics. *Physical Review E 51*, 4282–4286.

Helbing, D., P. Molnar, I. J. Farkas & K. Bolay (2001). Self-organizing pedestrian movement. *Environment and Planning B: Planning and Design 28*, 361–383.

Helbing, D. & M. Schreckenberg (1999). Cellular automata simulating experimental properties of traffic flow. *Physical Review E 59*, R2505–R2508.

Helbing, D. & T. Vicsek (1999). Optimal self-organization. *New Journal of Physics 1*, 13.1–13.7.

Henderson, L. F. (1971). The statistics of crowd fluids. *Nature 229*, 381–383.

Hillier, B. & J. Hanson (1984). *The Social Logic of Space*. Cambridge University Press, Cambridge.

Hinich, M. J. & M. C. Munger (1997). *Analytical Politics*. Cambridge University Press.

Hobbes, T. (1651). *Leviathan*. Ed. C. B. Macpherson. Penguin, London, 1985.

Hobsbawm, E. (1995). *Age of Extremes*. Abacus, London.

Hod, S. & E. Nakar (2002). Self-segregation versus clustering in the evolutionary minority game. *Physical Review Letters 88*, 238702.

Huang, Z.-F. & S. Solomon (2001). Finite market size as a source of extreme wealth inequality and market instability. *Physica A 294*, 503-13.

Huberman, B. A. & L. A. Adamic (1999). Growth dynamics of the World-Wide Web. *Nature 401*, 131.

Huberman, B. A. & R. M. Lukose (1997). Social dilemmas and Internet congestion. *Science 277*, 535–537.

Jencks, C. (1995). *The Architecture of the Jumping Universe*. Academic Editions, London.

Jeong, H., B. Tombor, R. Albert, Z. N. Oltvai & A.-L. Barabási (2000). The large-scale organization of metabolic networks. *Nature 407*, 651–654.

Kagan, R. (2002). Power and weakness. *Policy Review 113*, June & July.

Keen, S. (2001). *Debunking Economics: The Naked Emperor of the Social Sciences*. Pluto Press & Zed Books, Sydney & London.

Keen, S. (2003). Standing on the toes of pygmies: why econophysics must be careful of the economic foundations on which it builds. *Physica A 324*, 108–116.

Keen, S., J. Legge, G. Fishburn & R. Standish (2003). Aggregation problems in the non-interactive equilibrium theory of markets. Preprint.

Kerner, B. S. (1998). Experimental features of self-organization in traffic flow. *Physical Review Letters 81*, 3797–3800.

Kerner, B. S. (1999). The physics of traffic. *Physics World*, August, 25–30.

Kerner, B. S. (2003). Control of spatial-temporal congested traffic patterns at highway bottlenecks. Preprint, arxiv.org/abs/cond-mat/0309017.

Kerner, B. S., S. L. Klenov & D. E. Wolf (2002). Cellular automata approach to three-phase traffic theory. Preprint, arxiv.org/abs/cond-mat/0206370.

Kerner, B. S. & H. Rehborn (1996). Experimental features and characteristics of traffic jams. *Physical Review E 53*, R1297–R1300.

Kerner, B. S. & H. Rehborn (1997). Experimental properties of phase transitions in traffic flow. *Physical Review Letters 79*, 4030–4033.

Kirchner, A. & A. Schadschneider (2002). Simulation of evacuation processes using a bionics-inspired cellular automaton model for pedestrian dynamics. Preprint, arxiv.org/abs/cond-mat/0203461.

Kirman, A. P. (1994). Economics with interacting agents. Working paper 94-05-030, Santa Fe Institute, Santa Fe, N.M.

Kirman, A. (1996) Some observations on interactions in economics. Paper presented at Workshop on Structural Change, Manchester Metropolitan University, U.K., 20–21 May. Available at www.cpm.mmu.ac.uk/pub/workshop/kirman.html.

Klemm, K. & V. M. Eguíluz (2002). Growing scale-free networks with small-world behavior. *Physical Review E 65*, 057102.

Klemm, K., V. M. Eguíluz, R. Toral & M. San Miguel (2002). Global culture: a noise induced transition in finite systems. Preprint, arxiv.org/abs/cond-mat/0205188.

Knospe, W., L. Santen, A. Schadschneider & M. Schreckenberg (2001). Human behaviour as the origin of traffic phases. *Physical Review E, 65*, 015101.

Krugman, P. (1994). *Peddling Prosperity.* W. W. Norton, New York.

Krugman, P. (1996). *The Self-Organizing Economy.* Blackwell, Oxford.

Kurtz, S. (2002). The future of "history." *Policy Review 113*, June/July.

Laloux, L., M. Potters, R. Cont, J.-P. Aguilar & J.-P. Bouchard (1999). Are financial crashes predictable? *Europhysics Letters 45*, 1–5.

Landauer, R. (1975). Inadequacy of entropy and entropy derivatives in characterizing the steady state. *Physical Review A, 12*, 636.

Latora, V. & M. Marchioro (2001). Efficient behaviour of small-world networks. *Physical Review Letters 87*, 198701.

Lawrence, S. & C. L. Giles (1998). Searching the World Wide Web. *Science 280*, 98–100.

LeBaron, B. (2002). Short-memory traders and their impact on group learning in financial markets. *Proceedings of the National Academy of Sciences USA 99*, 7201–7206.

Lee, H. Y., H.-W. Lee & D. Kim (2000). Phase diagram of congested traffic flow: an empirical study. *Physical Review E 62*, 4737–4741.

Lewin, R. (1992). *Complexity.* Macmillan, New York.

Li, W. & X. Cai (2003). Statistical analysis of airport network of China. Preprint, arxiv.org/abs/cond-mat/0309236.

Liljeros, F., C. R. Edling, L. A. Nunes Amaral, H. E. Stanley & Y. Åberg (2001). The web of human sexual contacts. *Nature 411*, 907–908.

Lillo, F. & R. N. Mantegna (2000). Symmetry alteration of ensemble return distribution in crash and rally days of financial markets. *European Physical Journal B 15*, 603–606.

Lillo, F. & R. N. Mantegna (2000). Variety and volatility in financial markets. *Physical Review E 62*, 6126–6134.

Litan, R. E. & A. N. Rivlin (2000). The economy and the Internet: what lies ahead? Conference report no. 4, Brookings Institution, Washington, D.C.

Locke, J. (1689). *Two Treatises of Government.* Ed. M. Goldie, 1993. J. M. Dent, London.

Lombardo, M. P. (1985). Mutual restraint in tree swallows: a test of the TIT FOR TAT model of reciprocity. *Science 227*, 1363–1365.

Louzoun, Y., S. Solomon, J. Goldenberg & D. Mazursky (2002). The risk at being unfair: world-size global markets lead to economic instability. Preprint.

Lübeck, S., M. Schreckenberg & K. D. Usadel (1998). Density fluctuations and phase transition in the Nagel-Schreckenberg traffic flow model. *Physical Review E 57*, 1171–1174.

Lux, T. & M. Marchesi (1999). Scaling and criticality in a stochastic multi-agent model of a financial market. *Nature 397*, 498–500.

Makse, H. A., J. S. Andrade Jr., M. Batty, S. Havlin & H. E. Stanley (1998). Modeling urban growth patterns with correlated percolation. *Physical Review E 58*, 7054–7062.

Makse, H. A., S. Havlin & H. E. Stanley (1995). Modelling urban growth patterns. *Nature 377*, 608–612.

Mandelbrot, B. (1963). The variation of certain speculative prices. *Journal of Business 35*, 394–419.

Mandelbrot, B. (1977). *The Fractal Geometry of Nature*. W. H. Freeman, New York.

Mandelbrot, B. (1999). A multifractal walk down Wall Street. *Scientific American*, February, 70–73.

Mantegna, R. N. & H. E. Stanley (1995). Scaling behaviour in the dynamics of an economic index. *Nature 376*, 46–49.

Mantegna, R. N. & H. E. Stanley (1996). Turbulence and financial markets. *Nature 383*, 587–588.

Mantegna, R. N. & H. E. Stanley (1999). *An Introduction to Econophysics*. Cambridge University Press.

Mathias, N. & V. Gopal (2001). Small worlds: how and why. *Physical Review E 63*, 021117.

Matthews, R. (1999). Get connected. *New Scientist*, 4 December, 24–28.

May, R. (1987). More evolution of cooperation. *Nature 327*, 15–17.

May, R. M. & A. L. Lloyd (2001). Infection dynamics on scale-free networks. *Physical Review E 64*, 066112.

Merchant, C. (1990). *The Death of Nature: Women, Ecology, and the Scientific Revolution*. Harper, San Francisco.

Mikhailov, A. S. & D. H. Zanette (1999). Noise-induced breakdown of coherent collective motion in swarms. *Physical Review E 60*, 4571–4575.

Milgram, S. (1967). The small world problem. *Psychology Today 2*, 60–67.

Milinski, M. (1987). TIT FOR TAT in sticklebacks and the evolution of cooperation. *Nature 325*, 433–435.

Mill, J. S. (1859). *On Liberty*. Ed. G. Himmelfarb, 1985. Penguin, London.

Mill, J. S. (1872). *System of Logic*. Longmans, Green, Reader & Dyer, London.

Mill, J. S. & J. Bentham. *Utilitarianism and Other Essays*. Ed. A. Ryan, 1987. Penguin, London.

Mitzenmacher, M. (2003). A brief history of generative models for power law and lognormal distributions. Preprint, Harvard University.

Moore, C. & M. E. J. Newman (2000). Epidemics and percolation in small-world networks. *Physical Review E 61*, 5678–5682.

More, T. (1516). *Utopia*. Everyman's Library, London, 1992.

Mossa, S., M. Barthélémy, H. E. Stanley & L. A. Nuñes Amaral (2002). Truncation of power law behavior in "scale-free" network models due to information filtering. *Physical Review Letters 88*, 138701.

Muchnik, L. & S. Solomon (2003). Statistical mechanics of conventional traders may lead to non-conventional market behaviour. Preprint.

Mumford, L. (1938). *The Culture of Cities*. Secker & Warburg, London.

Naughton, J. (1999). *A Brief History of the Future*. Weidenfeld & Nicolson, London.

Néda, Z., E. Ravasz, Y. Brechet, T. Vicsek & A.-L. Barabási (2000). Self-organizing processes: the sound of many hands clapping. *Nature 403*, 849–850.

Neubert, L., L. Santen, A. Schadschneider & M. Schreckenberg (1999). Single-vehicle data of highway traffic: a statistical analysis. *Physical Review E 60*, 6480–6490.

Newman, M.E.J. (2000). Models of the small world: a review. Preprint, arxiv.org/abs/cond-mat/0001118.

Newman, M.E.J. (2001). Ego-centred networks and the ripple effect. Preprint, arxiv.org/abs/cond-mat/0111070.

Newman, M.E.J. (2001). The structure of scientific collaboration networks. *Proceedings of the National Academy of Sciences USA 98*, 404–409.

Newman, M.E.J., S. Forrest & J. Balthrop (2002). Email networks and the spread of computer viruses. *Physical Review E 66*, 035101.

Newman, M.E.J. & J. Park (2003). Why social networks are different from other types of networks. Preprint, arxiv.org/abs/cond-mat/0305612.

Nowak, M. & R. M. May (1992). Evolutionary games and spatial chaos. *Nature 359*, 826–829.

Nowak, M. & R. May (1993). The spatial dilemmas of evolution. *International Journal of Bifurcation and Chaos 3*, 35–78.

Nowak, M., R. M. May & K. Sigmund (1995). The arithmetics of mutual help. *Scientific American*, June, 50–55.

Nowak, M. & K. Sigmund (1992). Tit for tat in heterogeneous populations. *Nature 355*, 250–253.

Nowak, M. & K. Sigmund (1993). Chaos and the evolution of cooperation. *Proceedings of the National Academy of Sciences USA 90*, 5091–5094.

Nowak, M. & K. Sigmund (1993). A strategy of win-stay, lose-shift that outperforms tit-for-tat in the Prisoner's Dilemma game. *Nature 364*, 56–58.

Nussbaum, F. L. (1965). *The Triumph of Science and Reason 1660–1685*. Harper & Row, New York.

Olson, R. (1990). *Science Deified and Science Defied*, vol. 2. University of California Press, Berkeley, Calif.

Ormerod, P. (1998). *Butterfly Economics*. Faber & Faber, London.

Ormerod, P. (2000). The Keynesian micro-foundations of the business cycle: some implications of globalisation. Preprint.

Ormerod, P. (2002). The U.S. business cycle: power law scaling for interacting units with complex internal structure. *Physica A 314*, 774–785.

Ormerod, P. (2002). Sense on segregation. *Prospect*, February, 12–14.

Ormerod, P. & M. Campbell (2000). The evolution of family structures in a social context. Preprint.

Parrish, J. K. & L. Edelstein-Keshet (1999). Complexity, pattern, and evolutionary trade-offs in animal aggregation. *Science 284*, 99–101.

Pastor-Satorras, R. & A. Vespignani (2001). Epidemic spreading in scale-free networks. *Physical Review Letters 86*, 3200–3203.

Pastor-Satorras, R. & A. Vespignani (2001). Immunization of complex networks. Preprint, arxiv.org/abs/cond-mat/0107066.

Perez, G. J., G. Tapang, M. Lim & C. Saloma (2002). Streaming, disruptive interference and power-law behavior in the exit dynamics of confined pedestrians. *Physica A 312*, 609–618.

Peterson, I. (1996). The Gods of Sugarscape. *Science News 150*, 23 November.

Peterson, I. (1996). The shapes of cities. *Science News 149*, 8–9.

Plerou, V., P. Gopikrishnan, L. A. Nuñes Amaral, M. Meyer & H. E. Stanley (1999). Scaling of the distribution of price fluctuations of individual companies. *Physical Review E 60*, 6519–6529.

Popper, K. (1997). *The Lesson of This Century*. Routledge, London.

Porter, R. (2000). *Enlightenment*. Penguin, London.

Porter, T. M. (1986). *The Rise of Statistical Thinking 1820–1900*. Princeton University Press.

Potters, M., R. Cont & J.-P. Bouchard (1998). Financial markets as adaptive ecosystems. *Europhysics Letters 41*, 239–244.

Potts, W. K. (1984). The chorus-line hypothesis of manoeuvre coordination in avian flocks. *Nature 309*, 344–345.

Prigogine, I. (1980). *From Being to Becoming*. W. H. Freeman, New York.

Pustilnik, L. A. & G. Yom Din (2003). Influence of solar activity on state of wheat market in medieval England. Preprint, arxiv.org/abs/astro-ph/0312244.

Ravasz, E. & A.-L. Barabási (2002). Hierarchical organization in complex networks. Preprint, arxiv.org/abs/cond-mat/0206130.

Rawcliffe, C. (1999). *Medicine and Society in Later Medieval England*. Sandpiper, London.

Read, D. (2002). A multitrajectory, competition model of emergent complexity in human social organization. *Proceedings of the National Academy of Sciences USA 99*, 7251–7256.

Reynolds, C. W. (1987). Flocks, herds and schools: a distributed behavioral model. *Computer Graphics 21 (4)*, 25–34.

Reynolds, C. (1995). Boids. Web site at www.red3rd.com/cwr/boids.

Richardson, L. F. (1960). *Statistics of Deadly Quarrels*, ed. Q. Wright & C. C. Lienau. Boxwood Press, Pittsburgh.

Roberts, C. A., D. Stallman & J. A. Bieri (2002). Modeling complex human-environment interactions: the Grand Canyon river trip simulator. *Journal of Ecological Modelling 153*, 181–196.

Rowlinson, J. S., ed. (1988). *J. D. van der Waals: On the Continuity of the Gaseous and Liquid States. Studies in Statistical Mechanics* XIV. North-Holland, Amsterdam.

Russell, B. (1984). *A History of Western Philosophy*. Unwin Paperbacks, London.

Saloma, C., G. J. Perez, G. Tapang, M. Lim & C. Palmes-Saloma (2003). Self-organized queuing and scale-free behavior in real escape panic. *Proceedings of the National Academy of Sciences U.S.A. 100*, 11,947–11,952.

Schechter, B. (1999). Birds of a feather. *New Scientist*, 23 January, 30–33.

Schelling, T. C. (1978). *Micromotives and Macrobehavior*. W. W. Norton, New York.

Schneewind, J. B., ed. (1990). *Moral Philosophy from Montaigne to Kant*, vol. II. Cambridge University Press, Cambridge.

Schweitzer, F., ed. (2002). *Modelling Complexity in Economic and Social Systems*. World Scientific, Singapore.

Sen, P., et al. (2002). Small-world properties of the Indian Railway network. Preprint, arxiv.org/abs/cond-mat/0208535.

Serrano, M. A. & M. Boguñá (2003). Topology of the world trade web. *Physical Review E 68*, 015101 (R).

Shvetsov, V. & D. Helbing (1999). Macroscopic dynamics of multilane traffic. *Physical Review E 59*, 6328–6339.

Sigmund, K. (1993). *Games of Life.* Oxford University Press.

Sigmund, K., E. Fehr & M. A. Nowak (2002). The economics of fair play. *Scientific American*, January, 82–87.

Simon, H. A. (1996). *The Sciences of the Artificial*, 3rd ed. MIT Press, Cambridge, Mass.

Skyrms, B. (1996). *Evolution of the Social Contract.* Cambridge University Press.

Smith, A. (1776). *An Inquiry into the Nature and Causes of the Wealth of Nations*, ed. L. Dickey. Hackett, Indianapolis, 1993.

Solomon, S. (1999). Behaviorly realistic simulations of stock market traders with a soul. *Computer Physics Communications 121–2*, 161.

Sornette, D. (2003). Critical market crashes. Preprint, arxiv.org/abs/cond-mat/0301543.

Sornette, D. (2003). *Why Stock Markets Crash.* Princeton University Press, Princeton, N.J.

Soros, G. (2002). *George Soros on Globalization.* Perseus, New York.

Stanley, M.H.R., et al. (1996). Scaling behaviour in the growth of companies. *Nature 379*, 804–806.

Stix, G. (1998). A calculus of risk. *Scientific American*, May, 92–97.

Sutton, J. (1997). Gibrat's legacy. *Journal of Economic Literature XXXV*, 40–59.

Sznajd-Weron, K. & R. Weron (2002). How effective is advertising in duopoly markets? Preprint, arxiv.org/abs/cond-mat/0211058.

Thompson, D'A. W. (1942). *On Growth and Form: A New Edition.* Cambridge University Press. Reissued by Dover, New York, 1992.

Tjaden, B. & G. Wasson (1997). The Oracle of Bacon at Virginia. Available at http://www.cs.virginia.edu/oracle/.

Toner, J. & Y. Tu (1995). Long-range order in a two-dimensional dynamical XY model: how birds fly together. *Physical Review Letters 75*, 4326–4329.

Toner, J. & Y. Tu (1998). Flocks, herds, and schools: a quantitative theory of flocking. *Physical Review E 58*, 4828–4858.

Treiber, M. & D. Helbing (2002). Microsimulations of freeway traffic including control measures. Preprint, arxiv.org/abs/cond-mat/0210096.

Treiber, M., A. Hennecke & D. Helbing (2000). Congested traffic states in empirical observations and microscopic simulations. *Physical Review E 62*, 1805–1824.

Tuck, R. (2002). *Thomas Hobbes: A Very Short Introduction.* Oxford University Press, Oxford.

Vicsek, T., A. Czirók, E. Ben-Jacob, I. Cohen & O. Shochet (1995). Novel type of phase transition in a system of self-driven particles. *Physical Review Letters 75*, 1226–1229.

Vines, S. (2003). *Market Panic.* Profile, London.

Waldrop, M. M. (1994). *Complexity.* Penguin. London.

Waltz, K. N. (1954). *Man, the State, and War.* Columbia University Press, New York.

Ward, M. (2001). *Universality.* Macmillan, London.

Watts, D. J. (1999). *Small Worlds.* Princeton University Press, Princeton, N.J.

Watts, D. J. (2003). *Six Degrees: The Science of a Connected Age.* W. W. Norton, New York.

Watts, D. J., P. S. Dodds & M. Newman (2002). Identity and search in local networks. *Science 296*, 1302–1305.

Watts, D. J. & S. H. Strogatz (1998). Collective dynamics of "small-world" networks. *Nature 393*, 440–442.

Weidlich, W. (2000). *Sociodynamics: A Systematic Approach to Modeling in the Social Sciences.* Harwood Academic, Amsterdam.

Wilkinson, G. S. (1984). Reciprocal food sharing in the vampire bat. *Nature 308*, 181–184.

Wilson, E. O. (1998). *Consilience.* Knopf, New York.

Winch, P. (1958). *The Idea of a Social Science and its Relation to Philosophy.* Routledge, London.

Wolf, D. (1999). Cellular automata for traffic simulations. *Physica A 263*, 438–451.

Wu, J. & R. Axelrod (1995). How to cope with noise in the iterated Prisoner's Dilemma. *Journal of Conflict Resolution 39*, 183–189.

Wyart, M. & J.-P. Bouchard (2002). Statistical models for company growth. Preprint, arxiv.org/abs/cond-mat/0210479.

Yang, I., H. Jeong, B. Kahng & A.-L. Barabási (2003). Emerging behavior in electronic bidding. *Physical Review E 68*, 016102.

Yook, S.-H., H. Jeong & A.-L. Barabási (2002). Modeling the Internet's large-scale topology. *Proceedings of the National Academy of Sciences USA 99*, 13,382–13,386.

Zhou, W.-X. & D. Sornette (2003). 2000–2003 real estate bubble in the UK but not in the USA. *Physica A 329*, 249–263.

Zipf, G. K. (1965). *Human Behavior and the Principle of Least Effort.* Hafner, New York.

ACKNOWLEDGMENTS

Writing this book has often felt like a collaborative exercise, for which I am eager to spread some credit (while of course remaining accountable for all blame). For wise, perceptive, and tolerant suggestions and advice, and for the provision of materials and references, I am deeply indebted to many real experts in this arena, in particular Robert Axelrod, Robert Axtell, Albert-Laszló Barabási, Eshel Ben-Jacob, Rama Cont, Dirk Helbing, Steve Keen, Thomas Lux, Mitsugu Matsushita, Joe McCauley, Mark Newman, Paul Ormerod, Craig Reynolds, Sorin Solomon, Gene Stanley, Alessandro Vespignani and Tamás Vicsek. The support of my editors, John Glusman, Ravi Mirchandani, and Caroline Knight, has been vital, and many rough edges were knocked off the text by the careful attention of the copy editor, John Woodruff. I have been encouraged as ever by the good judgment of my agent, Peter Robinson, and the support of my wife, Julia.

Philip Ball
London, October 2003

INDEX

Abdus Salam International Centre for Theoretical Physics, 344
Achenwall, Gottfried, 53
Ackroyd, Peter, 148
Adamic, Lada, 384
Advanced Research Projects Agency (ARPA), 376
Afghanistan, war in, 339, 443
Aguilar, Jean-Pierre, 227, 234, 235
AIDS, 358, 394, 395
Aix-Marseilles, University of, 213
Akerlof, George, 211
Albania, 431
Alberick, Ricardo, 395
Albert, Eddie, 352–53, 360, 370
Albert, Prince Consort of Queen Victoria, 74
Albert, Réka, 381, 383, 385, 395, 397–99
Alembert, Jean le Rond d', 54
alliances, 270–94; business, 270–76, 282; international, 284–89, 291–93; political, 271, 284
America Online (AOL), 377, 384
Anaxagoras, 34
Anderson, Philip, 246
Andrews, Thomas, 83
Anglicans, 10, 28
Apelles, 405
Apple Computers, 270
Aquinas, St. Thomas, 461
Aristophanes, 405

Aristotle, 16, 34, 35, 48, 303, 461
ARPANET, 376, 379
Arrow, Kenneth, 305
Arthur, W. Brian, 220, 272n, 331–33
Ashworth, Tony, 414–15
Asimov, Isaac, 365
AT&T, 274
atoms, 34–35, 43; motion of, 36–40
attractive forces, 84–86, 132, 138
Aubrey, John, 15
Ausloos, Marcel, 234, 235
Austria, 1848 revolution in, 186
autocorrelation function, 200–201
automata, 17, 36, 125
Axelrod, Robert, 424, 431, 438, 439, 441; interaction-based model of cultural transmission of, 340–41, 345, 347, 348, 350, 351; landscape model of alliance formation of, 275, 276, 281–83, 285, 292, 293, 301; Prisoner's Dilemma tournaments of, 419–22, 430, 432, 433, 440, 447
Axtell, Robert, 223, 224, 258–62, 264, 265, 267–69, 313n, 348–50

Babbage, Charles, 72
Bachelier, Louis, 192–94, 196–98
Bacillus, 119–22, 127; *B. circulans*, 120; *B. subtilis*, 110, 112
Bacon, Francis, 14, 21, 35, 50

Bacon, Kevin, 352–53, 355, 359–60, 369–70

bacteria: motion of, 118–22, 127; nonequilibrium growth of, 110–14

Bak, Per, 237–43, 246

Balearic Isles, University of, 395

Bangladesh, 346

Barabási, Alber-László, 381, 385, 389–91, 394, 395, 397–400, 467

Baran, Paul, 377–79, 387

Barings Bank, 202

Barrett, William, 87

Batman Returns (movie), 125

Batty, Michael, 138, 145–46, 149–50

Beck, Harry, 361

Becker, Gary, 316, 323–26, 330

"Beethoven in Heaven" (Boltzmann), 48

Belgium, 58–59; Royal Observatory, 59

Bellamy, Edward, 451

Bell Laboratories, 274

Bénard, Henri, 104, 105

Ben-Jacob, Eshel, 119, 120, 127

Bennett, Scott, 283, 285

Bentham, Jeremy, 30, 65, 183, 262, 458

Bentley, Wilson A., 100, 114

Bernardes, Americo Tristão, 300

Berners-Lee, Tim, 376–77, 381

Bernoulli, Daniel, 35, 39, 62, 84

Bernoulli, Jacob, 62

Betamax video system, 272

Bhutan, 341

Bible, 52; Book of Job, 26

bifurcations, 106–9

Big Picture, The (movie), 352

Biham, Ofer, 175

biometrics, 70

Bismarck, Otto von, 305

Black, Duncan, 302

Black, Fischer, 201

Black, Joseph, 457n

Black-Scholes model, 201–3

Bogart, Humphrey, 352

Bolay, Kai, 134

Boltzmann, Ludwig, 33, 45–49, 69, 70, 78, 81, 96, 99, 130, 131, 205, 206

Bonaparte, Napoleon, 204, 226, 271, 449

Bonn, University of, 215, 217

Borda, Jean Charles de, 304

Borges, Jorge Luis, 108, 289

Bornholdt, Stefan, 393

Boscovich, Roger Joseph, 35–36

Boston University, 151, 197, 242, 260, 399

Bouchard, Jean-Philippe, 203, 248

Boulding, Kenneth, 204

Bowie, David, 133

Boyle, Robert, 37n, 40

Brahe, Tycho, 4

Brando, Marlon, 369

Brave New World (Huxley), 450–51

Brazil, elections in, 298, 299

Brinton, Thomas, 26

Britain, 291, 340, 341; crowd behavior in, 141, 145–46; divorce rate in, 327; elections in, 297, 301; in invasion of Iraq, 339; marriage age in, 324; Muslims in, 312; Parliament of, 10–11, 13, 16, 19, 28, 372, 407; political parties in, 284, 297, 391, 426; population growth in, 52; Roman, 252; supermarkets in, 250–51, 256; traffic in, 156; in World War I, 412, 413, 433; in World War II, 286–88; *see also* England

Brock, William, 215

Brookhaven National Laboratory, 237–38

Brookings Institution, 223, 258, 306, 348, 350

Brown, Robert, 43

Brownian motion, 43, 44, 111, 192

Buchanan, Mark, 243, 244

Buckle, Henry Thomas, 65–69, 74, 75

Bulgaria, 293

Burda, Zdzisław, 223

Burke, Edmund, 57–58, 222, 441

Burundi, 347

Bush, George W., 295–96

business cycle, 187–89; chaos in, 192; real, theory of, 210

Business Cycles (Schumpeter), 188n

Cagliari, University of, 218
Cagniard de la Tour, Charles, Baron, 82–83
California, University of, Los Angeles, 235
California Institute of Technology, 240
Calvin, John, 12
Cambodia, 346
Cambridge University, 12, 125, 159, 205
Campbell, Lewis, 67
Campbell, Michael, 317–19, 327, 330
Candolle, Alphonse de, 72
Cantillon, Richard, 456
capitalism, *see* market economy
Carey, John, 450
Carlyle, Thomas, 204
Carnegie Endowment for International Peace, 411
Carnival Review Group (Greater London Authority), 146
Carnot, Nicholas-Léonard-Sadi, 36–37, 39
cartels, 256
Carvalho, Rui, 153
Castells, Manuel, 374
catastrophe theory, 5
Catholics, 3, 20, 13, 52, 285, 324, 431
Cat's Cradle (Vonnegut), 80, 161
Ceará, Federal University of, 298
Central Intelligence Agency (CIA) Information Warfare Center, 388
CERN particle physics laboratory, 376–77
Challet, Damien, 333–34, 397
chaos theory, 5
Charlemagne, 438
Charles, Jacques, 39
Charles I, King of England, 11, 15, 19, 51, 458
Charles II, King of England, 10, 20, 51
Chechnya, 339
Chen, Donald, 149

Chen, Kan, 242
Chicago, University of, 241, 362
China, 430; ancient, 252
choice theory, 304–5
Christianity, 21, 28, 35, 289; fundamentalist, 154
chromodynamics, quantum, 96
Chuo University, 110
Churchill, Winston, 288
Cicero, 460, 461
Clark, Wesley, 379
Clausius, Rudolph, 38, 40, 45, 68, 70, 77, 99
Clinton, Bill, 387, 426
clustering, 364–68
coalitions, *see* alliances
Coase, Ronald, 262
Cobb, G. Belton, 413, 422
Coconut Grove nightclub fire, 141
Cohen, Michael, 350
Cold War, 378, 379, 403, 418, 430, 444–46, 448
Cologne, University of, 159, 248
Columbia University, 370
Commerce Department, U.S., 372
Commonwealth, Hobbesian, 26–27, 178
Communist Manifesto (Marx and Engels), 186
Communist Party, 189
complexity theory, 5, 126
compressibility, 228
computer viruses, 392–94
Comte, Auguste, 58, 65, 290, 459
Condorcet, Marie-Jean-Antoine-Nicolas de Caritat, Marquis de, 54–56, 71, 303, 304, 459
Congress, U.S., 407
Conservative Party, British, 284, 297, 426
Contrite Tit for Tat (CTFT) strategy, 431, 443
Conway, John Horton, 125–26
Cooper, Lord Ashley, 406
Co-operative Societies (Great Britain), 450
Cootner, Paul, 196
Copernicus, Nicolaus, 12

Cornell University, 233, 364
Cornhill Magazine, 74
correlated percolation, 151–52
Correlates of War project, 86
correlation function, 200
corresponding states, principle of, 94–95
Costa Filho, Raimundo, 298–300
Counter-Reformation, 12
Cournot, Antoine-Augustin, 71
Cours de philosophie positive (Comte), 58
crime, 316–22, 330, 331
Crimean War, 65*n*
Critical Infrastructure Protection, Commission of, 387
critical points, 83, 86–95, 227–37
Cromwell, Oliver, 10, 11, 28, 51, 284, 450
crowd movement, 131, 141–46
Cuban missile crisis, 445
cultural transmission, 337–47; models of, 347–51
Curie, Marie, 87
Curie, Pierre, 87, 215
Curie point, 87–88, 90, 92, 95, 108, 109, 164, 229, 230, 233
cyber-terrorism, 387–89
Cyples, William, 74
Czechoslovakia, 286, 339
Czirók, András, 127–28

Daimler-Benz research laboratories, 168
Dalzell, Stewart, 375–76
Darwin, Charles, 52, 69–70, 101, 154, 408
Darwinism, 30, 276, 404, 409, 433, 436, 464
Davenant, Charles, 222
Davies, Cecil, 447
Dean, James, 352
Death of Nature, The (Merchant), 30
decision making, 306–12; criminal, 316–22; in elections, 295–305; marital, 323–30; minority, 331–36; segregation and, 312–15
De Cive (Hobbes), 19, 20
Declaration of Independence, 57

deductive reasoning, 15
Deep Blue computer, 208
Defoe, Daniel, 180
de Gaulle, Charles, 288
De Legibus (Cicero), 460
Democratic Party, U.S., 296, 297
Democritus, 34–35
demographic statistics, 49–54
Dene Nation, 337
Denmark, 286, 327
Depardieu, Gérard, 369
De rerum natura (Lucretius), 35
Desaguliers, Jean Théophile, 56
Descartes, René, 12, 15–16, 18–20, 35
Descent of Man, The (Darwin), 69
determinism, 18, 36, 47, 289–90; balance between choice and, 349; entropy and, 77–78; statistics and, 73–76
Dezso, Zoltan, 394
Dickens, Charles, 67
Dictyostelium discoideum, 121–22
dielectric breakdown model (DBM), 150
diffusion, 43
diffusion limited aggregation (DLA), 111–12, 127, 149–52
Diggers, 11, 450
Digital Equipment Corporation, 274
Discourse on the Origin of Inequality (Rousseau), 418
Disraeli, Benjamin, 67
dissipative structures, 106
divergences, 228–30
Dr. Strangelove (movie), 430–31, 437, 446
Domesday Book, 52
Dostoevsky, Fyodor, 74–75, 317
Dow, Charles, 189
Downs, Anthony, 302
Dresher, Melvin, 416, 417
Dugas, Gaetan, 358
Duisburg, University of, 159, 171, 175
Duke University, 432
Durlauf, Steven, 215

Eastern Orthodox Church, 285
economics: uncertainty in, 211–14; rationality in, 206–11; *see also* market economy
economies of scale, 253, 262; increasing, 262
econophysics, 182, 209, 227
Edelstein-Keshet, Leah, 122–23, 147
Eden, M., 112
Edgeworth, Francis, 183, 205, 206, 208, 213
Egypt, ancient, 247, 248
Ehrenfest, Paul, 49
Einstein, Albert, 43, 192, 209, 354
elections, 295–305, 331
Elementary Principles in Statistical Mechanics (Gibbs), 99
Elements of Geography (Euclid), 14–15
"El Farol problem," 331–34, 336
Elizabeth I, Queen of England, 148
Elliot, Ralph, 188
emergence, 126
Emerson, Ralph Waldo, 23, 67, 181, 459
energy, conservation of, 38, 78
energy landscapes, 278–81
Engels, Friedrich, 186, 450
England, 3–4, 16, 28, 339, 407; during
    Civil War, 10–11, 13, 19, 178*n*, 405,
    450; Elizabethan, 12, 148; population of,
    52; *see also* Britain
Enlightenment, 12, 53, 54, 57, 58, 245,
    249, 402, 455, 457, 458; debates on
    international power politics during, 411;
    market economy of, 29, 179, 180; mechanistic philosophy of, 33 (*see also specific
    philosophers*)
Enron, 223
entropy, 38, 46, 77–78, 103
Éole Normale Supérieure, 192
Eötvös University, 142
Epicurus, 34, 35
epidemics, 322, 358–95
Epstein, Joshua, 223, 224, 306–10, 313*n*,
    348–50
Erdös, Paul, 353–54, 359, 361–62, 387,
    395, 397

error curve, 60–63, 68–69, 131, 193
*Esquisse d'un tableau historique des progrès
    de l'esprit humain* (Condorcet), 54–56
Essai sur la nature du commerce en général
    (Cantillon), 456
*Essai sur l'application de l'analyse à la
    probabilité des décisions rendues à la
    plurité des voix* (Condorcet), 54
Essay on the Principle of Population
    (Malthus), 52, 56
Estonia, 285, 286
Euclid, 14–16, 71
eugenics, 70
Euler, Leonhard, 353*n*
Euripedes, 14, 405
European Union, 339, 411
exponents, critical, 229–30, 233

Fable of the Bees (Mandeville), 453–54
Fama, Eugene, 197*n*
Farkas, Illés, 133*n*, 142
Farr, William, 72
Federal Bureau of Investigation (FBI), 388
Federal Reserve, 221
Fehr, Ernst, 426–27
"Felicific Calculus" (Bentham), 183
Ferguson, Adam, 458–60
Ferguson, Niall, 289–91
Few Good Men, A (movie), 360
Fibonacci sequence, 188–89
Filmer, Robert, 405
Finland, 286, 288
firms: alliances of, 270–76, 282; growth of,
    250–69
fractals, 110–12, 150, 196
Fisher, Irving, 182, 190, 192
Flood, Merrill, 416, 417
Florian, Razvan, 283*n*
Florida Supreme Court, 296
Flynn, Errol, 353
Fölmer, Hans, 215
Ford Motor Company, 267

Fourier, Joseph, 63

Fox, J. J., 71

France, 20, 48, 408; divorce rate in, 327; 1848 revolution in, 186; Napoleonic, 271; study of statistics in, 54–65; in World War I, 412; in World War II, 285–88

Francis, St., 447

Franz Ferdinand, Archduke of Austria-Hungary, 244

Free Democratic Party, German, 271

French Academy of Sciences, 304

French Revolution, 10, 54–56, 302

French Royal Observatory, 60

Freud, Sigmund, 49, 402–3

Fribourg, University of, 333

Gächter, Simon, 426–27

Galam, Serge, 283n

Galileo, 12, 16–17, 34, 35, 58

Galton, Francis, 70

Game of Life (Conway), 125–26

game theory, 215–16, 256, 318, 404–5, 416–17, 438, 454; see also Prisoner's Dilemma

Gandhi, Mohandas K., 447

gases, kinetic theory of, 39–47, 67–70, 81–84, 130–31, 193

Gassendi, Pierre, 20, 35

Gauss, Carl Friedrich, 61

Gaussian statistics, 196, 198, 209, 210, 218, 219, 232n, 236, 242, 248, 258, 298–99; in Black-Scholes theory, 202, 203; error curve and, 61, 193–95, 199; "noise" in, 197, 217, 220; power-law distributions versus, 246; networks and, 362, 384n, 397

Gauthier, David, 424

Gay-Lussac, Joseph-Louis, 40

Geller, Uri, 49–50

General Motors, 160, 267

general equilibrium theory, 191

Generous Tit for Tat (GTFT) strategy, 431, 435–37, 443

geometry, 14–16

Georg-August University, 131

Georgia, 339

Germany, 90; divorce rate in, 327; 1848 revolution in, 186; political parties in, 271; population growth in, 52; reunification of, 293; traffic in, 157–59, 168, 173, 176; in World War I, 413, 433; in World War II, 284–88, 447

Gibbs, Josiah Willard, 99, 101, 103, 106, 108, 206, 329

Gibrat, Robert, 257–60, 264

Giddings, Franklin, 464

Girondins, 302

Giuliani, Rudolph, 322

Gladwell, Malcolm, 81, 321–22

Goodbye to All That (Graves), 414

Goodstein, David, 49

Gopal, Venkatesh, 400–401

Gore, Al, 295–96

Grand Canyon National Park, 351

Granovetter, Mark, 355

Graunt, John, 51, 53

Graves, Robert, 414

Great Crash (October 1929), 181–82

Great Depression, 188

Greater London Authority, 146

Greece, 286, 287; ancient, 12, 34–35, 271, 405–6, 425

Green Acres (television series), 352

Grim Trigger strategy, 437

Grotius, Hugo, 22, 24

Guare, John, 357–58

Gutenberg, Beno, 240, 244

Haiti, 248

Halley, Edmond, 59

Hamilton, William, 245

Hang Seng Index, 199, 235

Hanson, Julienne, 139–40, 153

Hare, Maurice Evan, 75

Harrington, James, 178n
Harvard University, 126, 315, 356
Havlin, Schlomo, 151
Hayek, Friedrich von, 461–64
Hebrew University, 175, 223
Hegel, Friedrich, 432
Heilbroner, Robert, 180, 205–7
Heisenberg, Werner, 79, 354
Helbing, Dirk, 131–33, 134n, 136–38, 141, 142, 144, 145, 161, 170–77, 107
heliocentric theory, 12
Helmholtz, Hermann von, 39, 77
Henderson, L. F., 130–31
Hennecke, Ansgar, 172, 173
Hepburn, Audrey, 352
Heraclitus, 100
Herapath, John, 39
herding, 216, 301
Hereditary Genius (Galton), 70
Herschel, John, 64, 68, 83
Hillier, Bill, 139–40, 153
Hirn, Gustave-Adolphe, 85
Hiroshima, atomic bombing of, 412
history: counterfactual, 289–91; science of, 65–67
History of Civilization in England (Buckle), 66, 67
Hitler, Adolf, 284, 288, 461, 462
Hobbes, Francis, 14
Hobbes, Thomas, 6, 11–32, 51, 57, 118, 121, 229, 305, 416, 418, 434n, 443, 448, 454, 455, 464–65; economic theories and, 178–80, 183, 208, 456; historical setting for theories of, 11–13, 28, 265; on international conflict, 410, 411, 429; Internet as manifestation of Leviathan of, 373; mechanistic philosophy of, 12, 16–21, 34, 36, 50, 58, 132; model of society of, 23–32, 339, 403, 461; "State of Nature" of, 23–24, 129, 349, 406, 408, 423–24, 457
Hobsbaum, Eric, 288, 412

Hod, Shahar, 334–35
Hokkaido University, 114
Hölderlin, Friedrich, 451
Hollar, Wenceslas, 26
Holocaust, 412, 447
homme moyen, concept of, 64, 70
Hopkinson, John, 87
Hopper, Dennis, 369
Horney, Karen, 131–32
Household Words (periodical), 67
Huang, Zhi-Feng, 248
Huberman, Bernardo, 175, 384
Human Behavior and the Principle of Least Effort (Zipf), 245
Humboldt, Wilhelm von, 407
Hume, David, 28, 56–57
Humphreys, W. J., 100, 114
Hungary, 284, 286
Hunt, Frederick, 67
Hussein, Saddam, 433
Hutton, James, 457n
Huxley, Aldous, 450–51
Huxley, Julian, 450
hyperlinks, 376–77, 380, 381

IBM, 208, 274, 282
Ideas Toward a Definition of the Limits of State Action (Humboldt), 407
"I Love You" computer virus, 372
I'm a Fool (television drama), 352–53, 255
impossibility theorem (Arrow), 305
income distribution, 246–48
increasing returns of scale, 262
India, 346
Industrial Revolution, 53, 140, 183–84, 189, 253
Inégalités économiques (Gibrat), 257
inertia, law of, 16–17
Information Age, 374
Inland Revenue (U.K.), 247
Inquisition, 12, 17

interacting-agents models, 214–21, 313, 327

international alliances, 284–89, 291–93

International Court of Justice, 410

International Statistical Congress, 74

Internet, 158, 372–80, 385–91, 401; spread of computer viruses in, 392–94; *see also* World Wide Web

Internet Movie Database, 353

Iran, 293

Iraq, war in, 339, 433, 443

Ireland, marriage age in, 324

Iron Law of Wages (Ricardo), 184

Irving, Larry, 372

Ising, Ernst, 89–90

Ising models, 89–92, 96, 109, 230–31, 277–79, 281, 291, 441

Islam, 312, 411, 438

Israel, 293, 438, 443

Italy, 341; 2003 blackout in, 391; in World War II, 284, 286

Jacobins, 302

Jagiellonian University, 223

James II, King of England, 3, 51

Japan, 252, 341; art in, 338; crime in, 317; labor relations in, 253; marriage in, 324; social networks in, 356; weddings in, 337

Jefferson, Thomas, 57

Jencks, Charles, 82

Jeong, Hawoong, 381, 385

Jesuits, 12, 35

Jevons, Stanley, 190

Jews, 324

Johns Hopkins University, 355

Jones, David, 447

Jordan, 293

Juglar cycles, 189

Kagan, Robert, 411

Kahn, Hermann, 444

Kamerlingh Onnes, Heike, 94–95

Kant, Immanuel, 33, 52–53, 66, 73, 76, 244, 290, 404, 407, 411, 443

Kapitsa, Pyotr, 95

Kapteyn, Jacobus, 258

Karinthy, Frigyes, 357n

Karlsruhe University, 160

Kasparov, Garry, 208

Kay, John, 181, 211, 214, 268

Keltsch, Joachim, 136

Kelvin, Lord (William Thomson), 38–39, 77, 78

Kepler, Johannes, 4, 116

Kerner, Boris, 168, 170, 171, 172

Kett, Henry, 148

Keylock, Brian, 250, 269

Keynes, John Maynard, 191, 192n, 205, 206, 208, 212, 216–17, 221, 226

Khrushchev, Nikita, 445

Kiel, University of, 217, 393

kinetic theory of gases, 39–47, 67–70, 81–84, 130–31, 193

King's College (London), 87

Kinnock, Neil, 297, 301

Kirman, Alan, 213, 215–18

Kitchin cycles, 189

Klemm, Konstantin, 344n

Kochen, Manfred, 356

Kondratieff cycles, 189, 190

Kosovo, 339

Kraines, David, 432

Kraines, Vivian, 432

Kropotkin, Peter, 409, 428

Krugman, Paul, 182, 183, 188n, 206, 221, 224n

Kubrick, Stanley, 430, 446

Kuhn, Thomas, 82

Kuznets cycles, 89

labor, commodification of, 183–85

labor theory of value (Marx), 65, 254

Labour Party, British, 284, 297, 301

laissez-faire, principle of, 65, 66, 180, 222, 462

Lambert, John, 10–11
landscape model, 275–94, 301
Laplace, Pierre-Simon, 60–63, 68, 69n, 71, 77, 83, 84
Latvia, 286
Lavoisier, Antoine-Laurent, 101
"law of large numbers" (Poisson), 62, 63, 77–79, 249
"law of least action" (Hamilton), 245, 311
Laws of War and Peace (Grotius), 22
Lee, Bruce, 369
Lee, Christopher, 369
Leeuwenhoek, Antony van, 118–20
Leibniz, Gottfried, 19, 71
Leiden Physical Laboratory, 94
Lenz, Wilhelm, 88–90
Letters from the Underworld (Dostoevsky), 74–75
Leucippus, 34
Levelers, 11
Levi, Primo, 373
Leviathan (Hobbes), 12, 13, 19–21, 23–29, 34, 50, 178, 424, 450
Lévy, Paul, 196
Lévy flight, 195–99
Lewin, Kurt, 132
Li, Gong, 369
Liberal Party, British, 284
Lidl supermarket chain, 251
Lighthill, James, 159–60, 172
Lillo, Fabrizio, 199
Lind, Michael, 315
Lion King, The (movie), 125
Lithuania, 286
"live and let live" strategy, 413–15, 422
Locke, John, 29, 30, 403–11, 423, 429, 448, 461
London Business School, 181
London School of Economics, 259
London Stock Exchange, 215
London Underground, 361–62, 380
Longfellow, Henry Wadsworth, 64n
Longley, Paul, 150

Looking Backward (Bellamy), 451
Lorentz, Hendrik Antoon, 79
Los Alamos National Laboratory, 175
Louis XV, King of France, 57
Love Bug computer virus, 372, 393
Lucretius, 35
Luther, Martin, 12
Lux, Thomas, 217–19

Mach, Ernst, 43
MacIver, Robert M., 23
Mahler, Gustav, 49
Makse, Hernán, 151
Malik, Kenan, 464
Malthus, Thomas Robert, 52, 56, 72, 226, 408
Manchester University, 159
Mandelbrot, Benoit, 196–99, 244n, 246
Mandeville, Bernard, 453–54
Manhattan Project, 416
Mantegna, Rosario, 197–99
Man, the State, and War (Waltz), 4
Man Without Qualities, The (Musil), 49
Marchesi, Michele, 218, 219
market economy, 178–203, 221–25, 461–62; chaos theory as model for, 5; crime in, 316; critical points in, 234–37; "hidden hand" of, 180–81, 190n, 311, 453; inequalities in, 457, 459; interacting-agents models of, 214–21, 241; Marx's critique of, 183–86; mathematical approach to, 183; rise of, 178–79; self-organized criticality in, 241–42; statistics of fluctuations in, 186–203; *see also* firms
Markowitz, Harry, 211
marriage, 323–31
Marshall, Alfred, 205–7, 262
Marvel Comics, 395–96
Marx, Karl, 30, 52, 64–65, 183–86, 190, 191, 208, 253, 254, 437, 450, 459–60
Marxism, 290

Massachusetts Institute of Technology (MIT), 356; School of Management, 196

Mathematical Psychics (Edgeworth), 183, 205

Mathias, Nish, 400–401

Matisse, Henri, 338

Matsushita, Mitsugu, 110–13, 116, 119, 149

Maurois, André, 289

Maxwell, James Clerk, 77–78, 85n, 87, 96, 99, 127, 204–6, 329; kinetic theory of gases of, 39–47, 67–70, 81, 130, 131, 193

May, Robert, 440, 441

McNamara, Robert, 445

McNaughton, John, 445

"mean field" theories, 215–16, 233

mechanical philosophy, 18–20, 31, 33–47; of Hobbes, 12, 16–21, 34, 36, 50, 58, 132

mechanics, 16; Newtonian, 56–58, 60, 68, 78, 180, 245; quantum, 78–79, 95–96; social, 59–60; statistical, 44–47, 49, 50, 81, 87, 96, 98–99, 269

Mein Kampf (Hitler), 450

Merchant, Carolyn, 30–31

Meredith College, 432

Mersenne, Marin, 20

Merton, Robert, 201

metabolic networks, 389–91

metastability, 163–66

Michigan, University of, 275, 350

Micromotives and Macrobehavior (Schelling), 310

Microsoft, 270, 276; Internet Explorer, 377

Middle Ages, 180, 184, 253, 309, 456

Milgram, Stanley, 356–57, 363, 370

Mill, John Stuart, 30, 65, 208, 407, 452, 458, 461, 464

Moe, Richard, 157

Moelwyn-Hughes, Emyr Alun, 41–42

Moivre, Abraham de, 61–63, 193

Molnár, Péter, 132–33, 136, 138

Monbiot, George, 250, 251, 256

Monk, George, 11

monopolies, 255, 256

Montesquieu, Charles-Louis de Secondat, Baron de la Brède et de, 56, 453, 457

Moors, in Spain, 338

moral law, 458, 460

More, Thomas, 20–21, 451

Morgenstern, Oskar, 417

Mormons, 400

morphology diagrams, 113–16, 119

Morris, William, 451

motion: collective, 118–55; laws of, 16–18, 37, 39, 45–47, 54, 77, 95, 127, 131

Moynihan, Daniel Patrick, 313

Mumford, Lewis, 23, 149, 150, 455

Murdoch, Rupert, 256

Musil, Robert, 49

Muslims, *see* Islam

Mussolini, Benito, 461

Mutual Aid (Kropotkin), 409

Nagel, Kai, 159–61, 168, 172, 175

Nagel-Schreckenberg (NaSch) model, 159, 160, 164–69

Nakar, Ehud, 334–35

Nakaya, Ukichiro, 114

Napier, John, 17

Nash equilibria, 263–64, 277

National Aeronautics and Space Administration (NASA), 273

National Security Agency (NSA), 388

National Telecommunications and Information Administration, 373

National Trust for Historic Preservation, 157

natural selection, 70

Nazis, 286n, 287

Netherlands, 58; traffic in, 173

Netscape, 377

networks, 397–401; electronic, *see* Internet; World Wide Web; metabolic, 389–91; social, 352–71, 394–97

New America Foundation, 315

New Atlantis (Bacon), 21

New Lanark community, 450

Newman, Mark, 354n, 393
Newmarch, William, 66, 330
News from Nowhere (Morris), 451
News International group, 256
Newton, Isaac, 12, 35, 101, 159, 181, 192n,
   258, 457n; on gravitation, 3, 34, 76, 148;
   Halley and, 59; laws of motion of, 16, 37,
   39, 45–47, 54, 77, 95, 127, 131; Leibniz
   and, 19, 71; mechanics of, 56–58, 60, 68,
   78, 180, 245
Newtonian System of the World, The
   (Desaguliers), 56
New York City, crime rates in, 321–22
New York Stock Exchange, 215
Nicholson, Jack, 360
Niels Bohr Institute, 344n
Nietzsche, Friedrich, 67
Nightingale, Florence, 64, 72
Nikkei Index, 199
Nixon, Richard M., 313
Nobel Prize, 44, 87, 103, 201, 211, 233,
   324
"noise," 106, 127–28, 192, 193, 197
nonequilibrium systems: growth and form of,
   101–17, 151; motion of, 127–29
normal distribution, 61
North Atlantic Treaty Organization (NATO),
   284, 293, 430, 438
Northern Ireland, 339, 431, 443
Notre Dame University, 381, 383
Nowak, Martin, 433–36, 440, 441
Nuñes Amaral, Luis, 394
Nussbaum, Frederick, 27

Oakeshott, Michael, 289
Observations upon the Bills of Morality
   (Graunt), 51
Oceana (Harrington), 178n
oligopolies, 255–57
Olson, Richard, 28, 460
On Growth and Form (Thompson), 101
Onion, The, 295

On Liberty (Mill), 407
Ono, S., 91
Onsager, Lars, 90, 101–3, 233
On the Governance of Rulers (Aquinas),
   461
Open Group, 274
Open Software Foundation (OSF), 274–75,
   282
options pricing, theory of, 201–3
ordered graphs, 363–64; clustering in,
   364–67
Origin of Species, The (Darwin), 69, 101
Ormerod, Paul, 197, 219, 220, 312, 317–20,
   327–30, 459–60
Osborne, M.F.M., 190n
Osiander, Andreas, 12
Ostwald ripening, 275
Ottoman Empire, 438
Ouro Prêto, Federal University of, 300
Owen, Robert, 450
Owen, Wilfred, 447
Oxford University, 3, 14, 64, 433, 440

Paczuski, Maya, 161
Paine, Thomas, 55
Palestinians, 443
paradigm shifts, 82
Pareto, Vilfredo, 190, 246–48, 349
Pareto optimum, 190–91
Park, Robert F., 452–53
Parrish, Julia, 122–23, 147
Partial Tit for Tat strategy, 443
Pascal, Blaise, 17
Pasteur, Louis, 110
Pastor-Satorras, Romualdo, 393, 394
Patriarcha (Filmer), 405
Paul, St., 289
Pavlov strategy, 432, 435–37, 442
Peck, Gregory, 352
Peddling Prosperity (Klugman), 221
pedestrian movement, models of, 132–40;
   crowd safety and, 141–47

Peierls, Rudolf, 90n
Peirce, Charles, 69–70
Peloponnesian Wars, 406
Penn, Alan, 153
Pentagon, 388
percolation, correlated, 151–52
Pericles, 425
Perrin, John, 44
Petty, William, 3–4, 6, 51, 455
phase transitions, 81–100, 109, 128–29, 164; continuity and, 82–85; corresponding states and, 94–95; in criminal behavior, 320–22; in crowds, 131; Curie point and, 87–88, 90; Ising models of, 89–92
Philippines, University of the, 142n
Philosophical Radicals, 30
Plato, 13, 20, 21, 405, 406, 451
Platonism, 100, 115
Pleasence, Donald, 369, 370
Poincaré, Henri, 192
Poisson, Siméon-Denis, 60, 62, 63
Poland, 286–88
Political Arithmetick (Petty), 4
political parties, 301–3; alliances of, 271, 284; *see also specific parties*
Polynesians, 409
Pool, Ithiel de Sola, 355–56
Popper, Karl, 403, 425, 426, 444, 445, 448
Portugal, 286, 287
positivism, 58
Potts, Wayne, 123
power-law probability distributions, 236–44, 246, 247, 249, 398–400; of growth of firms, 260, 264–66; of voting, 299–300; of World Wide Web, 382–83, 384
Presbyterians, 10, 53
Prigogine, Ilya, 103, 106–8
Principia Mathematica (Newton), 16, 101, 457n
Prisoner's Dilemma, 417–48; strategies in, 420–23, 426, 427, 429–37, 439–40, 442–48

probability, 44–46, 53–54; error curve and, 60–63, 68–69; of market fluctuations, 195–96; in quantum mechanics, 79; *see also* power-law probability distributions
Procter and Gamble, 202
Project 2050, 350
Proportionate Growth, Law of (Gibrat), 258, 259
Protestants, 52, 431
Prussia, 271, 305
Ptolemy, 189
Pythagoreans, 249

Quakers, 244
quantum mechanics, 78–79, 95–96
Queen's College (Belfast), 83
Quesnay, François, 57
Quetelet, Adolphe, 58–60, 63–66, 68–70, 74, 131, 193, 246
Quincke, Georg Hermann, 104n

racial segregation, 312–15
RAND Corporation, 378, 416, 418, 444
random graphs, 359–64
randomness, 60–63, 67, 69
random walks, 42–43, 192–97
Rapoport, Anatol, 362, 420, 432
Rayleigh, John William Strutt, Lord, 104
Rayleigh-Bénard convection patterns, 106, 114
Reagan, Ronald, 352
"real business cycle" (RBC) theory, 210
recessions, 221–22, 224–25
Redner, Sidney, 242, 243
Reformation, 12, 22
Rehborn, Hubert, 168, 170
Renaissance, 12, 22, 64
Rényi, Alfred, 359, 361–62, 387, 397
Republic (Plato), 13, 20
Republican Party, U.S., 296, 297, 426
repulsion, molecular, 86, 132

Restoration, English, 10–11
Reynolds, Craig, 124, 126, 127
Ricardo, David, 184, 226
Richardson, Lewis Fry, 243–44
Richter, Charles, 240, 244
Rights of Man, The (Paine), 55
Road to Serfdom, The (Hayek), 461
Robespierre, Maximilien-François-Marie-
    Isidore de, 54–56, 302
Roman Holiday (movie), 352
Romania, 284, 286, 293
Romans, ancient, 27, 35, 252, 338, 405, 438
Romanticism, 23
Roosevelt, Franklin D., 288
Rousseau, Jean-Jacques, 29, 55, 408, 411,
    418, 457
Royal Society, 51, 118
Rudolf, Crown Prince of Austria, 49
Rumford, Count (Benjamin Thompson), 36
Russian Revolution, 185, 450
Rwanda, 347

Safeway supermarket chain, 250–51
Sainsbury's supermarket chain, 251, 256
Saint-Simon, Claude-Henri de Rouvroy,
    Comte de, 56
Salinger, Michael, 260
Sander, Len, 111, 114
Santa Fe Institute, 241, 331, 350, 393, 401
"satisficing," 257
Scheinkman, Jose, 241, 242
Schelling, Thomas, 310–15, 317, 327, 351,
    463
Schoenberg, Arnold, 49
Scholes, Myron, 201
Schreckenberg, Michael, 159–61, 168, 171,
    172, 175; *see also* Nagel-Schreckenberg
    (NaSch) model
Schrödinger, Erwin, 37n, 79, 354
Schumpeter, Joseph, 187–90, 208
Science & Finance consulting company, 203
Scotland, 10, 19

search engines, 377, 383–84
Secret Service, U.S., 388
segregation, 312–15
self-organized criticality (SOC), 236–43
Senate, U.S., 372
Senior, Nassau, 66
Serbia, 430, 443
Shannon, Claude, 78
Shaw, George Bernard, 323, 340
Shiller, Robert, 210, 217
Shirras, George Findlay, 248
Short, Thomas, 52
Sigmund, Karl, 433–36, 446
SimCity, 348
Simmel, Georg, 452
Simon, Herbert, 153–54, 351
Sinatra, Frank, 369–70
Sinclair, John, 53
Single Unix Specification, 274
Six Degrees of Separation (Guare), 357–58,
    369
Skyrms, Brian, 445
Slammer computer worm, 372
Smart Growth America, 149
Smith, Adam, 29, 57, 58, 178–81, 183,
    185, 208, 223, 254, 456–59, 461–62;
    "hidden hand" concept of, 180–81, 190n,
    311, 453; on labor, 184, 253, 323; laissez-
    faire advocated by, 65, 180, 222, 462
Smith, John Maynard, 432
Smoluchowski, Marian, 79
Snow, C. P., 38
Snow Crystals (Bentley and Humphreys),
    100, 114
Snyder, Carl, 248
Social Democratic Party, German, 271
social norms, 306–10, 324
Solaria networks, 365–66, 369
Solomon, Sorin, 223, 248
Sophocles, 405
Sornette, Didier, 235, 236
Soros, George, 213
South Korea, 372, 430

South Sea Bubble, 185

Soviet Union, 291, 292, 378, 430, 446; collapse of, 293, 339; in World War II, 286–88

Spain, 341; Moorish, 338

Spanish Inquisition, 17

Spence, Michael, 211

Spencer, Herbert, 70, 98, 151

spin glasses, 277–81

Spinoza, Baruch, 418

Spirit of the Law, The (Montesquieu), 56

Stalin, Joseph, 287, 288, 461

Stalinism, 189, 286*n*

Standard & Poor's (S&P) 500 Index, 194, 195, 198, 199, 220, 235, 242

Stand to Arms (Cobb), 413

Stanley, Gene, 151, 153, 197–98, 260, 394, 399, 400

"State of Nature": of Grotius, 22, 24; of Hobbes, 23–24, 129, 349, 406, 408, 423–24, 457

Statistical Account of Scotland (Sinclair), 53

Statistical Society of London, 66, 71, 72, 330

statistics, 6, 44–47, 49–53, 58, 60–79; of crowd motion, 130–31; error curve in, 60–63, 68–69; Gaussian, *see* Gaussian statistics; historical, 65–67; of market fluctuations, 197–200; political use of, 248; in science, 67–79 (*see also* mechanics, statistical)

Steiger, Rod, 369, 395

Stiglitz, Joseph, 211

stochastic processes, 61

Stockholm University, 394

stock market crash of 1929, *see* Great Crash

Strogatz, Steven, 364–68, 382–83, 385, 389, 393, 395, 400

Stuttgart University, 132, 136, 172

Sugarscape, 223, 348–51

Sun Also Rises, The (movie), 353

Sun Microsystems, 274, 276

superconductivity, 95

Supreme Court, U.S., 296, 407

susceptible-infected-susceptible (SIS) model, 392–93

Süssmilch, Johann Peter, 52, 73

Sutherland, Donald, 369

Sutton, John, 259

Sweden, 286

Swift, Jonathan, 57, 455

Switzerland, 168, 286

Sydney, University of, 130

Symbolics computer company, 124

Syria, 293

System of Logic, A (Mill), 65

Szilard, Leo, 410

Tableau économique (Quesnay), 57

Tailandier, A., 72

Tait, Peter Guthrie, 69

Tang, Chao, 237–40, 243

Tate Gallery, 138–39

technological change, 185, 190

Tennyson, Alfred, Lord, 77, 243, 408

terrorism, cyber-, 387–89

Tesco supermarket chain, 256

Thatcher, Margaret, 337

Theory of Games and Economic Behavior (von Neumann and Morgenstern), 417

thermodynamics, 99–100; laws of, 37–39, 45–47, 77–78, 99, 102; nonequilibrium, 101–9

Thirty Years' War, 11, 52

Thom, René, 5, 81

Thompson, Benjamin, *see* Rumford, Count

Thompson, D'Arcy Wentworth, 101, 104

Thomson, William, *see* Kelvin, Lord

Thoreau, Henry David, 98

Thucydides, 276, 405

Timely Comics, 395

tipping points, 81–83, 321, 322

Tit for Tat (TFT) strategies, 420–23, 426, 427, 429–37, 439–40, 442–48

Tit for Two Tats strategy, 431–32, 443

Tokyo Stock Exchange, 215

Tolstoy, Leo, 75
Toronto, University of, 420
traffic, 156–77; disturbed flow of, 170–73; forms of, 167–70; measurements of, 158–59; metastability of, 163–66
Traité élémentaire de chimie (Lavoisier), 101
Transportation Analysis and Simulation System, 175
Treatise of Human Nature (Hume), 56
Treiber, Martin, 172, 173, 176, 177
Trent, Council of, 12
Trevor-Roper, Hugh, 290
Tucker, Josiah, 180
Turkey, 293
Twain, Mark, 67
Two Cultures, The (Snow), 38
Two Treatises of Government (Locke), 403, 405

Ubiquity (Buchanan), 243
Ulam, Stanislaw, 125
Uncertainty Principle (Heisenberg), 79
uniform resource locator (URL) links, 380
United Nations, 339, 410
United States, 271, 304, 340, 346, 411; Air Force, 388; August 2003 blackout in, 391; during Civil War, 10, 272; during Cold War, 378; computer industry in, 274; Constitution of, 296; crowd behavior in, 141; e-commerce in, 380; economy of, 187–88, 198, 221; government of, 407 (*see also specific governmental bodies*); income distribution in, 246; Iraq invaded by, 339; marketplace choice in, 270; traffic in, 156, 157, 159, 167, 175–76; 2000 presidential election in, 295–96, 298; urban grid structure in, 149
universality, 92, 234
Universal Declaration of Human Rights, 23
University College (London), 138, 139, 153
Unix International Incorporation (UII), 274–75, 282

urban planning, 147–55; pedestrian movement in, 138–40
Utah State University, 123
utilitarianism, 30, 183, 262, 458
utopias, 20–23, 29–32, 54–56, 450–51

van der Waals, Johannes Diderik, 83–87, 93, 94, 99, 132, 227–30, 232, 233, 276, 291, 328
van Gogh, Vincent, 338
Veblen, Thorstein, 212, 213
Versailles Treaty, 244
Verschaffelt, Jules, 230
Vespignani, Alessandro, 344, 393, 394
VHS video system, 272
Vicsek, Tamás, 127–28, 133n, 141, 142, 144, 467
Victorian era, 140, 291, 337
Vienna, University of, 433
Vietnam, 252
Vietnam War, 304, 445
Virginia, University of, 353n
Virtual History (Ferguson), 289
viruses, computer, 392–94
Voltaire, 54, 450
Vonnegut, Kurt, 80, 102, 163
von Neumann, John, 125, 416–17
voting, 295–305, 331

Wagons Roll at Night, The (movie), 352
Walras, Léon, 191
Waltz, Kenneth, 4, 284
War and Peace (Tolstoy), 75
Warsaw Pact, 284, 293
Washington, George, 271
Watergate scandal, 426
Waterloo, Battle of, 271
Watts, Duncan, 357n, 370–71, 382–83, 385, 389, 393, 395, 400
Wealth of Nations (Smith), 29, 57, 178, 180, 222, 453, 456

Weiss, Pierre, 87–88, 215, 233
Wellington, Arthur Wellesley, First Duke of, 226
Wells, H. G., 410
Whewell, William, 72
Whitman, Gerald, 159, 172
Wiedemann, Rainer, 160
Wiesenfeld, Kurt, 237–40, 243
Wilde, Oscar, 340n
William III, King of England, 3
Wilson, Edward O., 126, 315, 423
Wilson, Kenneth, 233
Winstanley, Gerrard, 450
Witten, Tom, 111, 114
Wittgenstein, Ludwig, 49
Woodford, Michael, 241, 242
WorldCom, 223
World Resources Institute, 350
World War I, 244, 412–15, 422, 433

World War II, 188, 244, 284–89, 412, 416, 438, 447
World Wide Web, 372, 374, 376, 377, 379–85, 397–98

Xerox Research Center, 175, 384
X/Open Company, 274

Yale University, 99, 182
Yugoslavia, 286, 287; breakup of, 304, 338

Zaire, 248
Zhang, Yi-Cheng, 333–34, 397
Zhou, Wei-Xing, 236n
Zipf, George Kingsley, 245–46, 249, 311, 398